INTRODUCTION TO
Scheduling

Chapman & Hall/CRC
Computational Science Series

SERIES EDITOR

Horst Simon
Associate Laboratory Director, Computing Sciences
Lawrence Berkeley National Laboratory
Berkeley, California, U.S.A.

AIMS AND SCOPE

This series aims to capture new developments and applications in the field of computational science through the publication of a broad range of textbooks, reference works, and handbooks. Books in this series will provide introductory as well as advanced material on mathematical, statistical, and computational methods and techniques, and will present researchers with the latest theories and experimentation. The scope of the series includes, but is not limited to, titles in the areas of scientific computing, parallel and distributed computing, high performance computing, grid computing, cluster computing, heterogeneous computing, quantum computing, and their applications in scientific disciplines such as astrophysics, aeronautics, biology, chemistry, climate modeling, combustion, cosmology, earthquake prediction, imaging, materials, neuroscience, oil exploration, and weather forecasting.

PUBLISHED TITLES

PETASCALE COMPUTING: Algorithms and Applications
Edited by David A. Bader

PROCESS ALGEBRA FOR PARALLEL AND DISTRIBUTED PROCESSING
Edited by Michael Alexander and William Gardner

GRID COMPUTING: TECHNIQUES AND APPLICATIONS
Barry Wilkinson

INTRODUCTION TO CONCURRENCY IN PROGRAMMING LANGUAGES
Matthew J. Sottile, Timothy G. Mattson, and Craig E Rasmussen

INTRODUCTION TO SCHEDULING
Yves Robert and Frédéric Vivien

INTRODUCTION TO
Scheduling

EDITED BY
YVES ROBERT
FRÉDÉRIC VIVIEN

CRC Press
Taylor & Francis Group
Boca Raton London New York

CRC Press is an imprint of the
Taylor & Francis Group an **informa** business
A CHAPMAN & HALL BOOK

CRC Press
Taylor & Francis Group
6000 Broken Sound Parkway NW, Suite 300
Boca Raton, FL 33487-2742

First issued in paperback 2017

© 2010 by Taylor and Francis Group, LLC
CRC Press is an imprint of Taylor & Francis Group, an Informa business

No claim to original U.S. Government works

ISBN 13: 978-1-138-11772-3 (pbk)
ISBN 13: 978-1-4200-7273-0 (hbk)

Library of Congress Cataloging-in-Publication Data

Introduction to scheduling / Yves Robert, Frederic Vivien.
     p. cm. -- (Chapman & Hall/CRC computational science series)
   Includes bibliographical references and index.
   ISBN 978-1-4200-7273-0 (hardcover : alk. paper)
   1. Electronic data processing--Distributed processing. 2. Computer scheduling--Mathematical models. 3. Multiprocessors. 4. Computational grids (Computer systems) I. Robert, Yves, 1938- II. Vivien, Frédéric. III. Title. IV. Series.

QA76.9.D5I673 2009
004'.36--dc22
                                                          2009032786

Visit the Taylor & Francis Web site at
http://www.taylorandfrancis.com

and the CRC Press Web site at
http://www.crcpress.com

Contents

Preface

Aim and Scope

The objective of this book is twofold:

- To introduce the basic concepts, methods, and fundamental results of scheduling theory.
- To expose recent developments in the area.

In each chapter, emphasis is given to self-contained, easy-to-follow, but still rigorous presentations. The book is full of examples, beautiful theorems, and pedagogical proofs. It also contains an in-depth coverage of some key application fields. Altogether, this book provides a comprehensive overview of important results in the field, organized as a handy reference for students and researchers.

There are numerous books and proceedings devoted to scheduling techniques, among others [1, 2, 3, 4]. However, we believe that this book is unique, because it is meant to be accessible to a large audience while covering both the foundations and modern developments.

Book Overview

For the sake of presentation, we have organized the chapters into five main categories: *Foundations*, *Online and Job Scheduling*, *Cyclic Scheduling*, *Advanced Topics*, and *Platform Models*. However, each chapter can be read independently, or following the pointers to relevant material in other chapters whenever needed.

Foundations

The first two chapters present fundamental concepts. The purpose of Chapter 1 is to introduce and classify scheduling problems and their complexity. Furthermore, the notations used in the book are introduced. Firstly, the resource constrained project scheduling problem is defined and several applications of this problem are discussed. An important subclass is the class of machine scheduling problems. For a classification of these problems, the well-known three field notation is introduced. Secondly, the basic definitions of the theory of computational complexity are given and applied to machine scheduling problems. It is shown how the three field notation can be used to identify, for different classes of machine scheduling problems, the hardest problems known to be polynomially solvable, the easiest NP-hard problems, and the minimal and maximal open problems.

Because many scheduling problems are NP-hard (hence it is quite unlikely that we will ever be able to find efficient polynomial-time algorithms for their solution), it is worth looking for algorithms that always return a feasible solution whose measure is not too far from the optimum. Chapter 2 is devoted to the study of such *approximation algorithms*. The fundamentals of this approach are first recorded via a classification of optimization problems according to their polynomial-time approximability. Some examples are then presented in order to show successful techniques for the design of efficient approximation algorithms, and some basic ideas of methods for proving lower bounds on polynomial-time inapproximability are provided.

Online and Job Scheduling

The next two chapters introduce the reader to two important related topics, *online* scheduling and *job* scheduling. Traditional scheduling theory assumes that an algorithm, when computing a schedule, has complete knowledge of the entire input. However, this assumption is often unrealistic in practical applications. In an online setting, jobs arrive over time; whenever a new job arrives scheduling decisions must be made without knowledge of any future jobs. On arrival of a job, its complete characteristics such as the processing time may or may not be known in advance. We seek scheduling algorithms that work well in online scenarios and have a proven good performance. In this context, an online algorithm A is called c-competitive if, for all inputs, the objective function value of a schedule computed by A is at most a factor of c away from that of an optimal schedule. In Chapter 3, we study a number of classical problems as well as fresh problems of current interest. As for classical problems, we consider, for instance, the famous makespan minimization problem in which jobs have to be scheduled on parallel machines so as to minimize the makespan of the resulting schedule. We analyze Graham's List scheduling algorithm, that is 2-competitive, as well as more recent algorithms that improve upon the factor of 2. With respect to recent advances, we study energy-efficient scheduling algorithms. Here jobs must be scheduled on a variable speed processor so as to optimize certain objective functions.

In Chapter 4, we target job scheduling problems that comprise independent and possibly parallel jobs. Total (weighted) completion or response times and machine utilization are the most common objective functions, although complex objectives defined by system owners have recently received increasing interest. The machine models are usually restricted to parallel identical machines but may consider communication delays and machine subsets. The schedules may either be preemptive or non-preemptive. Occasionally, some other constraints, like migration with penalties, are considered as well. Since basically all interesting deterministic job scheduling problems are NP-hard, many theoretical results yield approximation factors. As in practice most job scheduling problems include unknown processing times and submission over time many online results are also given in the literature. However, in general, only simple algorithms, like various forms of list scheduling, are analyzed as complex approaches with potentially good competitive or approximation factors are of limited practical interest. In this chapter, we present several important results and explain how they are related to each other. Furthermore, we describe various methods to prove competitive or approximation factors of simple job scheduling algorithms.

Cyclic Scheduling

Cyclic scheduling problems occur when a set of tasks is to be repeated many times. Chapter 5 begins with a short introduction to applications which motivated the study of cyclic scheduling problems. Then we introduce the basic model for precedence relations between tasks in this context, called uniform precedences. Two main kinds of studies have been devoted to cyclic scheduling problems in this context. The first one aims at analyzing the behavior of the earliest schedule without any resource constraints. The second approach aims at defining efficient periodic schedules in contexts where resources are to be taken into account. We discuss the notion of periodicity, the complexity of these problems, and exhibit some special cases which can be polynomially solved. In the general case, we present an approach taken from the field of *software pipelining*. We also give a list of resource contexts in which such problems have been studied (parallel or dedicated processors, registers) with different optimization criteria.

The major objective of Chapter 6 is to present a practical application of the results of Chapter 5 for the synthesis of embedded systems. The first part of this chapter is devoted to the formulation of the practical problem using Generalized Timed Event Graphs (in short GEGT), a subclass of Petri Nets. The second part of this chapter is devoted to the presentation of some mathematical tools to study the properties of a GEGT. It has been shown that pseudo-polynomial algorithms can be developed to evaluate the liveness or the throughput of a GEGT. More recently, a nice simplification of the GEGT was introduced leading to a simple sufficient condition of liveness. Some new results on the evaluation of the throughput will also be presented.

Chapter 7 deals with *steady-state* scheduling, an approach that also builds upon cyclic scheduling techniques, but in the context of large scale platforms. Steady-state scheduling consists in optimizing the number of tasks that can be processed per time unit when the number of tasks becomes arbitrarily large. In this chapter, we concentrate on bag-of-tasks applications and collective communications (broadcast and multicast) and we prove that efficient schedules can be derived in the context of steady-state scheduling, under realistic communication models that take into account both the heterogeneity of the resources and the contentions in the communication network.

Advanced Topics

Chapter 8 focuses on scheduling large and compute-intensive applications on parallel resources, typically organized as a master-worker platform. We assume that we can arbitrarily split the total work, or load, into chunks of arbitrary sizes, and that we can distribute these chunks to an arbitrary number of workers. The job has to be perfectly parallel, without any dependence between sub-tasks. In practice, this model, known as the the *Divisible Load* model, is a reasonable relaxation of an application made up of a large number of identical, fine-grain parallel computations; instead of scheduling an integer number of tasks on each resource, we allow for a rational number of tasks. The chapter provides several examples of divisible load computations, and discusses several extensions to assess both the power and the limitations of the divisible model.

Chapter 9 considers multi-objective scheduling, i.e., scheduling with the goal of simultaneously optimizing several objectives. In a multi-objective setting, one can explicitly model, optimize, and trade off between various performance measures. The chapter provides three motivating problems to illustrate different approaches. In *MaxAndSum*, a scheduler optimizes both the makespan and the sum of completion times. In *EfficientReliable*, the goal is to find the trade-off between the makespan and the reliability when scheduling on failing processors. Finally, in *TwoAgentMinSum*, a processor must be shared fairly between jobs produced by two independent users. In this latter example, we propose an axiomatic way to characterize schedules that are *fair* to individual users.

Chapter 10 shows how to compare the performance of stochastic task resource systems. Here, task resource systems are modeled by dynamic systems whose inputs are the task and resource data, given under the form of random processes, and whose outputs are the quantities of interest (response times, makespan). The approach presented here is based on stochastic orders. Several typical examples (some simple, others rather sophisticated) are used to show how orders can help the system designer to compare several options but also to compute bounds on the performance of a given system.

Platform Models

Chapter 11, the last chapter of the book, aims at assessing the impact of platform models on scheduling techniques. First, we survey a wide and representative panel of platform models and topologies used in the literature. Then we explain why such models have been used in the past and why they may be ill-adapted to current applications and current platforms. Next we present through a few simple examples the impact of these models on the complexity of the corresponding problems. Lastly we present a few promising network models for modern platforms, discussing their realism as well as their tractability.

References

[1] J. Blazewicz, K. H. Ecker, E. Pesch, G. Schmidt, and J. Weglarz, editors. *Handbook on Scheduling: From Theory to Applications*. Springer, 2007.

[2] P. Chrétienne, E. G. Coffman Jr., J. K. Lenstra, and Z. Liu, editors. *Scheduling Theory and its Applications*. John Wiley & Sons, 1995.

[3] J. Y.-T. Leung, editor. *Handbook of Scheduling: Algorithms, Models, and Performance Analysis*. Chapman & Hall/CRC, 2004.

[4] B. A. Shirazi, A. R. Hurson, and K. M. Kavi. *Scheduling and load balancing in parallel and distributed systems*. IEEE Computer Science Press, 1995.

Acknowledgments

This book originates from the invited talks that were given during the 35th French Spring School in Theoretical Computer Science (*EPIT'2007*), which took place in June 2007 at La Villa Clythia, Fréjus, in southern France. We thank *CNRS*—the French Council for National Research—and INRIA—the French National Institute for Research in Computer Science and Control—whose support made the school possible.

We also thank the whole editorial team at CRC Press, whose encouragement and friendly pressure helped us finalize this project (almost in time).

Finally, we sincerely thank all chapter authors for their contributions. It was really nice working with this great coterie of French and German colleagues!

Yves Robert and Frédéric Vivien
Lyon and Honolulu
Contact: {Yves.Robert|Frederic.Vivien}@ens-lyon.fr

Contributors

Susanne Albers
Humboldt University Berlin
Berlin, Germany

Olivier Beaumont
INRIA
Bordeaux, France

Peter Brucker
University of Osnabrück
Osnabrück, Germany

Pierre-François Dutot
Université de Grenoble
Grenoble, France

Lionel Eyraud-Dubois
INRIA
Université de Bordeaux
Bordeaux, France

Matthieu Gallet
École normale supérieure de Lyon
Université de Lyon
Lyon, France

Bruno Gaujal
INRIA
Université de Grenoble
Grenoble, France

Rodolphe Giroudeau
LIRMM
Montpellier, France

Claire Hanen
Université Paris Ouest
          Nanterre-La Défense
Nanterre, France

Sigrid Knust
Institute of Mathematics
Technical University of Clausthal
Clausthal, Germany

Jean-Claude König
LIRMM
Montpellier, France

Arnaud Legrand
CNRS
Université de Grenoble
Grenoble, France

Loris Marchal
CNRS
Université de Lyon
Lyon, France

Olivier Marchetti
Université Pierre et Marie Curie
          Paris 6
Paris, France

Alix Munier-Kordon
Université Pierre et Marie Curie
    Paris 6
Paris, France

Yves Robert
École normale supérieure de Lyon
Institut Universitaire de France
Université de Lyon
Lyon, France

Krzysztof Rzadca
Polish-Japanese Institute of
    Information Technology
Warsaw, Poland

Erik Saule
Université de Grenoble
Grenoble, France

Uwe Schwiegelshohn
Technische Universität Dortmund
Dortmund, Germany

Denis Trystram
Université de Grenoble
Grenoble, France

Jean-Marc Vincent
INRIA
Université de Grenoble
Grenoble, France

Frédéric Vivien
INRIA
Université de Lyon
Lyon, France

Chapter 1

On the Complexity of Scheduling

Peter Brucker

University of Osnabrück

Sigrid Knust

Technical University of Clausthal

Abstract   This chapter presents a survey on scheduling models (project scheduling and processor scheduling) and their computational complexity.

1.1 Introduction

Scheduling is concerned with the allocation of limited resources to activities over time. Activities may be tasks in computer environments, steps of a construction project, operations in a production process, lectures at the university, etc. Resources may be processors, workers, machines, lecturers, and so on. General scheduling models will be introduced and specific applications like project or processor scheduling will be discussed.

Methods for solving scheduling problems depend on the computational complexity. For machine (processor) scheduling problems a classification scheme has been introduced. After describing it we survey some polynomially solvable and NP-hard scheduling problems.

1.2   Scheduling Models

The so-called resource-constrained project scheduling problem (RCPSP, see also Brucker and Knust [9]) is one of the basic complex scheduling problems. In the following we introduce this problem and some of its generalizations. Processor (machine) scheduling problems, that may be considered as special cases, are introduced in Section 1.3.

The resource-constrained project scheduling problem (RCPSP) is a very general scheduling problem that may be used to model many applications in practice (e.g., a production process, a software project, a school timetable, the construction of a house, or the renovation of an airport). The objective is to schedule some activities over time such that scarce resource capacities are respected and a certain objective function is optimized. Examples for resources may be processors, machines, people, or rooms, which are only available with limited capacities. As objective functions the project duration, the deviation from deadlines, or costs concerning resources may be minimized.

The **resource-constrained project scheduling problem (RCPSP)** may be formulated as follows. Given are n **activities** (jobs) $j = 1, \ldots, n$ and r **renewable resources** $k = 1, \ldots, r$. A constant amount of R_k units of resource k is available at any time. Activity j must be processed for p_j time units. During this time period a constant amount of r_{jk} units of resource k is occupied. All data are assumed to be integers. A resource is called **disjunctive** if $R_k = 1$, otherwise it is called **cumulative**. If resource k is disjunctive, two activities i, j with $r_{ik} = r_{jk} = 1$ cannot be processed simultaneously.

Furthermore, **precedence constraints** are defined between some activities. They are given by relations $i \to j$, where $i \to j$ means that activity j cannot start before activity i is completed.

The objective is to determine starting times S_j for the activities $j = 1, \ldots, n$ in such a way that

- at each time t the total resource demand is less than or equal to the resource availability of each resource $k = 1, \ldots, r$,

- the given precedence constraints are fulfilled, i.e., $S_i + p_i \leq S_j$ if $i \to j$, and

- the **makespan** $C_{\max} = \max\limits_{j=1}^{n} \{C_j\}$ is minimized, where $C_j := S_j + p_j$ is assumed to be the completion time of activity i.

The fact that an activity which starts at time S_j finishes at time $C_j = S_j + p_j$ implies that activities are not preempted. We may relax this condition by allowing **preemption** (activity splitting). In this case the processing of any activity may be interrupted and resumed later.

It is often useful to add a unique **dummy starting activity** 0 and a unique **dummy termination activity** $n + 1$, which indicate the start and the end of the project, respectively. The dummy activities need no resources and have processing time 0. In order to impose $0 \to j \to n+1$ for all activities $j = 1,\ldots,n$ we set $0 \to j$ for all activities j without any predecessor and $j \to n+1$ for all activities j without any successor. Then S_0 is the starting time of the project and $S_{n+1} - S_0$ may be interpreted as the makespan of the project. Usually we set $S_0 := 0$.

If preemption is not allowed, the vector $S = (S_j)$ defines a **schedule** of the project. S is called **feasible** if all resource and precedence constraints are fulfilled.

Example 1 *Consider a project with $n = 4$ activities, $r = 2$ resources with capacities $R_1 = 5, R_2 = 7$, a precedence relation $2 \to 3$ and the following data:*

j	1	2	3	4
p_j	4	3	5	8
r_{j1}	2	1	2	2
r_{j2}	3	5	2	4

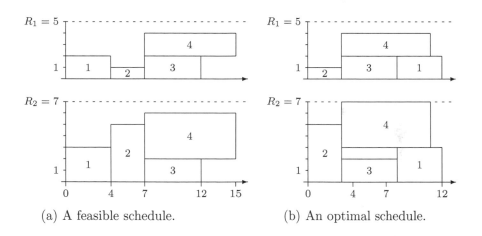

(a) A feasible schedule.   (b) An optimal schedule.

FIGURE 1.1: Two feasible schedules for Example 1.

In Figure 1.1(a) a so-called **Gantt chart** of a feasible schedule with $C_{\max} = 15$ is drawn. This schedule does not minimize the makespan, since by moving activity 1 to the right, a shorter schedule is obtained. An optimal schedule with makespan $C_{\max} = 12$ is shown in (b).

A precedence relation $i \to j$ with the meaning $S_i + p_i \le S_j$ may be generalized by a start-start relation of the form

$$S_i + d_{ij} \le S_j \tag{1.1}$$

with an arbitrary integer number $d_{ij} \in \mathbb{Z}$. The interpretation of relation (1.1) depends on the sign of d_{ij}:

- If $d_{ij} \geq 0$, then activity j cannot start before d_{ij} time units after the start of activity i. This means that activity j does not start before activity i and d_{ij} is a minimal distance (time-lag) between both starting times (cf. Figure 1.2(a)).

- If $d_{ij} < 0$, then the earliest start of activity j is $-d_{ij}$ time units before the start of activity i, i.e., activity i cannot start more than $-d_{ij}$ time units later than activity j. If $S_j \leq S_i$, this means that $|d_{ij}|$ is a maximal distance between both starting times (cf. Figure 1.2(b)).

(a) Positive time-lag.   (b) Negative time-lag.

FIGURE 1.2: Positive and negative time-lags.

If $d_{ij} > 0$ holds, the value is also called a **positive time-lag**; if $d_{ij} < 0$, it is called a **negative time-lag**.

Relations (1.1) are very general timing relations between activities. For example, (1.1) with $d_{ij} = p_i$ is equivalent to the precedence relation $i \rightarrow j$. More generally, besides start-start relations, finish-start, finish-finish, or start-finish relations may be considered. But if no preemption is allowed, any type of these relations can be transformed to any other type. For example, finish-start relations $C_i + l_{ij} \leq S_j$ with finish-start time-lags l_{ij} can be transformed into start-start relations $S_i + p_i + l_{ij} \leq S_j$ by setting $d_{ij} := p_i + l_{ij}$.

Generalized precedence relations may, for example, be used in order to model certain timing restrictions for a chemical process. If $S_i + p_i + l_{ij} \leq S_j$ and $S_j - u_{ij} - p_i \leq S_i$ with $0 \leq l_{ij} \leq u_{ij}$ are required, then the time between the completion time of activity i and the starting time of activity j must be at least l_{ij} but no more than u_{ij}. This includes the special case $0 \leq l_{ij} = u_{ij}$ where activity j must start exactly l_{ij} time units after the completion of activity i. If $l_{ij} = u_{ij} = 0$, then activity j must start immediately after activity i finishes (**no-wait constraint**).

Also release times r_j and deadlines d_j of activities j can be modeled by relations of the form (1.1). While a **release time** r_j is an earliest starting time for activity j, a **deadline** d_j is a latest completion time for j. To model release times we add the restrictions $S_0 + r_j \leq S_j$. To model deadlines we add the restrictions $S_j - (d_j - p_j) \leq S_0$. In both cases we assume that $S_0 = 0$.

If $r_j \leq d_j$ for a release time r_j and a deadline d_j, then the interval $[r_j, d_j]$ is called a **time window** for activity j. Activity j must be processed completely within its time window.

In processor scheduling sometimes also so-called **communication delays** $c_{ij} > 0$ are considered. If two activities i, j with $i \rightarrow j$ are processed on the same processor, no communication overhead arises and the constraint is interpreted as a usual precedence relation $S_i + p_i \leq S_j$. If, on the other hand, i, j are processed on different processors, a communication delay occurs which means that $S_i + p_i + c_{ij} \leq S_j$ has to be satisfied.

In the **multi-mode case** a set $\mathcal{M}_j$ of so-called modes (processing alternatives) is associated with each activity j. The processing time of activity j in mode m is given by p_{jm} and the per period usage of a renewable resource k is given by r_{jkm}. One has to assign a mode to each activity and to schedule the activities in the assigned modes. Multiple modes may, for example, be used in order to model a situation in which an activity can be processed quickly by many workers or more slowly with less people.

In the multi-mode situation, besides renewable resources like processors, machines, or people we may also have so-called **non-renewable resources** like money or energy. While renewable resources are available with a constant amount in each time period again, the availability of non-renewable resources is limited for the whole time horizon of the project. This means that non-renewable resources are consumed, i.e., when an activity j is processed, the available amount R_k of a non-renewable resource k is decreased by r_{jk}.

Non-renewable resources are only important in connection with multi-mode problems, because in the single-mode case the resource requirements of non-renewable resources are schedule independent (the available non-renewable resources may be sufficient or not). On the other hand, in the multi-mode case the resource requirements of non-renewable resources depend on the choice of modes (i.e., feasible and infeasible mode assignments may exist).

Besides the objective of minimizing the makespan $C_{\max} := \max\limits_{j=1}^{n} \{C_j\}$ one may consider other objective functions $f(C_1, \ldots, C_n)$ depending on the completion times of the activities. Examples are the **total flow time** $\sum\limits_{j=1}^{n} C_j$ or more generally the **weighted (total) flow time** $\sum\limits_{j=1}^{n} w_j C_j$ with non-negative weights $w_j \geq 0$.

Other objective functions depend on **due dates** d_j that are associated with the activities. With the **lateness** $L_j := C_j - d_j$, the **tardiness** $T_j := \max\{0, C_j - d_j\}$, and the **unit penalty** $U_j := \begin{cases} 0, \text{ if } C_j \leq d_j \\ 1, \text{ otherwise} \end{cases}$ the following objective functions are common:

the **maximum lateness**	$L_{\max} := \max\limits_{j=1}^{n} L_j$
the **total tardiness**	$\sum\limits_{j=1}^{n} T_j$
the **total weighted tardiness**	$\sum\limits_{j=1}^{n} w_j T_j$
the **number of late activities**	$\sum\limits_{j=1}^{n} U_j$
the **weighted number of late activities**	$\sum\limits_{j=1}^{n} w_j U_j.$

All these objective functions f are monotone non-decreasing in the completion times C_j, i.e., they satisfy $f(C_1, \ldots, C_n) \le f(C'_1, \ldots, C'_n)$ for completion time vectors C, C' with $C_j \le C'_j$ for all $j = 1, \ldots, n$. They are also called **regular**.

1.3 Processor (Machine) Scheduling

In this section we introduce some important classes of processor (machine) scheduling problems (see also Błażewicz et al. [7], Brucker [8]).

Single processor scheduling

In the simplest scheduling model we are given n jobs $J_1, \ldots, J_n$ with processing times p_j $(j = 1, \ldots, n)$ which have to be processed on a single processor. Additionally, precedence constraints may be given. Such a problem can be modeled as an RCPSP with one renewable resource with unit capacity where all jobs require one unit of the resource.

Example 2 *Consider an instance with $n = 5$ jobs, processing times $p_1 = 3$, $p_2 = 2$, $p_3 = 4$, $p_4 = 2$, $p_5 = 3$ and precedence constraints $1 \to 3$, $2 \to 4$, $4 \to 5$. A feasible schedule for this instance with makespan $C_{\max} = 14$ is shown in Figure 1.3.*

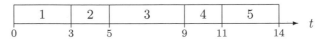

FIGURE 1.3: Single processor schedule.

Parallel processor scheduling

Instead of a single processor we may have m processors $P_1, \ldots, P_m$ on which the jobs have to be processed. If the processors are **identical**, the processing time p_j of job J_j does not depend on the processor on which J_j is processed. This problem corresponds to an RCPSP with one cumulative resource with capacity m where all jobs require one unit of the resource.

Example 3 *Consider an instance with $n = 8$ jobs, $m = 3$ processors, processing times $p_1 = 1$, $p_2 = 3$, $p_3 = 4$, $p_4 = 2$, $p_5 = 2$, $p_6 = 3$, $p_7 = 1$, $p_8 = 5$, and precedence constraints as shown in the left part of Figure 1.4. A feasible schedule with makespan $C_{\max} = 9$ is shown in the right part of the figure.*

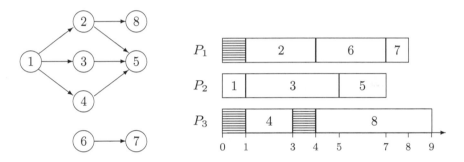

FIGURE 1.4: Schedule for identical parallel processors.

By contrast with the case with identical processors, for so-called **unrelated** processors the processing time p_{jk} of job J_j depends on the processor P_k ($k = 1, \ldots, m$) on which J_j is processed. The processors are called **uniform** if $p_{jk} = p_j/s_k$ where s_k may be interpreted as the speed of processor P_k. Problems with unrelated processors can be modeled as a multi-mode RCPSP with m renewable resources where each resource has unit capacity. Each job J_j has m modes corresponding to the processors on which J_j may be processed. If job J_j is processed in mode k, then p_{jk} is its processing time and J_j uses only one unit of resource k (processor P_k).

A further generalization are scheduling problems with **multi-purpose processors** (flexible machines). In this situation we associate with each job J_j a subset of processors $\mu_j \subseteq \{P_1, \ldots, P_m\}$ indicating that J_j can be executed by any processor of this set. If job J_j is processed on processor P_k, then its processing time is equal to p_{jk} (or simply to p_j if the processing time does not depend on the assigned processor). This problem can be formulated as above as a multi-mode RCPSP with m renewable resources where each job J_j has only $|\mu_j|$ modes corresponding to the processors on which J_j may be processed.

Multi-processor task scheduling

In a **multi-processor task scheduling problem** we have n jobs $J_1, \ldots, J_n$ and m processors $P_1, \ldots, P_m$. Associated with each job J_j is a processing time p_j and a subset of processors $\mu_j \subseteq \{P_1, \ldots, P_m\}$. During its processing job J_j occupies each of the processors in μ_j. Finally, precedence constraints may be given between certain jobs.

This problem can be formulated as an RCPSP with m renewable resources where each resource has unit capacity. Furthermore, job J_j requires one unit of all processors $P_k \in \mu_j$ simultaneously.

Example 4 *Consider the following instance with $n = 5$ jobs and $m = 3$ processors:*

j	1	2	3	4	5
μ_j	$\{P_1, P_2\}$	$\{P_2, P_3\}$	$\{P_1, P_2\}$	$\{P_3\}$	$\{P_1, P_2, P_3\}$
p_j	1	2	2	3	1

In Figure 1.5 a feasible schedule with makespan $C_{\max} = 7$ for this instance is shown. It does not minimize the makespan since by processing job J_1 together with job J_4 we can get a schedule with $C_{\max} = 6$.

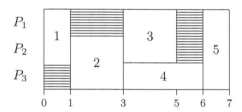

FIGURE 1.5: Feasible schedule for a multi-processor task problem.

In another variant of multi-processor task problems all processors are assumed to be identical and each job can be processed by any processor. With each job J_j an integer number $size_j$ is associated with the meaning that during its processing J_j needs $size_j$ processors simultaneously.

Shop scheduling

In so-called shop scheduling problems the jobs consist of several operations which have to be processed on different processors. In **general-shop scheduling problems** we have jobs $J_1, \ldots, J_n$ and m processors $P_1, \ldots, P_m$. Job J_j consists of n_j operations $O_{1j}, \ldots, O_{n_j,j}$. Two operations of the same job cannot be processed at the same time and a processor can process at most one operation at any time. Operation O_{ij} must be processed for p_{ij} time units on a dedicated processor from the set $\{P_1, \ldots, P_m\}$. Furthermore, precedence constraints may be given between arbitrary operations.

Such a general-shop scheduling problem can be modeled as an RCPSP with $m + n$ renewable resources where each resource has unit capacity. While the resources $k = 1, \ldots, m$ correspond to the processors, the resources $m + j$ ($j = 1, \ldots, n$) are needed to model the fact that different operations of job J_j cannot be processed at the same time. Furthermore, we have $\sum_{j=1}^{n} n_j$ activities O_{ij}, where operation O_{ij} needs one unit of the "processor resource" corresponding to the dedicated processor of O_{ij} and one unit of the "job resource" $m + j$.

Important special cases of the general-shop scheduling problem are job-shop, flow-shop, and open-shop problems, which will be discussed next.

A **job-shop problem** is a general-shop scheduling problem with chain precedences of the form

$$O_{1j} \to O_{2j} \to \ldots \to O_{n_j,j}$$

for $j = 1, \ldots, n$ (i.e., there are no precedences between operations of different jobs and the precedences between operations of the same job build a chain). Note that for a job-shop problem no "job resource" is needed, since all operations of the same job are linked by a precedence relation (and thus cannot be processed simultaneously).

A **flow-shop problem** is a special job-shop problem with $n_j = m$ operations for $j = 1, \ldots, n$ where operation O_{ij} must be processed on P_i. In a so-called **permutation flow-shop problem** the jobs have to be processed in the same order on all processors.

Example 5 *In Figure 1.6 a feasible schedule for a permutation flow-shop problem with $n = 4$ jobs and $m = 3$ processors is shown.*

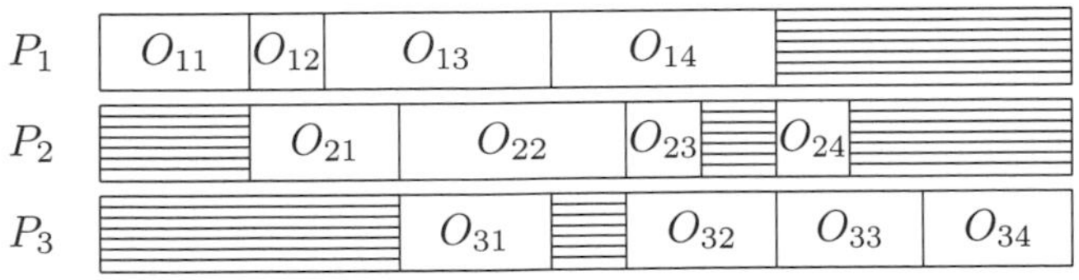

FIGURE 1.6: Permutation schedule for a flow-shop problem.

An **open-shop problem** is like a flow-shop problem but without any precedence relations between the operations. Thus, it also has to be decided in which order the operations of a job are processed.

Example 6 *In Figure 1.7 a feasible schedule for an open-shop problem with* $n = 2$ *jobs and* $m = 3$ *processors is shown. In this schedule the operations of job* J_1 *are processed in the order* $O_{11} \rightarrow O_{31} \rightarrow O_{21}$, *the operations of job* J_2 *are processed in the order* $O_{32} \rightarrow O_{22} \rightarrow O_{12}$.

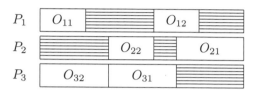

FIGURE 1.7: Feasible schedule for an open-shop problem.

A classification scheme

Classes of scheduling problems are often specified in terms of a three-field classification $\alpha|\beta|\gamma$ introduced by Graham et al. [12]. The first entry α specifies the **processor environment**, β specifies **job characteristics**, and γ denotes an **optimality criterion** (objective function).

The first entry α specifies the processor environment, where 1 stands for a single processor, P denotes parallel identical, Q uniform, and R unrelated processors. For open-shop, flow-shop, and job-shop problems the symbols O, F, and J are used. If the number m of processors is fixed (i.e., not part of the input), this number m is added to the α-field (e.g., $P3$ specifies a problem with three identical processors, Pm denotes a parallel processor problem with a fixed number m of processors, and $F2$ denotes a two-processor flow-shop problem).

The β-field specifies job characteristics concerning processing times, release dates, precedences, etc. The entry $p_j = p$ $(p_{ij} = p)$ denotes that all jobs (operations) have constant processing times and $p_j = 1$ denotes that they are all equal to one. Furthermore, r_j indicates release dates for the jobs and $pmtn$ denotes that preemption of the jobs is allowed. If precedence relations are given, the entry *prec* is used (or *chains, tree, sp-graph* if the precedences have a special structure). Additionally, in a multi-processor task environment with parallel processors, the entry $size_j$ says that a job needs several processors simultaneously. Communication delays are indicated by an entry c_{ij} (or $c_{ij} = c$, $c_{ij} = 1$ for constant and unit communication delays, respectively).

Finally, γ specifies the objective function. It is set to $C_{\max}, L_{\max}, \sum C_j$,

$\sum w_j C_j$, $\sum U_j$, $\sum w_j U_j$, $\sum T_j$, $\sum w_j T_j$ to denote one of the objective functions introduced in Section 1.2.

In order to illustrate the $\alpha|\beta|\gamma$-notation three examples are presented:

- $1 \mid r_j; pmtn \mid L_{\max}$ denotes a problem with a single processor, where the jobs have release dates, preemption is allowed, and the maximum lateness has to be minimized.

- $P2 \mid p_j = p; r_j; tree \mid C_{\max}$ denotes a parallel processor problem with two processors, where the jobs have release dates and the same processing time p, precedence constraints in form of a tree are given and the makespan has to be minimized.

- $Jm \mid p_{ij} = 1 \mid \sum w_j U_j$ denotes a job-shop problem with a fixed number of processors, where all operations have unit processing times and the weighted number of late jobs has to be minimized.

1.4 Easy and Hard Problems

When we consider scheduling problems (or more generally combinatorial optimization problems) an important issue is the complexity of a problem. For a new problem we often first try to develop an algorithm that solves the problem in an efficient way. If we cannot find such an algorithm, it may be helpful to prove that the problem is NP-hard, which implies that most likely no efficient algorithm exists. In this section we review the most important aspects of complexity theory (cf. also Garey and Johnson [11]) which allow us to decide whether a problem is "easy" or "hard".

A computational problem can be viewed as a function h which maps each input x in some given domain to an output $h(x)$ in some given range. Of interest are algorithms which solve computational problems, i.e., compute $h(x)$ for each input x. The length $|x|$ of an input x is defined as the length of some encoding of x. The efficiency of an algorithm may be measured by its running time, i.e., the number of steps it needs for a certain input. Since it is reasonable that this number is larger for a larger input, an efficiency function should depend on the input size $|x|$. The size of an input for a computer program is usually defined by the length of a **binary encoding** of the input. For example, in a binary encoding an integer number a is represented as a binary number using $\log_2 a$ bits, an array with m numbers needs $m \log_2 a$ bits when a is the largest number in the array. On the other hand, in a so-called **unary encoding** a number a is represented by a bits.

Since in most cases it is very difficult to determine the average running time of an algorithm for an input with size n, often the worst-case running time of an algorithm is studied. For this purpose, the function $T(n)$ may be defined as an upper bound on the running time of the algorithm on any input x with size $|x| = n$. Since often it is also difficult to give a precise description of $T(n)$, we only consider the growth rate of $T(n)$ which is determined by its asymptotic order. We say that a function $T(n)$ is $O(g(n))$ if constants $c > 0, n_0 \in \mathbb{N}$ exist such that $T(n) \leq cg(n)$ for all $n \geq n_0$. For example, the function $T_1(n) = 37n^3 + 4n^2 + n$ is $O(n^3)$, the function $T_2(n) = 2^n + n^{100} + 4$ is $O(2^n)$.

If the running time of an algorithm is bounded by a polynomial function in n, i.e., if it is $O(n^k)$ for some constant $k \in \mathbb{N}$, the algorithm is called a **polynomial-time** algorithm. If the running time is bounded by a polynomial function in the input size of a unary encoding, an algorithm is called **pseudo-polynomial**. For example, an algorithm with running time $O(n^2 a)$ is pseudo-polynomial if the input size of a binary encoding is $O(n \log a)$.

As an example for a polynomial-time algorithm we consider the scheduling problem $1 || \sum w_j C_j$.

Example 7 *The input x of problem $1 || \sum w_j C_j$ consists of the number n of jobs and two n-dimensional vectors (p_j) and (w_j). Thus, the input length $|x|$ in a binary encoding is given by $|x| = n \log_2 a$ where $a := \max\limits_{j=1}^{n}\{p_j, w_j\}$ denotes the largest numeric value in the input. The output $h(x)$ for this problem is a schedule S minimizing $\sum\limits_{j=1}^{n} w_j C_j$. It can be represented by a vector of the completion times C_j for all jobs J_j. The following algorithm of Smith [27], also called Smith's weighted shortest processing time first (WSPT) rule, calculates optimal C_j-values:*

Algorithm $1 || \sum w_j C_j$
1. Enumerate the jobs such that $w_1/p_1 \geq w_2/p_2 \geq \ldots \geq w_n/p_n$;
2. $C_0 := 0$;
3. FOR $j := 1$ TO n DO
   $C_j := C_{j-1} + p_j$;

The number of computational steps in this algorithm can be bounded as follows. In Step 1 the jobs have to be sorted, which takes $O(n \log n)$ steps. Furthermore, Step 3 can be done in $O(n)$ time. Thus, $T(n) \in O(n \log n)$. If n is replaced by the input length $|x| = n \log_2 a \geq n$, the bound is still valid.

Finally, we show that each schedule constructed by the algorithm is optimal. Clearly, in an optimal schedule starting at time zero the jobs are scheduled one after the other without idle times on the machine. It has to be shown that scheduling the jobs in an order of non-increasing ratios $\frac{w_j}{p_j}$ is optimal. This can be done by proving that if in a schedule S a job J_i is scheduled immediately before a job J_j with $\frac{w_i}{p_i} \leq \frac{w_j}{p_j}$ (or, equivalently, $p_j w_i \leq p_i w_j$), swapping J_i and J_j does not increase the objective value. Let S_i be the starting time of job J_i in S. Then, the contribution of the jobs J_i, J_j in S to the objective value $\sum w_j C_j$ is $c(S) := (S_i + p_i)w_i + (S_i + p_i + p_j)w_j$. Let S' be the schedule derived from S by swapping J_i and J_j. The contribution of the jobs J_i, J_j in S' to the objective value is $c(S') := (S_i + p_j)w_j + (S_i + p_j + p_i)w_i$, which is not greater than $c(S)$.

To see that a schedule constructed by the algorithm is optimal assume that the jobs are enumerated as in Step 1 and consider an arbitrary optimal schedule S. If in S job J_1 is not scheduled as the first job, J_1 can be swapped with its predecessors until it is scheduled in the first position. Due to $\frac{w_1}{p_1} \geq \frac{w_j}{p_j}$ for all jobs J_j, the objective value does not increase. Next, job J_2 will be moved to the second position, etc. Finally, the schedule constructed by the algorithm is reached, which has the same objective value as the optimal schedule S.

Since exponential functions grow much faster than polynomial functions, exponential-time algorithms cannot be used for larger problems. If we compare two algorithms with running times $O(n^3)$ and $O(2^n)$ under the assumption that a computer can perform 10^9 steps per second, we get the following result: while for $n = 1000$ the first algorithm is still finished after one second, the second algorithm needs already 18 minutes for $n = 40$, 36 years for $n = 60$ and even 374 centuries for $n = 70$.

For this reason a problem is classified as "easy" (tractable) if it can be solved by a polynomial-time algorithm. In order to classify a problem as "hard" (intractable), the notion of NP-hardness was developed which will be discussed next.

A problem is called a **decision problem** if its answer is only "yes" or "no". An instance for which the answer is "yes" is also called a "yes"-instance, an instance for which the answer is "no" a "no"-instance. A famous example for a decision problem is the so-called **partition problem**: Given n positive integers $a_1, \ldots, a_n$ and the value $b := \frac{1}{2} \sum_{i=1}^{n} a_i$, is there a subset $A \subset \{1, \ldots, n\}$ such that $\sum_{i \in A} a_i = \sum_{i \notin A} a_i = b$?

On the other hand, most scheduling problems belong to the class of **optimization problems**, where the answer is a feasible solution which minimizes (or maximizes) a given objective function c. For each scheduling problem a corresponding decision problem may be defined by asking whether there exists a feasible schedule S with $c(S) \leq y$ for a given threshold value y.

The class of all polynomially solvable decision problems is denoted by $\mathcal{P}$. Another important complexity class is the set $\mathcal{NP}$ which contains all decision problems that are nondeterministically polynomially solvable. An equivalent definition of this class is based on so-called certificates which can be verified in polynomial time.

A **certificate** for a "yes"-instance of a decision problem is a piece of information which proves that a "yes"-answer for this instance exists. For example, for the partition problem a set A with $\sum_{i \in A} a_i = b$ is a certificate for a "yes"-instance. For a decision problem corresponding to a scheduling problem a feasible schedule S with $c(S) \leq y$ provides a certificate for a "yes"-instance.

If such a certificate is short and can be checked in polynomial time, the corresponding decision problem belongs to the class $\mathcal{NP}$. More precisely, the set $\mathcal{NP}$ contains all decision problems where each "yes"-instance I has a certificate which

- has a length polynomially bounded in the input size of I, and

- can be verified by a polynomial-time algorithm.

Note that there is an important asymmetry between "yes"- and "no"-instances. While a "yes"-answer can easily be certified by a single feasible schedule S with $c(S) \leq y$, a "no"-answer cannot be certified by a short piece of information.

It is easy to see that every polynomially solvable decision problem belongs to $\mathcal{NP}$ (the output of the polynomial-time algorithm may be used as a certificate which can also be verified by the algorithm in polynomial time). Thus, $\mathcal{P} \subseteq \mathcal{NP}$. One of the most challenging open problems in theoretical computer science today is the question whether $\mathcal{P} = \mathcal{NP}$ holds. Although it is generally conjectured that $\mathcal{P} \neq \mathcal{NP}$ holds, up to now nobody was able to prove this conjecture.

The central issue for the definition of NP-hardness is the notion of a polynomial **reduction**. A decision problem P is said to be **polynomially reducible** to another decision problem Q (denoted by $P \propto Q$) if a polynomial-time computable function g exists which transforms every instance I for P into an instance $I' = g(I)$ for Q such that I is a "yes"-instance for P if and only if I' is a "yes"-instance for Q.

A polynomial reduction $P \propto Q$ between two decision problems P, Q implies that Q is at least as difficult as P. This means that if Q is polynomially solvable, then also P is polynomially solvable. Equivalently, if P is not polynomially solvable, then also Q cannot be solved in polynomial time. Furthermore, reducibility is transitive, i.e., if $P \propto Q$ and $Q \propto R$ for three decision problems P, Q, R, then also $P \propto R$.

A decision problem P is called **NP-complete** if

(i) P belongs to the class $\mathcal{NP}$, and

(ii) every other decision problem $Q \in \mathcal{NP}$ is polynomially reducible to P.

A problem P is called **NP-hard** if only (ii) holds. Especially, an optimization problem is NP-hard if the corresponding decision problem is NP-complete.

If any single NP-complete problem could be solved in polynomial time, then all problems in $\mathcal{NP}$ would also be polynomially solvable, i.e., we would have $\mathcal{P} = \mathcal{NP}$. Since nobody has found a polynomial-time algorithm for any NP-complete problem yet, most likely no polynomial-time algorithm exists for these problems.

An extension of NP-completeness is the concept of strongly NP-complete problems. If a problem P is **strongly NP-complete**, then it is unlikely that it can be solved by a pseudo-polynomial algorithm, since this would imply $\mathcal{P} = \mathcal{NP}$.

In 1971 Steve Cook was the first to prove that NP-complete problems exist. He showed that the so-called **satisfiability problem** (in which it has to be decided whether a given boolean formula can be satisfied or not) is NP-complete. This result together with the transitivity of reducibility implies that for an NP-completeness proof of a decision problem $P \in \mathcal{NP}$ it is sufficient to show that

(ii') **one** known NP-complete problem $Q \in \mathcal{NP}$ is polynomially reducible to P

(since then we have $R \propto P$ for **all** problems $R \in \mathcal{NP}$ due to $R \propto Q \propto P$).

In this way the partition problem and other fundamental combinatorial problems have been shown to be NP-complete by reductions from the satisfiability problem. Usually, it is easy to show that P belongs to the class $\mathcal{NP}$. Thus, the main task in proving NP-completeness of a problem P is to find a known NP-complete problem Q which can be reduced to P.

The following example shows a reduction from the partition problem to the three-machine flow-shop scheduling problem $F3||C_{\max}$.

Example 8 *The partition problem is reducible to the decision version of problem $F3||C_{\max}$. To prove this reduction, consider a three-machine flow-shop problem with $n + 1$ jobs $J_1, \ldots, J_{n+1}$, where the processing times of the operations are given by*

$$p_{1j} := 0, \quad p_{2j} := a_j, \quad p_{3j} := 0 \quad for \ j = 1, \ldots, n;$$
$$p_{1,n+1} = p_{2,n+1} = p_{3,n+1} := b = \tfrac{1}{2} \sum_{j=1}^{n} a_j.$$

and the threshold for the corresponding decision problem is set to $y := 3b$. Obviously, the input of the flow-shop problem can be computed in polynomial time from the input of the partition problem.

If the partition problem has a solution, then an index set $A \subseteq \{1, \ldots, n\}$ exists such that $\sum_{j \in A} a_j = b$. In this case the schedule shown in Figure 1.8 solves the decision version of problem $F3||C_{\max}$.

If, on the other hand, the flow-shop problem has a solution with $C_{\max} \leq 3b$, then job $n + 1$ must be scheduled as shown in Figure 1.8. Furthermore, the set $A = \{j \mid job \ J_j \ finishes \ not \ later \ than \ b\}$ solves the partition problem.

Many reductions exist between scheduling problems. For example, problems with unit processing times $p_j = 1$ are a special case of constant processing times $p_j = p$ which can be reduced to the general case of arbitrary processing times p_j. Problems without release dates may be reduced to problems with arbitrary release dates by setting $r_j = 0$ for all j. Furthermore, problem $1||\sum C_j$ reduces to $1||\sum w_j C_j$ because any instance of $1||\sum C_j$ defines also

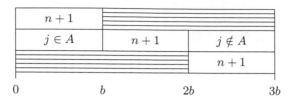

FIGURE 1.8: Solution of $F3||C_{\max}$.

an instance of $1||\sum w_j C_j$ by setting $w_j := 1$ for all $j = 1,\ldots,n$. These reductions are called elementary.

Other elementary reductions for objective functions can be found in Figure 1.9. In it we use the results that $\sum f_j$ reduces to $\sum w_j f_j$ by setting $w_j = 1$ for all j. Furthermore, $C_{\max}, \sum C_j$, and $\sum w_j C_j$ reduce to $L_{\max}, \sum T_j$, and $\sum w_j T_j$, respectively, by setting $d_j = 0$ for all j.

Finally, for any threshold value y we have

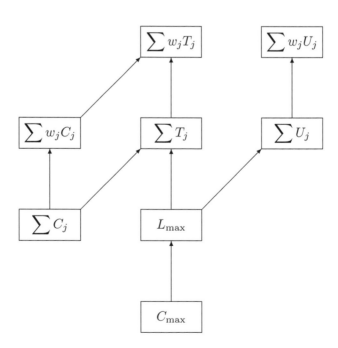

FIGURE 1.9: Elementary reductions for objective functions.

$L_{\max} \leq y \Leftrightarrow C_j - d_j \leq y$ for all j
$\qquad \Leftrightarrow C_j - (d_j + y) \leq 0$ for all j
$\qquad \Leftrightarrow \max\{0, C_j - (d_j + y)\} \leq 0$ for all j
$\qquad \Leftrightarrow \sum T_j = \sum \max\{0, C_j - (d_j + y)\} \leq 0$
$\qquad \Leftrightarrow \sum U_j \leq 0.$

Thus, $L_{\max}$ reduces to $\sum T_j$ and $\sum U_j$.

For a single processor in the case of a regular objective function it can be shown that if no release dates are given or if all processing times are equal to one, allowing preemption does not lead to better solutions (see e.g., Brucker [8]). Thus, for a single processor we get reductions between characteristics in the β-field as shown in Figure 1.10. Note that the symbol ∘ denotes that the corresponding entry is empty, i.e., no precedence constraints in Figure 1.10(a); no release dates, no premption allowed, and arbitrary processing times in (b).

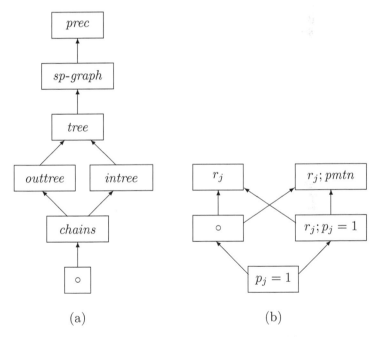

(a)   (b)

FIGURE 1.10: Reductions for single processor problems.

1.5   Complexity Classification of Scheduling Problems

In this section all known complexity results for the class of single processor problems are presented. We distinguish between "easy" and "hard" problems which correspond to problems which are polynomially solvable and NP-hard, respectively.

Of interest are the hardest problems in the class which have been shown to be polynomially solvable, the simplest problems which are already known to be NP-hard, as well as the simplest and hardest problems for which the complexity status is unknown.

Based on the elementary reductions (see Figures 1.9 and 1.10), known polynomial-time algorithms and known NP-hardness results in Table 1.1 all complexity results for problems with a single processor are summarized. We distinguish the following subclasses:

- Maximal polynomially solvable problems: A problem P is maximal polynomially solvable if no elementary reduction from P to another problem Q exists, where Q has been shown to be polynomially solvable. For example, problem $1||\sum w_j C_j$ is not maximal polynomially solvable since its generalization $1|sp\text{-}graph|\sum w_j C_j$ is still polynomially solvable.

- Minimal NP-hard problems: A problem P is minimal NP-hard if no elementary reduction from another problem Q to P exists, where Q has been shown to be NP-hard. For example, problem $1|r_j|\sum w_j T_j$ is not minimal NP-hard since its special case $1|r_j|\sum C_j$ is already NP-hard. Problems that are strongly NP-hard are marked with $*$.

- Minimal and maximal open problems.

Results and corresponding literature references for other classes of processor scheduling problems (parallel processor problems, shop problems, multiprocessor task problems, and several generalized problems) can be found at our website:

    http://www.mathematik.uni-osnabrueck.de/research/OR/class/.

maximal polynomially solvable:	
$1\|prec; r_j\|C_{\max}$	Lawler [15]
$1\|prec; p_j = p; r_j\|L_{\max}$	Simons [25]
$1\|prec; r_j; pmtn\|L_{\max}$	Błażewicz [6], Baker et al. [2]
$1\|prec; p_j = p; r_j\|\sum C_j$	Simons [26]
$1\|prec; pmtn; p_j = p; r_j\|\sum C_j$	Baptiste et al. [5]
$1\|r_j; pmtn\|\sum C_j$	Baker [1]
$1\|p_j = p; r_j\|\sum w_j C_j$	Baptiste [4]
$1\|sp\text{-}graph\|\sum w_j C_j$	Lawler [17]
$1\|r_j; pmtn\|\sum U_j$	Lawler [18]
$1\|p_j = p; r_j\|\sum w_j U_j$	Baptiste [3]
$1\|pmtn; p_j = p; r_j\|\sum w_j U_j$	Baptiste [3]
$1\|p_j = p; r_j\|\sum T_j$	Baptiste [4]
$1\|pmtn; p_j = p; r_j\|\sum T_j$	Tian et al. [28]
$1\|p_j = 1; r_j\|\sum w_j T_j$	Assignment problem

	minimal NP-hard:	
*	$1\|r_j\|L_{\max}$	Lenstra et al. [23]
*	$1\|chains; r_j; pmtn\|\sum C_j$	Lenstra [20]
*	$1\|prec\|\sum C_j$	Lawler [17], Lenstra and Rinnooy Kan [21]
*	$1\|r_j\|\sum C_j$	Lenstra et al. [23]
*	$1\|chains; p_j = 1; r_j\|\sum w_j C_j$	Lenstra and Rinnooy Kan [22]
*	$1\|prec; p_j = 1\|\sum w_j C_j$	Lawler [17], Lenstra and Rinnooy Kan [21]
*	$1\|r_j; pmtn\|\sum w_j C_j$	Labetoulle et al. [14]
*	$1\|chains; p_j = 1\|\sum U_j$	Lenstra and Rinnooy Kan [22]
	$1\|\|\sum w_j U_j$	Lawler and Moore [19], Karp [13]
	$1\|\|\sum T_j$	Lawler [16], Du and Leung [10]
*	$1\|chains; p_j = 1\|\sum T_j$	Leung and Young [24]
*	$1\|\|\sum w_j T_j$	Lawler [16], Lenstra et al. [23]

minimal open:	maximal open:
$1\|pmtn; p_j = p; r_j\|\sum w_j C_j$	$1\|pmtn; p_j = p; r_j\|\sum w_j T_j$
$1\|p_j = p; r_j\|\sum w_j T_j$	$1\|p_j = p; r_j\|\sum w_j T_j$

Table 1.1: Complexity results for problems with a single processor.

References

[1] K. Baker. *Introduction to Sequencing and Scheduling.* John Wiley & Sons, New York, 1974.

[2] K. Baker, E. Lawler, J. Lenstra, and A. Rinnooy Kan. Preemptive scheduling of a single machine to minimize maximum cost subject to release dates and precedence constraints. *Operations Research*, 31:381–386, 1983.

[3] P. Baptiste. Polynomial time algorithms for minimizing the weighted number of late jobs on a single machine with equal processing times. *Journal of Scheduling*, 2:245–252, 1999.

[4] P. Baptiste. Scheduling equal-length jobs on identical parallel machines. *Discrete Applied Mathematics*, 103(1):21–32, 2000.

[5] P. Baptiste, P. Brucker, S. Knust, and V. Timkovsky. Ten notes on equal-execution-time scheduling. *4OR*, 2:111–127, 2004.

[6] J. Błażewicz. Scheduling dependent tasks with different arrival times to meet deadlines. In *Model. Perform. Eval. Comput. Syst., Proc. int. Workshop, Stresa 1976*, pages 57–65. North Holland, Amsterdam, 1976.

[7] J. Błażewicz, K. Ecker, E. Pesch, G. Schmidt, and J. Węglarz. *Scheduling Computer and Manufacturing Processes.* Springer, Berlin, 2nd edition, 2001.

[8] P. Brucker. *Scheduling Algorithms.* Springer, Berlin, 5th edition, 2007.

[9] P. Brucker and S. Knust. *Complex Scheduling.* Springer, Berlin, 2006.

[10] J. Du and J.-T. Leung. Minimizing total tardiness on one machine is NP-hard. *Mathematics of Operations Research*, 15(3):483–495, 1990.

[11] M. R. Garey and D. S. Johnson. *Computers and Intractability. A Guide to the Theory of NP-Completeness.* W. H. Freeman & Co, San Francisco, 1979.

[12] R. Graham, E. Lawler, J. Lenstra, and A. Rinnooy Kan. Optimization and approximation in deterministic sequencing and scheduling: A survey. *Annals of Discrete Mathematics*, 5:287–326, 1979.

[13] R. Karp. Reducibility among combinatorial problems. In R. E. Miller and J. W. Thatcher, editors, *Complexity of computer computations (Proc. Sympos., IBM Thomas J. Watson Res. Center, Yorktown Heights, N.Y., 1972)*, pages 85–103. Plenum, New York, 1972.

[14] J. Labetoulle, E. Lawler, J. Lenstra, and A. Rinnooy Kan. Preemptive scheduling of uniform machines subject to release dates. In *Progress in combinatorial optimization (Waterloo, Ont., 1982)*, pages 245–261. Academic Press, Toronto, Ont., 1984.

[15] E. Lawler. Optimal sequencing of a single machine subject to precedence constraints. *Management Science*, 19:544–546, 1973.

[16] E. Lawler. A "pseudopolynomial" algorithm for sequencing jobs to minimize total tardines. *Annals of Discrete Mathematics*, 1:331–342, 1977.

[17] E. Lawler. Sequencing jobs to minimize total weighted completion time subject to precedence constraints. *Annals of Discrete Mathematics*, 2:75–90, 1978.

[18] E. Lawler. A dynamic programming algorithm for preemptive scheduling of a single machine to minimize the number of late jobs. *Annals of Operations Research*, 26(1-4):125–133, 1990.

[19] E. Lawler and J. Moore. A functional equation and its application to resource allocation and sequencing problems. *Management Science*, 16:77–84, 1969.

[20] J. Lenstra. Not published.

[21] J. Lenstra and A. Rinnooy Kan. Complexity of scheduling under precedence constraints. *Operations Research*, 26(1):22–35, 1978.

[22] J. Lenstra and A. Rinnooy Kan. Complexity results for scheduling chains on a single machine. *European Journal of Operational Research*, 4(4):270–275, 1980.

[23] J. Lenstra, A. Rinnooy Kan, and P. Brucker. Complexity of machine scheduling problems. *Annals of Discrete Mathematics*, 1:343–362, 1977.

[24] J.-T. Leung and G. Young. Minimizing total tardiness on a single machine with precedence constraints. *ORSA Journal on Computing*, 2(4):346–352, 1990.

[25] B. Simons. A fast algorithm for single processor scheduling. In *19th Annual Symposium on Foundations of Computer Science (Ann Arbor, Mich., 1978)*, pages 246–252. IEEE, Long Beach, Calif., 1978.

[26] B. Simons. Multiprocessor scheduling of unit-time jobs with arbitrary release times and deadlines. *SIAM Journal on Computing*, 12(2):294–299, 1983.

[27] W. Smith. Various optimizers for single-stage production. *Naval Research Logistics Quarterly*, 3:59–66, 1956.

[28] Z. Tian, C. Ng, and T. Cheng. An $O(n^2)$ algorithm for scheduling equal-length preemptive jobs on a single machine to minimize total tardiness. *Journal of Scheduling*, 9(4):343–364, 2006.

Chapter 2

Approximation Algorithms for Scheduling Problems

Jean-Claude König

LIRMM

Rodolphe Giroudeau

LIRMM

2.1 Introduction

In this chapter we show how to develop and analyze polynomial-time approximation algorithms with performance guarantees for scheduling problems. For a problem which is classified as NP-complete, several approaches may be used in order to propose an optimal solution or a solution close to the optimum. When looking at an optimal solution, we may recall the three classical exact methods: branch and bound, branch and cut, and dynamic programming.

In the following, we are interested in polynomial-time approximation algorithms that return a feasible solution whose measure is not too far from the optimum. Moreover, we want the gap between the two solutions to be bounded by a constant $\delta \in \mathbb{R}$. Also, approximation algorithms, designed to obtain a feasible solution, should admit a low complexity. The theory of approximation is needed to refine the classification given by complexity theory, which introduces NP-hardness (completeness) as a concept for proving the intractability of computing problems. Indeed, in complexity theory, problems are classified according to the difficulty to evaluate/determine an optimal solution, whereas the theory of approximation studies the behavior of polynomial-time algorithms from the view point of searching for approximated solutions. Approximation algorithms seem to be the most successful approach for solving hard optimization problems.

We will first recall the fundamentals of the concept of approximation of optimization problems, including the classification of the optimization problems according to their polynomial-time approximability. Second, we will present some examples of approximation algorithms in order to show successful techniques for designing efficient approximation algorithms. Moreover, we will give some basic ideas of methods to establish lower bounds on polynomial-time inapproximability.

DEFINITION 2.1 Performance ratio   *For a minimization problem, the performance ratio ρ^h of an algorithm h is defined as $\rho^h = max_I \frac{\mathcal{K}(I)}{\mathcal{K}^{opt}(I)}$ where*

- *I is an instance,*
- *$\mathcal{K}^h(I)$ (resp. $\mathcal{K}^{opt}(I)$) is the measure of the solution given by heuristic h (resp. an optimal) for instance I.*

Remarks:

- *We always have $\rho^h \geqslant 1$.*
- *For a maximization problem, the* max *operator is replaced by a* min *operator in the definition and the performance ratio is then always not greater than 1. In this chapter we only consider minimization problems.*

2.1.1   Approximation Algorithms

Let P be an NP-hard problem. By the theory of approximation we know that, whatever the value $\rho \geqslant 1$, either there exists a polynomial-time ρ-approximation solving P, or there does not exist any such polynomial-time ρ-approximation. We thus consider the two sets R_P and T_P defined as follows:

- $T_P = \{\rho \mid \nexists$ a polynomial-time ρ-approximation$\}$

- $R_P = \{\rho \mid \exists$ a polynomial-time ρ-approximation$\}$

Note that both sets are potentially unbounded, and $R_P \cup T_P = \mathbb{R}$. With this definition, in the case $T_P = \{1\}$ there exists a polynomial-time approximation scheme (PTAS) for problem P. Indeed, we have $R_P =]1, \infty[$. On the contrary, if $T_P = \mathbb{R}$ then there is no polynomial-time approximation with performance guarantee ($P \notin \mathcal{APX}$). In general, we have $T_p = [1, \rho_P)$ and $R_P = (\rho_P, \infty[$. Note that the value of ρ_P is unknown. The aim is to find on the first hand the smallest value ρ_{sup}^P such that $\rho_{sup}^P \geqslant \rho_P$ by developing a ρ_{sup}^P−approximation algorithm for the problem P. On the other hand, we try to find the largest value ρ_{inf}^P such that $\rho_{inf}^P \leqslant \rho_P$ by showing that the existence of polynomial-time ρ-approximation algorithm would imply, for instance, $\mathcal{NP} = \mathcal{P}$ in complexity theory.

Note that we know the exact value of ρ_P for a problem P if $\rho_{inf}^P = \rho_{sup}^P$. Note also that if $\exists c > 0$, such that $\rho_{inf}^P \geqslant 1 + 1/c$ then $P \notin PT\mathcal{AS}$, and if

$\exists \rho_{sup}^P$ then $P \in \mathcal{APX}$.

2.1.2 Definitions

In this section we give some basic results in complexity theory and approximation with performance guarantee. The two classical methods to obtain a lower bound for nonapproximation algorithms are given by the following results, the "Impossibility Theorem" [7] and the gap technique (see [1] for more details). We denote by A an approximation algorithm.

THEOREM 2.1 Impossibility Theorem
Consider a combinatorial optimization problem for which all feasible solutions have non-negative integer objective function values (in particular scheduling problems). Let c be a fixed positive integer. Suppose that the problem of deciding if there exists a feasible solution of value at most c is NP-complete. Then, for any $\rho < (c+1)/c$ there does not exist a polynomial-time ρ-approximation algorithm unless $\mathcal{P} = \mathcal{NP}$. In particular, there does not exist a polynomial-time approximation scheme.

THEOREM 2.2 Gap Theorem
Let Q' be an NP-complete decision problem and let Q be an NPO [1] minimization problem. Let us suppose that there exist two polynomial-time computable functions $f : I_{Q'} \rightarrow I_Q$ and $d : I_{Q'} \rightarrow \mathbb{N}$ and a constant gap > 0 such that, for any instance x of Q',

$$\begin{cases} \text{OPT}(f(x)) = d(x) \ \text{if } x \text{ is a positive instance} \\ \text{OPT}(f(x)) \geqslant d(x)(1 + gap) \ \text{otherwise} \end{cases}$$

where OPT associates to each instance of I_Q its measure under an optimal solution. Then no polynomial-time ρ-approximation algorithm for Q with $\rho < 1 + gap$ can exist, unless $\mathcal{P} = \mathcal{NP}$.

The rest of the chapter is organized as follows. In the next section we propose two absolute approximation algorithms, one for the edge-coloring problem, and one for the k-center problem. In Section 2.2 we develop a PTAS for the problem of scheduling a set of independent tasks on two processors. In Section 2.3 we introduce precedence constraints. In the last section we present the classical scheduling communication delay model. For this model several problems are presented, and several relaxation of instances are considered (task duplication, unbounded number of processors, etc.).

---

[1]$\mathcal{NPO}$ is the class of optimization problems such that the set of instances I_P of a problem $P \in \mathcal{NPO}$ is recognizable in polynomial time, the measure function is computable in polynomial time, and the value of a feasible solution is solvable in polynomial time.

2.1.3 Absolute Approximation Algorithms (Case $\rho_{inf} = \rho_{sup}$)

2.1.3.1 The Edge-Coloring Problem

Instance: Graph $G = (V, E)$

Question: Find an edge coloring of E , i.e., a partition of E into disjoint sets $E_1, E_2, \ldots, E_k$ such that, for $1 \leqslant i \leqslant k$, no two edges in E_i share a common endpoint in G (each set E_i induces a matching of G). The aim is to find the minimum value of k. This minimum value is called the edge chromatic number of G.

THEOREM 2.3

The problem to decide whether a given simple graph has an edge chromatic number equal to 3 is NP-complete, even when the graph has a maximum degree of 3 (see [11]).

COROLLARY 2.1

It is impossible to find a ρ-polynomial-time approximation algorithm with $\rho < 4/3$.

PROOF It is sufficient to use the Impossibility Theorem. Since for $k \geqslant 3$ the problem is NP-complete we have $\rho \leqslant \frac{k+1}{k} \leqslant 1 + \frac{1}{k} \leqslant 4/3$. Note that for $k \leqslant 2$ the problem to decide whether a simple graph has an edge-chromatic number equal to 2 is polynomial. ▯

THEOREM 2.4 Vizing's Theorem

Let G be an undirected simple graph with maximum degree k. Then G has an edge-coloring with at most $k + 1$ colors, and such a coloring can be found in polynomial time (see [22]).

Vizing's Theorem implies an absolute approximation algorithm for the edge coloring problem.

2.1.3.2 The k-Center Problem

The k-center problem is a classical combinatorial optimization problem. In the same way as previously, this problem admits an absolute approximation algorithm. Let first recall the problem:

Instance: A set V of n sites, a matrix D where d_{ij} is the distance between sites i and j, an integer k, and $\forall i, j, l \in V \times V \times V : d_{ij} \leqslant d_{il} + d_{lj}$.

Question: Find a set $S \subseteq V$, with $|S| = k$, minimizing $\max_v \{dist(v, S)\}$ where $dist(i, S) = \min_{j \in S}(d_{ij})$.

LEMMA 2.1
Assuming $\mathcal{P} \neq \mathcal{NP}$, there is no polynomial-time ρ-approximation with $\rho < 2$.

PROOF   If such an algorithm existed, it would be used to solve the problem of existence of dominating sets[2] of size k. Indeed, consider a graph $G = (V, E)$, and construct an instance of the k-center problem by considering a graph $G' = (V, E')$ with $d_{ij} = 1$ if $\{i, j\} \in E$, and $d_{ij} = 2$ otherwise. It is clear that the graph G' admits a k-center with maximum distance to the centers equal to 1, if and only if G admits a dominating set of size k.   ☐

DEFINITION 2.2   *We call $G_d = (V, E_d)$ the graph where $\{i, j\} \in E_d$ if and only if $d \geqslant d_{ij}$.*

DEFINITION 2.3   *Let $G = (V, E)$ be an undirected graph. We call $G^2 = (V, E')$ the graph such that $\{i, j\} \in E'$ if and only if $\{i, j\} \in E$ or $\exists k$ such that $\{i, k\} \in E$ and $\{k, j\} \in E$.*

Algorithm 2.1 2-approximation algorithm for the k-center problem

---

Sort d_{ij}: $d_1 < d_2 < \ldots < d_m$ (m denotes the number of possible values for the distances)
$i := 0$;
repeat
$\quad i := i + 1$; $d := d_i$;
$\quad$ Construct G_d^2;
$\quad$ Find a maximal independent set S of G_d^2, that is, a maximal set of pairwise nonadjacent vertices.
until $k \geqslant |S|$
Return S as the solution

LEMMA 2.2
Let i_0 be the value of the last i in the loop in Algorithm 2.1, then the distance between any vertex and S is at most $2d_{i_0}$.

PROOF   Let i_0 be the value of the last i in the loop. Thus, the solution

---

[2]**Instance:** $G = (V, E)$; **Question:** Find a minimum dominating set for G, i.e., a subset $V' \subseteq V$ such that for all $u \in V - V'$ there is a $v \in V'$ for which $(u, v) \in E$

is $d = 2d_{i_0}$. Indeed, S is an independent set, and so, any vertex in the graph $G^2_{d_{i_0}}$ admits a neighbor in the set S.   ▯

THEOREM 2.5
Algorithm 2.1 is a 2-approximation algorithm, and this is a best possible result.

PROOF   In order to prove that $d^* \geqslant d_{i_0}$, where d^* denotes an optimal solution, it is sufficient to prove $d^* > d_{i_0-1}$. This can be proved by contradiction since the graph $G^2_{d_{i_0-1}}$ admits an independent set S' of size at least $(k+1)$.

We thus suppose that $d^* \leqslant d_{i_0-1}$. Let C be a smallest k'-center, denoted by $C = \{x_1, \ldots, x_{k'}\}$ such that, for any $i \in V$, $dist(i, C) \leqslant d^*$. Then, for any $x \in S'$, there exists a vertex $f(x) \in C$ such that $d_{xf(x)} \leqslant d^*$. Let us suppose that there exist $(x, y) \in S' \times S'$, $x \neq y$, such that $f(x) = f(y)$. Then there exist $z = f(x)$ such that $\{z, y\} \in G_{d^*} \subset G_{d_{i_0-1}}$ and $\{x, z\} \in G_{d^*} \subset G_{d_{i_0-1}}$ and so $\{x, y\} \in G^2_{d_{i_0-1}}$. This is impossible since S' is an independent set of $G^2_{d_{i_0-1}}$. Therefore, $\forall x \in S', \forall y \in S', x \neq y \Rightarrow f(x) \neq f(y)$. Consequently, $|C| \geqslant |S'| \geqslant k+1$, and C cannot be a k-center.

We conclude using Lemma 2.1.   ▯

2.2   A Fully Polynomial-Time Approximation Scheme

In this section, we propose a polynomial-time approximation algorithm for the following problem, denoted π:

Instance: An independent set of tasks (i.e., without any precedence constraint), each task i having a processing time p_i.

Question: Determine a minimum length schedule of these tasks on two processors.

THEOREM 2.6
Problem π is NP-complete.

PROOF   The partition problem is a particular case problem π.   ▯

THEOREM 2.7
There exists a polynomial-time approximation scheme (PTAS) for the above mentioned problem.

Preliminary result. We consider the intermediate problem called the sum problem:

Instance: Let $E = [e_1, e_2, \ldots, e_n]$ be a list of n integers and let t be an integer.

Question: Find $E' \subseteq E$ such that $\sum_{i \in E'} i$ is maximum and $\sum_{i \in E'} i \leqslant t$.

Algorithm 2.2 A pseudo-polynomial-time algorithm for the sum problem

$L_0 = \{0\}$
$L_1 = \{0, e_1\}$
$L_i = L_{i-1} \cup \{x \mid \exists y \in L_{i-1}, x = y + e_i, x \leqslant t\}$
The solution is $y^* = \max(L_n)$

LEMMA 2.3

There exists a pseudo-polynomial-time algorithm which solves the sum problem in $O(n \times \min(t, 2^n))$.

PROOF In an iteration i, the computation of list L_i is proportional to the size of L_{i-1}. List L_i is at worst twice as long as list L_{i-1}. Therefore, the complexity of algorithm 2.2 is $O(\sum_i^n |L_i|) \leqslant O(n \times \max_i |L_i|) \leqslant O(n \times \min(t, 2^n))$. ☐

How can we reduce the size of the lists? An idea is to filter the list. Suppose that two integers 45 and 47 are in a list L_i: the sums that can be obtained using 47 are close to those obtained using 45. So, 45 can represent 47. Let $Filter(L, \delta)$ be the function that suppresses from list L any integer y such that there exists in L an integer z satisfying $(1 - \delta) \times y \leqslant z < y$.

Example: If $L = \{0, 1, 2, 4, 5, 11, 12, 20, 21, 23, 24, 25, 26\}$ and $\delta = 0.2$, then $Filter(L, \delta) = \{0, 1, 2, 4, 11, 20, 26\}$.

A filter will be used at each iteration of Algorithm 2.2. If $L_i' = Filter(L, \delta) = \{0, z_1, z_2, \ldots, z_{m+1}\}$ then, by construction, $z_i/z_{i-1} \geqslant 1/(1-\delta)$, $t \geqslant z_{m+1}$, and $z_1 \geqslant 1$. So $t \geqslant z_{m+1}/z_1 = \prod_{i=2,\ldots,m+1} z_i/z_{i-1} \geqslant (1/(1-\delta))^m$. Let $\delta = \frac{\epsilon}{n}$. Finally, we have $m < n \log(t)/\epsilon$ since $-\log(1-x) > x$ and the size of the lists is in $O((n \log t)/\epsilon)$. The problem becomes polynomial.

LEMMA 2.4

The error is slight. The ratio between the obtained solution and an optimal one is less than ϵ.

PROOF By induction, we can prove that: $\forall y \in L_i, \exists z \in L_i'$ such that

$z \geqslant (1 - \delta)^i \times y$. This result is true for $i = n$ and $y = y^*$. So, we have $\exists z \in L'_n$, $z \geqslant (1 - \delta)^n y^*$. In conclusion, if $y' \geqslant z$ is an obtained solution, $y'/y^* \geqslant (1 - \epsilon/n)^n \geqslant 1 - \epsilon$ which induces that $\frac{y^*}{y'} \leqslant 1 + 2\epsilon$ if $\epsilon \leqslant \frac{1}{2}$. ⬜

PROOF of Theorem 2.7   We consider the reduction from the scheduling problem to the sum problem that takes $e_i = p_i$ and $t = B/2$, where B is the sum of the processing times: $B = \sum_{i=1}^{n} p_i$.

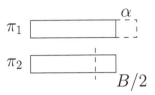

FIGURE 2.1: Application to the scheduling problem.

We denote by S' the solution of the sum problem given by Algorithm 2.3. Solution S' is equal to $S' = B/2 - \alpha$. Thus, we can have a solution S equal to $B/2 + \alpha$ for the scheduling problem (S is obtained by executing on one processor exactly the tasks corresponding to integers in the solution S', and by executing the other tasks on the other processor).

Reciprocally, from an optimal solution $S^* = B/2 + \beta$ for the scheduling problem, we can obtain an optimal solution $S'^* = B/2 - \beta$ for the sum problem, with $\beta \leqslant \alpha$. It is important to note that $\frac{S}{S^*} < \frac{S'^*}{S'}$. Therefore since there exists a polynomial-time approximation scheme for the sum problem, we can conclude that there exists a polynomial-time approximation scheme for the scheduling problem. ⬜

Note that this approach can be extended to any constant number of processors.

Algorithm 2.3 $(1 + \epsilon)$-approximation algorithm for the sum problem ($\epsilon < 1/2$)

$L'_0 = \{0\}$
$L'_1 = \{0, e_1\}$
$L'_i = Filter(\{x | \exists y \in L'_{i-1}, x = y + e_i, x \leqslant t\}, \delta)$ with $\delta = \epsilon/n$
The solution is $y' = max(L'_n)$

2.3   Introduction of Precedence Constraints

In this section, we introduce precedence constraints between tasks. $J_i \rightarrow J_j$ means that task J_j depends on task J_i and thus induces that $t_j \geqslant t_i + p_i$ where t_i (resp. p_i) is the starting time of the task i (resp. the processing time).

2.3.1   Unbounded Number of Processors

When the number of processors is unbounded the problem of scheduling a set of n tasks under precedence constraints is polynomial. It is sufficient to use the classical algorithm introduced by Bellman and the two techniques widely used in project management: *CPM* (Critical Path Method) and *PERT* (Project/Program Evaluation and Review Technique). See for instance [5, Section 7.3]

2.3.2   Bounded Number of Processors

Let us now study the scheduling problem with bounded resources.

THEOREM 2.8
The problem to decide if an instance of $P|prec; p_i = 1|C_{max}$ problem possesses a schedule of length three is NP-complete (see [16]).

THEOREM 2.9
For every list scheduling [3] *LS we have $\frac{C_{max}^{LS}}{C_{max}^*} \leqslant 2 - \frac{1}{m}$ and the bound is tight (see [10]).*

PROOF   Let C_{max}^* be the optimal length of a schedule.

- We have two lower bounds for C_{max}^*: $C_{max}^* \geqslant \frac{T_{seq}}{m}$ (where $T_{seq} = \sum_{i=1}^n p_i$) and $C_{max}^* \geqslant \sum_{k=1}^l p_{i_k}$ where a task $i_k, k = 1 \ldots l$ induced a path.

- Let P be a path constructed in the following way: Let i_1 be a task such that $C_{max}^{LS} = C_{i_1} = t_{i_1} + p_{i_1}$. Let i_2 be a task such that $i_2 = j$ such that $j \in \Gamma^-(i_1) \wedge C_j = \max_{i \in \Gamma^-(i_1)} C_i$, where $\Gamma^-(i_1)$ designates the set of predecessors of i_1. In the similar way, we have $i_k = j$ such that $j \in \Gamma^-(i_{k-1}) \wedge C_j = \max_{i \in \Gamma^-(i_{k-1})} C_i$.

---

[3] Given a priority list of the tasks, a *list scheduling* algorithm always selects the first available processor to process the first available task in the list.

- By the list principle it is clear that all the processors are busy between the completion time of i_1 and the the starting time of i_2. In the same way, there is no idle time between the completion time of i_{k-1} and the starting time i_k. Thus, at each time, there is at least one processor busy and so the number of idle slots is at most $(m-1)$ and the total idle time over all the processors is $T_{idle} \leqslant (m-1) \sum_{p=1}^{k} p_{i_p}$.

Then, we obtain $C_{max}^{LS} \leqslant \frac{\sum_{i=1}^{n} p_i}{m} + \frac{(m-1)}{m} \sum_{p=1}^{k} p_{i_p} \leqslant (2 - 1/m) C_{max}^*$

To show that the bound is tight consider the following instance where K is an arbitrarily large integer. There are $2m+1$ tasks J_i, whose processing times are $p_i = K(m-1)$ for $1 \leqslant i \leqslant m-1$, $p_m = 1$, $p_i = K$ for $m+1 \leqslant i \leqslant 2m$, and $p_{2m+1} = K(m-1)$. There are precedence edges from J_m to J_i and from J_i to J_{2m+1} for $m+1 \leqslant i \leqslant 2m$. There are exactly m entry vertices, so any list schedule will start tasks J_1 to J_m at time 0. At time 1, the execution of J_m is complete and the free processor (the one that executed J_m) will be successively assigned $m-1$ of the m free tasks $T_{m+1}, T_{m+2}, \ldots, T_{mp}$. Note that this processor starts the execution of the last of its $m-1$ tasks at time $1 + K(m-2)$ and terminates it at time $1 + K(m-1)$. Therefore, the remaining m-th task will be executed at time $K(m-1)$ by another processor. Only at time $K(m-1) + K = Km$ will task J_{2m+1} be free, which leads to a length $K(2m-1)$ for any list schedule.

However, the graph can be scheduled in only $Km+1$ time units. The key is to deliberately keep $m-1$ processors idle while executing task J_p at time 0 (which is forbidden in a list schedule). Then, at time 1, each processor executes one of the m tasks $T_{m+1}, T_{m+2}, \ldots, T_{2m}$. At time $1 + K$, one processor starts executing T_{2m+1} while the other $m-1$ processors execute tasks $T_1, T_2, \ldots, T_{m-1}$. This defines a schedule with a makespan equal to $1 + K + K(m-1) = Km + 1$, which is optimal because it is equal to the weight of the path $J_m \to J_{m+1} \to J_{2m+1}$. Hence, we obtain the performance ratio

$$\frac{K(2m-1)}{Km+1} = 2 - \frac{1}{m} - \frac{2m-1}{m(Km+1)} = 2 - \frac{1}{m} - \epsilon(K),$$

where $\lim_{K \to +\infty} \epsilon(K) = 0$.   ⬜

2.4   Introduction of Communication Delays

2.4.1   Introduction

In this section we study the classical scheduling problem with communication delays. Formally, to each arc $(i, j) \in E$ is associated a value c_{ij}, which represents the potential communication delay between the tasks i and j. Let

π_i denote the processor on which task i is scheduled. Then, if $(i, j) \in E$, task j depends on task i. Furthermore, if $\pi_i = \pi_j$ task j must not be scheduled before task i completes $(t_i + p_i \leqslant t_j)$. If $\pi_i \neq \pi_j$, task j cannot be scheduled before the communication delay elapsed since the completion of task i $(t_i + p_i + c_{ij} \leqslant t_j)$.

In this section we focus on Unit Execution Time-Unit Communication Time problems (UET-UCT), that is, on instances where all computations and communication times are equal to one.

2.4.2 Unbounded Number of Processors

2.4.2.1 Complexity Results

THEOREM 2.10
The problem of deciding whether an instance of $P_\infty | prec, p_i = 1, c_{ij} = 1 | C_{max}$ has a schedule of length five is polynomial [21].

THEOREM 2.11
The problem of deciding whether an instance of $P_\infty | prec, p_i = 1, c_{ij} = 1 | C_{max}$ has a schedule of length six is NP-complete [21].

COROLLARY 2.2
There is no polynomial-time approximation algorithm for $P_\infty | prec; c_{ij} = 1; p_i = 1 | C_{max}$ with performance bound smaller than 7/6 unless $\mathcal{P} = \mathcal{NP}$ (see [21] and the Impossibility Theorem).

THEOREM 2.12
There is no polynomial-time approximation algorithm for $P_\infty | prec; c_{ij} = 1; p_i = 1 | \sum_j C_j$ with performance bound smaller than 9/8 unless $\mathcal{P} = \mathcal{NP}$.

PROOF   We proceed by contradiction. Let us assume that there exists a polynomial-time approximation algorithm, denoted A, with a performance guarantee $\rho < 9/8$.

Let I be an instance of the problem $P_\infty | prec; c_{ij} = 1; p_i = 1 | C_{max}$. Let n be the number of tasks in instance I. Let $f(I) = I'$ be an instance of the problem $P_\infty | prec; c_{ij} = 1; p_i = 1 | \sum_j C_j$ constructed by adding x new tasks to instance I, with $x > \frac{9 + 6\rho n}{9 - 8\rho}$. In I', we add a precedence constraint from any of the tasks corresponding to a task in I (the so-called old tasks) to any of the x new tasks.

We now show that Algorithm A, of performance guarantee $\rho < 1 + \frac{1}{8}$, can then be used to decide the existence of a schedule of length at most 6 for I. This will be in contradiction with Theorem 2.11.

Let $A(I')$ (resp. $A^*(I')$) be the result computed by A (resp. an optimal result) on instance I'. We have two cases to consider.

1. $A(I') < 8\rho x + 6\rho n$. Then $A^*(I') < 8\rho x + 6\rho n$ (because for any instance J, $A^*(J) \leqslant A(J)$). We can then show that, in an optimal schedule, the last of the old tasks had been completed by time 6. Indeed, let us suppose one task i among the n old tasks is completed no sooner than $t = 7$ in an optimal schedule. Among the x new tasks, only the task which is executed on the same processor as i can be completed before the time $t = 9$. So $A^*(I') > 9(x - 1)$. Combining the two inequalities on $A^*(I')$ we conclude that $x < \frac{9 + 6\rho n}{9 - 8\rho}$. This contradicts the definition of x. Therefore such a task i does not exist and there exists a schedule of length at most 6 for instance I.

2. $A(I') \geqslant 8\rho x + 6\rho n$. Then $A^*(I') \geqslant 8x + 6n$ because algorithm A is a polynomial-time approximation algorithm with performance guarantee ρ. Then we can show that there is no schedule of length at most 6 for instance I. Indeed, if there existed such a schedule, by executing the x tasks at time $t = 8$ we would obtain a schedule with a sum of completion times strictly smaller than $8x + 6n$ (at least one of the old tasks can be completed before time 6). This would be a contradiction since $A^*(I') \geqslant 8x + 6n$.

Therefore, if there was a polynomial-time approximation algorithm with performance guarantee strictly smaller than $1 + \frac{1}{8}$, it could be used to distinguish in polynomial time the positive instances from negative ones for the problem $P_\infty | prec; c_{ij} = 1; p_i = 1 | C_{max} = 6$. Thus it would provide a polynomial-time algorithm for an NP-hard problem. Consequently, the problem $P_\infty | prec; c_{ij} = 1; p_i = 1 | \sum C_j$ has no ρ-approximation with $\rho < 1 + \frac{1}{8}$.
☐

2.4.2.2   Polynomial-Time Approximation Algorithm

Good approximation algorithms seem to be very difficult to design, since the compromise between parallelism and communication delay is not easy to handle. In this section we present an approximation algorithm with a performance ratio bounded by 4/3 (see [14]) for the problem $P_\infty | prec; c_{ij} = 1; p_i = 1 | C_{max}$. This algorithm is based on a formulation on an integer linear program (ILP). A feasible schedule is obtained by a relaxation and rounding procedure. Note that there exists a straightforward 2-approximation algorithm: the tasks without predecessors are executed at $t = 0$, the tasks admitting predecessors scheduled at $t = 0$ are executed at $t = 2$, and so on.

We denote the tasks without predecessor (resp. successor) by Z (resp. U). We call *source* every task belonging to Z. We model the scheduling problem by a set of equations defined using the vector of starting times: $(t_1, \ldots, t_n)$. For every arc $(i, j) \in E$, we introduce a variable $x_{ij} \in \{0, 1\}$ which indicates the presence or not of a communication delay. We then have the following constraints: $\forall (i,j) \in E, t_i + p_i + x_{ij} \leqslant t_j$.

In every feasible schedule, every task $i \in V - U$ has at most one successor, w.l.o.g. call it $j \in \Gamma^+(i)$, that can be processed by the same processor as i and at time $t_j = t_i + p_i$. The other successors of i, if any, satisfy: $\forall k \in \Gamma^+(i) - \{j\}, t_k \geqslant t_i + p_i + 1$. Consequently, we add the constraints:

$$\sum_{j \in \Gamma^+(i)} x_{ij} \geqslant |\Gamma^+(i)| - 1.$$

Similarly, every task i of $V - Z$ has at most one predecessor, w.l.o.g. call it $j \in \Gamma^-(i)$, that can be performed by the same processor as i and at times t_j satisfying $t_i - (t_j + p_j) < 1$. So, we add the following constraints:

$$\sum_{j \in \Gamma^-(i)} x_{ji} \geqslant |\Gamma^-(i)| - 1.$$

If we denote by C_{max} the makespan of the schedule, $\forall i \in V, t_i + p_i \leqslant C_{max}$. Thus, we obtain the following *ILP* (denoted by Π):

$$(\Pi) \begin{cases} \min C_{max} \\ \forall (i,j) \in E, \ x_{ij} \in \{0,1\} \\ \forall i \in V, \quad t_i \geqslant 0 \\ \forall (i,j) \in E, \ t_i + p_i + x_{ij} \leqslant t_j \\ \forall i \in V - U, \ \sum_{j \in \Gamma^+(i)} x_{ij} \geqslant |\Gamma^+(i)| - 1 \\ \forall i \in V - Z, \ \sum_{j \in \Gamma^-(i)} x_{ji} \geqslant |\Gamma^-(i)| - 1 \\ \forall i \in V, \quad t_i + p_i \leqslant C_{max} \end{cases}$$

Let Π^{inf} denote the linear program corresponding to Π in which we relax the integrity constraints $x_{ij} \in \{0,1\}$ by replacing them with $x_{ij} \in [0,1]$. As there is a polynomial number of variables and of constraints, this linear program can be solved in polynomial time. A solution of Π^{inf} assigns to every arc $(i,j) \in E$ a value $x_{ij} = e_{ij}$ with $0 \leqslant e_{ij} \leqslant 1$. This solution defines a lower bound Θ^{inf} on the value of C_{max}.

LEMMA 2.5
Θ^{inf} *is a lower bound on the value of any optimal solution for* $P_\infty|prec; c_{ij} = 1; p_i \geqslant 1|C_{max}$.

PROOF This is true since any optimal feasible solution of the scheduling problem must satisfy all the constraints of the integer linear program Π and as any solution of Π is a solution of Π^{inf}. ▯

In the following, we call an arc $(i,j) \in E$ a *0-arc* (resp. *1-arc*) if $x_{ij} = 0$ (resp. $x_{ij} = 1$).

Algorithm 2.4 *Rounding algorithm and construction of the schedule*

Step 1: Rounding

Let e_{ij} be the value of the arc $(i, j) \in E$ given by Π^{inf}

Let $x_{ij} = 0$ if $e_{ij} < 0.5$ and $x_{ij} = 1$ otherwise

Step 2: Computation of the starting times

if $i \in Z$ **then**

$\quad t_i = 0$

else

$\quad t_i = \max\{t_j + p_j + x_{ji} \mid j \in \Gamma^-(i) \text{ and } (j, i) \in E\}$

end if

Step 3: Construction of the schedule

Let $G' = (V; E')$ where $E' = E \backslash \{(i, j) \in E | x_{ij} = 1\}$ {G' is generated by the 0-*arcs.*}

Map each connected component of G' on a different processor.

Each task is executed at its starting time.

LEMMA 2.6

Every task $i \in V$ has at most one successor (resp. one predecessor) such that $e_{ij} < 0.5$ (resp. $e_{ji} < 0.5$).

PROOF We consider a task $i \in V$ and its successors $j_1, \ldots, j_k$ ordered such that $e_{i,j_1} \leqslant e_{i,j_2} \leqslant \ldots \leqslant e_{i,j_k}$. We know that $\sum_{l=1}^{k} e_{i,j_l} \geqslant k - 1$. Then, $2e_{i,j_2} \geqslant e_{i,j_2} + e_{i,j_1} \geqslant k - 1 - \sum_{l=3}^{k} e_{i,j_l}$. Since $e_{i,j_l} \in [0, 1]$, $\sum_{l=3}^{k} e_{i,j_l} \leqslant k - 2$. Thus, $2e_{i,j_2} \geqslant 1$. Therefore for any $l \in \{2, \ldots, k\}$ we have $e_{ij} \geqslant 0.5$. We use the same arguments for the predecessors. □

LEMMA 2.7

Algorithm 2.4 provides a feasible schedule.

PROOF It is clear that each task i admits at most one incoming (resp. outcoming) 0-arc. □

THEOREM 2.13

The relative performance ρ^h of the heuristic is bounded above by $\frac{4}{3}$ and the bound is tight (see [14]).

PROOF Let $x_1 \to x_2 \to \ldots \to x_{k+1}$ be any path constituted by $(k + 1)$ tasks. We denote by x the number of arcs in this path that, in the considered solution of the linear program, have a weight smaller than $\frac{1}{2}$. Then, this path has $k - x$ ares whose weight in that solution is at least $\frac{1}{2}$. So the length of this path, under that solution, is at least $k + 1 + \frac{1}{2}(k - x) = \frac{3}{2}k - \frac{1}{2}x + 1$. Moreover, by the rounding procedure, the length of this path is at most $2k - x + 1$. Thus, for any given path of value p^* before rounding and value p after rounding, having x arc with weights smaller than $\frac{1}{2}$, we have $\frac{p}{p^*} \leqslant \frac{2k-x+1}{3/2k-1/2x+1} < 4/3$. This is, in particular, true for any critical path. Hence the upper bound on the performance ratio. In [14], the authors propose an instance on which the bound is reached. $\qquad\qquad\qquad\qquad\qquad\qquad\qquad\qquad\qquad\qquad\qquad\qquad\square$

2.4.3 Limited Number of Processors

2.4.3.1 Complexity Results

THEOREM 2.14
The problem of deciding whether an instance of $P|prec, c_{ij} = 1, p_i = 1|C_{max}$ has a schedule of length three is polynomial [16].

THEOREM 2.15
The problem of deciding whether an instance of $P|prec, c_{ij} = 1, p_i = 1|C_{max}$ has a schedule of length four is NP-complete [21].

PROOF The problem $P|prec, c_{ij} = 1, p_i = 1|C_{max}$ clearly belongs to $\mathcal{NP}$. The proof of NP-completeness is based on the polynomial-time reduction CLIQUE $\propto P|prec, c_{ij} = 1, p_i = 1|C_{max}$, and is illustrated by Figure 2.3. Let π^* be an instance of the CLIQUE problem. π^* is defined by a graph $G = (V, E)$ and an integer k. The question is: does the graph $G = (V, E)$ contain a clique of size k? In the following, let $l = \frac{k(k-1)}{2}$ be the number of edges of a clique of size k, let $m' = \max\{|V| + l - k, |E| - l\}$, and let $m = 2(m' + 1)$.

From π^* we build an instance π of $P|prec, c_{ij} = 1, p_i = 1|C_{max}$ as follows:
- There are m processors.
- For each node $v \in V$ we introduce the tasks T_v and K_v and the precedence constraint $T_v \to K_v$.
- For each $e = (u, v) \in E$ we create a task L_e and the two precedence constraints: $T_v \to L_e$ and $T_u \to L_e$.
- Finally, we add four other sets of tasks: $X = \{X_x\}_{1 \leqslant x \leqslant m-l-|V|+k}$, $Y = \{Y_y\}_{1 \leqslant y \leqslant m-|E|+l}$, $U = \{U_u\}_{1 \leqslant u \leqslant m-k}$, and $W = \{W_w\}_{1 \leqslant w \leqslant m-|V|}$. For any task X_x, any task Y_y, any task U_u, and any task W_w, we add the precedence constraints: $U_u \to X_x$, $U_u \to Y_y$, and $W_w \to Y_y$.

We now show that the graph G admits a clique of size k if only if there exists a schedule of length 4.

- We suppose that G admits a clique of size k. Then to build a schedule of length 4 we can map the tasks according to the schedule given by Figure 2.2. In this figure, T_{clique} exactly corresponds to the set of the tasks T_v which correspond to the nodes of the clique of G, etc. One can check that all m processors are busy during each of the four time-steps.

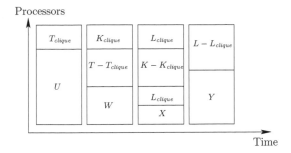

FIGURE 2.2: Gantt chart for a schedule built from a clique in the proof of Theorem 2.15.

- We now suppose that there exists a schedule of length 4. We must thus prove that G admits a clique of size k.

  All tasks in U must be completed at 1. If this was not the case, there would exist a task U_u completed no earlier than at time 2. Then at most one task in X and Y could start at time 2, the only one scheduled on the same processor than U_u. Then, $|X| + |Y| - 1$ tasks would have to start no earlier than at time 3. This is impossible as $|X| + |Y| - 1 \geqslant m + 1$. As all tasks in U are completed at time 1, no task in X can be started earlier than at time 2.

  All the tasks in Y must be started at time 3. Indeed, in order to be able to start processing any task of Y before time 3, all tasks in U and W would have to be completed no later than time 1. However, $|U| + |W| \geqslant m + 2$. So any task of Y depends on two tasks completed no earlier than at time 2, and thus can not be started earlier than at time 3. As all the tasks in Y are started at time 3, all tasks in W must be completed at time 2.

  As any task in L depends on two tasks in T, no task in L can start earlier than at time 2.

  As there are m processors and $4m$ tasks, the schedule contains no idle times. Therefore, besides all the tasks in U and W, there should be $k + |V|$ tasks from T and K completed by time 2. As any task in K must be preceded by the corresponding task in T, no more than $|V|$ tasks from T and K can be processed between times 1 and 2. Apart from tasks from T and K only the $m - |V|$ tasks of W can be scheduled during this interval. Therefore, all tasks of W are started at time 1, k

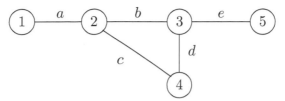

$$|V| = 5, \ |E| = 5, \ l = 3, \ k = 3, \ m' = 5, \text{ and } m = 12.$$

A dotted arc from one set to another one means that there is an arc from each element of the first set to any element of the second.

π^0	T_2	K_2	L_b	L_e
π^1	T_3	K_3	L_c	L_a
π^2	T_4	K_4	L_d	Y_1
π^3	U_1	T_1	K_1	Y_2
π^4	U_2	T_5	K_5	Y_3
π^5	U_3	W_1	X_1	Y_4
π^6	U_4	W_2	X_2	Y_5
π^7	U_5	W_3	X_3	Y_6
π^8	U_6	W_4	X_4	Y_7
π^9	U_7	W_5	X_5	Y_8
π^{10}	U_8	W_6	X_6	Y_9
π^{11}	U_9	W_7	X_7	Y_{10}

Gantt chart

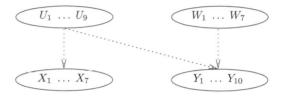

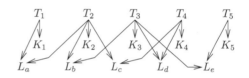

FIGURE 2.3: Example of polynomial-time reduction CLIQUE $\propto P|prec; c_{ij} = 1; p_i = 1|C_{max}$ with $m = 12$ processors.

tasks from T are completed by time 1, and the corresponding k tasks of K and the remaining tasks of T are started at time 1.

The tasks started at time 3 are tasks from X, remaining tasks from K, tasks from L which have both predecessors completed by time 1, and tasks from L which one predecessor completed at time 1 and one predecessor completed at time 2. There are m of these tasks as the schedule has no idle times. Note that both the tasks from K and those from L with one predecessor completed at time 2 should be executed on the same processor as their predecessor. There are thus at most $|V| - k$ of these tasks. Also $|X| = m - |V| - l + k$. Therefore, there are at least $m - (|V| - k) - (m - |V| - l + k) = l$ tasks in l which have both predecessors completed by time 1. This implies that the k tasks of T scheduled at time 0 correspond to the vertices of a k-clique of G.

□

THEOREM 2.16

Deciding whether an instance of $P|prec, c_{ij} = 1, p_i = 1, dup|C_{max}$ has a schedule of length four is an NP-complete problem [21]. (dup stands for the possibility to duplicate tasks.)

PROOF   In the proof of Theorem 2.15 there are m processors and $4m$ tasks for a schedule of length 4: there is no room for duplication.   ⬜

COROLLARY 2.3

The problem $P|prec, c_{ij} = 1, p_i = 1|C_{max}$ admits no polynomial-time approximation algorithm with a performance bound smaller than $5/4$ unless $\mathcal{P} = \mathcal{NP}$.

THEOREM 2.17

The problem $P|prec; c_{ij} = c = 1; p_i = 1|\sum_j C_j$ admits no polynomial-time approximation algorithm with a performance bound smaller than $11/10$ unless $\mathcal{P} = \mathcal{NP}$.

PROOF

Let I be an instance of the problem $P|prec; c_{ij} = 1; p_i = 1|C_{max}$, with m processors. Without loss of generality, we assume that I has $4m$ tasks (cf. in the proof of Theorem 2.15 all instances have $4m$ tasks for m processors). Let $I' = f(I)$ be an instance of the problem $P|prec; c_{ij} = 1; p_i = 1|\sum_j C_j$ constructed by adding $4m$ new tasks to instance I. We add a precedence constraint from each task corresponding to a task of I (the so-called old tasks) to each of the new $4m$ tasks.

Let $C^*(z)$ be the optimal value of the total job completion time minimization on an instance z. We want to conclude using the Gap theorem (Theorem 2.2). Therefore, we want to explicit a function $g : I \to \mathbb{N}$ and an integer gap such that:

$$\begin{cases} C^*(f(I) = I') = g(I) \text{ if } I \text{ is a positive instance,} \\ C^*(f(I) = I') \geqslant g(I)(1 + gap) \text{ otherwise.} \end{cases}$$

We have two cases to consider, depending whether I is a positive instance.

1. I is a positive instance. So there is a schedule of I whose makespan is 4. Under this schedule the sum of the completion times of *old tasks* of I' is equal to $10m$. None of the new tasks can be scheduled earlier than at time 5. These new tasks can be executed in any order starting at times 5 and thus, in an optimal schedule for the sum of completion times, m tasks are executed starting at time 5, m at time 6, m at time 7 and m at time 8. Therefore, when I is a positive instance, $C^*(f(I) = I') = 40m$.

2. Suppose I is a negative instance. In other words, I has no schedule of length 4. Then, in any schedule of I there is at least one task that is

completed no earlier than at time 5. Thus, in any schedule of I', even an optimal one, there is at least one old task that is completed no earlier than at time 5, no new task can start earlier than at time 5, and at most one new task can be completed at time 6 (the one being on the same processor than the only old task completed at time 5 is there is only one such task). Therefore, we have: $C^*(f(I) = I') \geqslant m * 1 + m * 2 + m * 3 + (m-1) * 4 + 5 + 6 + m * 7 + m * 8 + m * 9 + (m-1) * 10 = 44m - 3$.

Let $g(I) = 40m$. We then have established that:

$$\begin{cases} C^*(f(I) = I') = g(I) \text{ if } I \text{ is a positive instance,} \\ C^*(f(I) = I') \geqslant g(I)(1 + \frac{1}{10} - \frac{3}{40m}) \text{ otherwise.} \end{cases}$$

Then, for any $\frac{1}{10} > \epsilon > 0$, and for $m_0 = \lceil \frac{3}{40\epsilon} \rceil$, using Theorems 2.15 and 2.2, we established that, unless $\mathcal{P} = \mathcal{NP}$, problem $P|prec; c_{ij} = c = 1; p_i = 1| \sum_j C_j$ does not admit any polynomial-time approximation algorithm with a performance bound smaller than $11/10 - \epsilon$ for instances having at least m_0 processors. Consequently, $P|prec; c_{ij} = 1; p_i = 1| \sum C_j$ does not have a ρ-approximation with $\rho < 1 + \frac{1}{10}$. □

2.4.3.2 Polynomial-Time Approximation Algorithm

In this section, we present a simple algorithm that builds, for problem $P_\infty|prec, c_{ij} = 1, p_i = 1|C_{max}$, a schedule σ_m using m machines from a schedule σ_∞ using an unbounded number of processors. The validity of this algorithm is based on the fact there is at most one matching between the tasks executed at time t_i and the tasks processed at time $t_i + 1$ (this property is called Brent's Lemma [4]).

Algorithm 2.5 Schedule using m machines built from a schedule σ_∞ using an unbounded number

---
    for $i = 0$ to $C_{max}^\infty - 1$ **do**
        Let X_i be the set of tasks executed at time i under σ_∞.
        The X_i tasks are executed in $\lceil \frac{|X_i|}{m} \rceil$ units of time.
    end for

THEOREM 2.18
From any polynomial time approximation algorithm h^* with performance guarantee ρ_∞ for problem $P_\infty|prec, c_{ij} = 1, p_i = 1|C_{max}$, Algorithm 2.5 builds a polynomial-time approximation algorithm h with performance guarantee $\rho_m = (1 + \rho_\infty)$ for problem $P|prec, c_{ij} = 1, p_i = 1|C_{max}$.

PROOF Let $C_{max}^{h^*, UET-UCT, \infty}$ (respectively $C_{max}^{h, UET-UCT, m}$) be the length of the schedule produced by h^* (resp. by h). Let $C_{max}^{opt,UET-UCT,\infty}$ (resp. $C_{max}^{opt,UET-UCT,m}$) be the optimal length of the schedule on an unbounded number of processors (resp. on m processors). Let n be the number of tasks in the schedule. By definition, we have $C_{max}^{opt,UET-UCT,\infty} \leqslant C_{max}^{opt,UET-UCT,m}$ and $C_{max}^{h^*,UET-UCT,\infty} \leqslant \rho_\infty C_{max}^{h,UET-UCT,\infty}$. Therefore,

$$
\begin{aligned}
C_{max}^{h,UET-UCT,m} &\leqslant \sum_{i=0}^{\left(C_{max}^{h^*,UET-UCT,\infty}\right)-1} \left\lceil \frac{|X_i|}{m} \right\rceil \\
&\leqslant \sum_{i=0}^{\left(C_{max}^{h^*,UET-UCT,\infty}\right)-1} \left(\left\lfloor \frac{|X_i|}{m} \right\rfloor + 1 \right) \\
&\leqslant \left(\sum_{i=0}^{\left(C_{max}^{h^*,UET-UCT,\infty}\right)-1} \frac{|X_i|}{m} \right) + C_{max}^{h^*,UET-UCT,\infty} \\
&\leqslant C_{max}^{opt,UET-UCT,m} + C_{max}^{h^*,UET-UCT,\infty} \\
&\leqslant C_{max}^{opt,UET-UCT,m} + \rho_\infty C_{max}^{opt,UET-UCT,m}
\end{aligned}
$$

Therefore, $\rho_m \leqslant (1 + \rho_\infty)$. $\qquad\qquad\qquad\qquad\qquad\qquad\qquad\qquad\qquad$ □

2.4.4 Introduction of Duplication

The principle of duplication was first introduced by Papadimitriou and Yannakakis [15] for the problem with large communication delays, in order to minimize the impact of the communication delays. In this section, we consider the following conditions:

- there is at least one copy (i, π, t) [4] of each task J_i;

- at any time, each processor executes at most one copy;

- if $(J_i, J_k) \in E$ then each copy (k, π, t) has a supplier (j, π', t') such that $t \geqslant t' + p_j$ if $\pi = \pi'$ and $t \geqslant t' + p_j + c_{jk}$ otherwise.

In [6], the authors propose a polynomial-time algorithm with complexity $O(n^2)$ to solve the problem of scheduling with duplication a set of tasks on an unbounded number of processors with small communication delays. The algorithm has two steps: the first step (Algorithm 2.6) computes the tasks' release times; the second step uses critical trees defined by the first step to produce an optimal schedule in which all the tasks and their copies are executed at their release times.

---

[4]The triple means that a copy of task J_i is executed at time t on processor π.

Algorithm 2.6 Computing release dates and earliest schedule

---

for $i := 1$ to n **do**
  if $PRED(i) = \varnothing$ **then**
    $b_i := 0$
  else
    $C := \max\{b_k + p_k + c_{ki} \mid k \in PRED(i)\}$;
    Let s_i be such that: $b_{s_i} + p_{s_i} + c_{s_i i} = C$;
    $b_i := \max\{b_{s_i} + p_{s_i}, max\{b_k + p_k + c_{ki} \mid k \in PRED(i) \setminus \{s_i\}\}\}$.
  end if
end for
The graph formed by the arcs $\{i, s_i\}$ is an union of outtrees.
Each path is executed on a separate processor. Note that any task common to several paths must be duplicated. The number of paths is bound by n.

Without lost of generality, all copies of a task i have the same starting time, denoted by t_i. An arc $(i, j) \in E$ is a critical arc if $b_i + p_i + c_{ij} > b_j$. From this definition, it is clear that if (i, j) is a critical arc, then in any as-soon-as-possible schedule each copy of task j must be preceded by a copy of task i on the same processor. In order to construct an earliest schedule, each critical path is allocated on a separate processor, and each copy is executed at its release date.

THEOREM 2.19
Let b_i be the starting time computed by Algorithm 2.6. In no feasible schedule can the release date of a task i be less than b_i. Each sub-graph is a spanning forest. The whole procedure builds a feasible schedule and the overall complexity is $O(n^2)$.

The algorithm is illustrated with the example of Figure 2.4. Figure 2.5 presents the critical sub-graph and Figure 2.6 an optimal schedule.

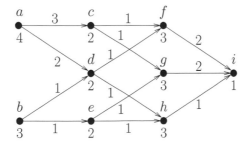

FIGURE 2.4: The problem P_0.

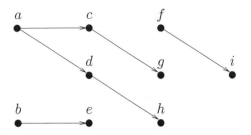

FIGURE 2.5: The critical sub-graph of P_0.

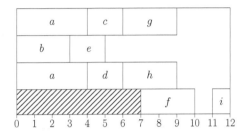

FIGURE 2.6: An earliest schedule of P_0.

2.4.5 Large Communication Delays

Scheduling in presence of large communication delays is one of the most difficult problems in scheduling theory. In order to reduce the influence of communication delays, it is possible to execute on the processor a task i is mapped on a subset, S, of its successors (i.e., $S \subseteq \Gamma^+(i)$). The difficulty is then to find the cardinality of S and an optimal order of the tasks of set S (and, so, of the starting times of these tasks).

2.4.5.1 Complexity Results

We consider the problem of scheduling a precedence graph with large communication delays and unit execution times (*UET-LCT*) on a restricted number of processors. Bampis et al. [2] proved that problem $P|prec; c_{ij} = c \geqslant 2; p_i = 1|C_{max}$ is polynomial for $C_{\max} = c + 1$, polynomial for $C_{\max} = c + 2$ and $c = 2$, and NP-complete for $C_{\max} = c + 3$. The algorithm for the $c = 2$ and $C_{\max} = c + 2$ case cannot be extended for $c \geqslant 3$. The NP-completeness proof of Bampis et al. is based on a reduction from the NP-complete problem *Balanced Bipartite Complete Graph (BBCG)* [8, 18]. Thus, Bampis et al. [2] proved that problem $P|prec; c_{ij} = c \geqslant 2; p_i = 1|C_{max}$ does not have a polynomial-time approximation algorithm with a performance ratio better than $(1 + \frac{1}{c+3})$, unless $\mathcal{P} = \mathcal{NP}$.

THEOREM 2.20

The problem of deciding whether an instance of $P_\infty|prec; c_{ij} = c; p_i = 1|C_{max}$ has a schedule of length smaller than or equal to $(c + 4)$ is NP-complete for $c \geqslant 3$ [9].

2.4.5.2 Polynomial-Time Approximation Algorithm

In this section, we present a polynomial-time approximation algorithm with performance guarantee for the problem $P_\infty|prec; c_{ij} = c \geqslant 2; p_i = 1|C_{max}$.

Notation: We denote by σ_∞ the *UET-UCT* schedule and by $\sigma_\infty^{(c)}$ the *UET-LCT* schedule. Moreover, we denote by t_i (resp. $t_i^{(c)}$) the starting time of the task i under schedule σ_∞ (resp. under schedule $\sigma_\infty^{(c)}$).

Principle: We keep the task assignment given by σ_∞, a "good" feasible schedule on an unrestricted number of processors. We then proceed to an expansion of the makespan while preserving communication delays ($t_j^{(c)} \geqslant t_i^{(c)} + 1 + c$) for any two tasks i and j such that $(i, j) \in E$ and the two tasks are processed on two different processors.

Consider a precedence graph $G = (V, E)$. We determine a feasible schedule σ_∞, for the $UET - UCT$ model, using a (4/3)-approximation algorithm proposed by Munier and König [14]. To any node $i \in V$ this algorithm associates a couple (t_i, π) where t_i is the starting time of task i under schedule σ_∞ and where π is the processor on which task i is processed.

We then define schedule $\sigma_\infty^{(c)}$ by determining, for any $\forall i \in V$ a couple $(t_i^{(c)}, \pi')$ as follows: $\pi = \pi'$ and starting time $t_i^{(c)} = d \times t_i = \frac{(c+1)}{2}t_i$. The justification of the expansion coefficient is given below. An illustration of the expansion is given in Figure 2.7.

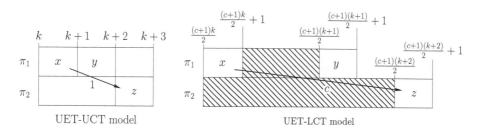

FIGURE 2.7: Illustration of the notion of makespan expansion (arrows explicit communication delays).

LEMMA 2.8
The expansion coefficient is $d = \frac{(c+1)}{2}$.

PROOF　Let there be two tasks i and j, such that $(i,j) \in E$, and that are processed on two different processors under the feasible schedule σ_∞. We are interested in obtaining a coefficient d such that $t_i^{(c)} = d \times t_i$ and $t_j^{(c)} = d \times t_j$. After expansion, in order to respect the precedence constraints and communication delays, we must have $t_j^{(c)} \geqslant t_i^{(c)} + 1 + c$, and so $d \times t_j - d \times t_i \geqslant c + 1$, $d \geqslant \frac{c+1}{t_j - t_i}$, $d \geqslant \frac{c+1}{2}$. It is sufficient to choose $d = \frac{(c+1)}{2}$.　　□

LEMMA 2.9
An expansion algorithm gives a feasible schedule for the problem denoted by $P_\infty | prec; c_{ij} = c \geqslant 2; p_i = 1 | C_{\max}$.

PROOF　It is sufficient to check that the solution given by an expansion algorithm produces a feasible schedule for the *UET-LCT* model. Let i and j be two tasks such that $(i,j) \in E$. We use π^i (respectively π^j) to denote the processor on which task i (respectively task j) is executed under schedule σ_∞. Moreover, π'^i (respectively π'^j) denotes the processor on which task i (respectively task j) is executed under schedule $\sigma_\infty^{(c)}$. Thus,

- If $\pi^i = \pi^j$ then $\pi'^i = \pi'^j$. Since the solution given by Munier and König [14] gives a feasible schedule on the model UET-UCT, we have $t_i + 1 \leqslant t_j$. Then: $\frac{c+1}{2}(t_j - t_i) \geqslant t_j - t_i \geqslant 1$.

- If $\pi^i \neq \pi^j$ then $\pi'^i \neq \pi'^j$. We have $t_i + 1 + 1 \leqslant t_j$. Then: $\frac{c+1}{2}(t_j - t_i) \geqslant \frac{c+1}{2} \times 2 = c + 1$.

　　□

THEOREM 2.21
An expansion algorithm gives a $\frac{2(c+1)}{3}$ – approximation algorithm for the problem $P_\infty | prec; c_{ij} = c \geqslant 2; p_i = 1 | C_{\max}$.

PROOF　We denote by C_{max}^h the makespan of the schedule computed by Munier and König's algorithm and C_{max}^{opt} the optimal makespan for the unit communication time problem. In the same way we denote by $C_{max}^{h^*}$ the makespan of the schedule computed by our algorithm and $C_{max}^{opt,c}$ the optimal makespan for the problem with communication delays of size c.

We know that $C_{max}^h \leqslant \frac{4}{3}C_{max}^{opt}$. Thus, we obtain $\frac{C_{max}^{h^*}}{C_{max}^{opt,c}} = \frac{\frac{(c+1)}{2}C_{max}^h}{C_{max}^{opt,c}} \leqslant \frac{\frac{(c+1)}{2}C_{max}^h}{C_{max}^{opt}} \leqslant \frac{\frac{(c+1)}{2}\frac{4}{3}C_{max}^{opt}}{C_{max}^{opt}} \leqslant \frac{2(c+1)}{3}$.　　□

2.5 Conclusion

In this chapter, we proposed some classical polynomial-time approximation algorithms for NP-complete problems. We designed two absolute approximation algorithms for the edge-coloring and k-center problems. Afterward, we focused on developing some efficient polynomial-time approximation algorithms for some classic scheduling problems (without communication delays, in presence of precedence constraints, and subject to communication delays) with makespan minimization as objective function. For this criteria, we derived a lower bound for any polynomial-time approximation algorithm using the Impossibility Theorem. We also proposed some inapproximability results using the Gap Theorem for the total job completion time minimization problem. Figure 2.8 gives a classification for scheduling in presence of communication delays with makespan minimization as objective function. For further material, we point readers to Vazirani's book [20] for polynomial-time approximation algorithms, and to the *Handbook on Scheduling* [3] for scheduling problems in general.

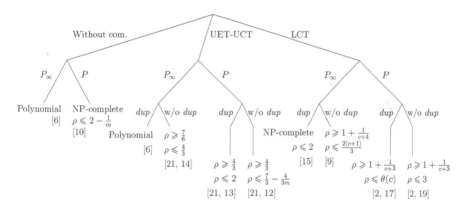

FIGURE 2.8: Main results in complexity and approximation for the minimization of the makespan for homogeneous model.

References

[1] G. Ausiello, P. Crescenzi, G. Gambosi, V. Kann, A. Marchetti-Spaccamela, and M. Protasi. *Complexity and Approximation: Combinatorial Optimization Problems and Their Approximability Properties*, chapter 3, pages 100–102. Springer-Verlag, 1999.

[2] E. Bampis, A. Giannakos, and J.-C. König. On the complexity of scheduling with large communication delays. *European Journal of Operational Research*, 94(2):252–260, 1996.

[3] J. Błażewicz, K. Ecker, E. Pesch, G. Schmidt, and J. Węglarz. *Handbook on Scheduling: From Theory to Applications*. International Handbooks on Information Systems. Springer, 2007.

[4] R. Brent. The parallel evaluation of general arithmetic expression. *Journal of the ACM*, 21(2):201–206, 1974.

[5] H. Casanova, A. Legrand, and Y. Robert. *Parallel Algorithms*. Chapman & Hall/CRC Press, 2008.

[6] P. Chrétienne and J. Colin. C.P.M. scheduling with small interprocessor communication delays. *Operations Research*, 39(3):680–684, 1991.

[7] P. Chrétienne and C. Picouleau. Scheduling with communication delays: A survey. In P. Chrétienne, E. C. Jr., J. Lenstra, and Z. Liu, editors, *Scheduling Theory and Its Applications*. John Wiley & Sons, 1995.

[8] M. R. Garey and D. S. Johnson. *Computers and Intractability. A Guide to the Theory of NP-Completeness*. W. H. Freeman & Co, San Francisco, 1979.

[9] R. Giroudeau, J.-C. König, F. K. Moulai, and J. Palaysi. Complexity and approximation for precedence constrained scheduling problems with large communication delays. *Theoretical Computer Science*, 401(1-3):107–119, 2008.

[10] R. Graham, E. Lawler, J. Lenstra, and A. Rinnooy Kan. Optimization and approximation in deterministic sequencing and scheduling: A survey. *Annals of Discrete Mathematics*, 5:287–326, 1979.

[11] I. Hoyler. The NP-completeness of edge-coloring. *SIAM Journal on Computing*, 10:718–720, 1981.

[12] A. Munier and C. Hanen. An approximation algorithm for scheduling unitary tasks on m processors with communication delays. Private communication, 1996.

[13] A. Munier and C. Hanen. Using duplication for scheduling unitary tasks on m processors with unit communication delays. *Theoretical Computer Science*, 178(1-2):119–127, 1997.

[14] A. Munier and J. König. A heuristic for a scheduling problem with communication delays. *Operations Research*, 45(1):145–148, 1997.

[15] C. Papadimitriou and M. Yannakakis. Towards an architecture-independent analysis of parallel algorithms. *SIAM Journal on Computing*, 19(2):322–328, 1990.

[16] C. Picouleau. New complexity results on scheduling with small communication delays. *Discrete Applied Mathematics*, 60:331–342, 1995.

[17] C. Rapine. *Algorithmes d'approximation garantie pour l'ordonnancement de tâches, Application au domaine du calcul parallèle.* Thèse de doctorat, Institut National Polytechnique de Grenoble, 1999.

[18] R. Saad. Scheduling with communication delays. *Journal of Combinatorial Mathematics and Combinatorial Computing*, 18:214–224, 1995.

[19] R. Thurimella and Y. Yesha. A scheduling principle for precedence graphs with communication delay. In *International Conference on Parallel Processing*, volume 3, pages 229–236, 1992.

[20] V. V. Vazirani. *Approximation Algorithms*. Springer, 2nd edition, 2002.

[21] B. Veltman. Multiprocessor Scheduling with Communications Delays. Ph.D. thesis, CWI-Amsterdam, 1993.

[22] V. Vizing. On an estimate of the chromatic class of a p-graph. *Diskret Analiz*, 3:23–30, 1964.

Chapter 3

Online Scheduling

Susanne Albers

Humboldt University Berlin

3.1 Introduction

Many scheduling problems that arise in practice are inherently *online* in nature. In these settings, scheduling decisions must be made without complete information about the entire problem instance. This lack of information may stem from various sources: (1) Jobs arrive one by one as a list or even as an input stream over time. Scheduling decisions must always be made without knowledge of any future jobs. (2) The processing times of jobs are unknown initially and during run time. They become known only when jobs actually finish. (3) Machine breakdown and maintenance intervals are unknown.

Despite the handicap of not knowing the entire input we seek algorithms that compute good schedules. Obviously, computing optimal solutions is, in general, impossible and we have to resort to approximations. Today, the standard means to evaluate algorithms working online is *competitive analysis* [48]. Here an online algorithm A is compared to an optimal offline algorithm OPT that knows the entire input in advance. Consider a given scheduling problem and suppose we wish to minimize an objective function. For any input I, let $A(I)$ be the objective function value achieved by A on I and let $OPT(I)$ be the value of an optional solution for I. Online algorithm A is called *c-competitive* if there exists a constant b such that, for all problem inputs I, inequality $A(I) \leqslant c \cdot OPT(I) + b$ holds. The constant b must be independent of the input. An analogous definition can be set up for maximization problems. Note that competitive analysis is a strong worst-case performance measure; no probabilistic assumptions are made about the input.

Online scheduling algorithms were already investigated in the 1960s but an extensive in-depth study has only started 10 to 15 years ago, after the concept of competitive analysis had formally been introduced. By now there exists a rich body of literature on online scheduling. The investigated problem settings, as in standard scheduling, address various machine models (identical, related, or unrelated machines), different processing formats (preemptive or non-preemptive scheduling) and various objective functions (such as makespan, (weighted) sum of completion times, (weighted) sum of flow times etc.).

In this chapter, due to space constraints, we can only present a selection of the known results. In the first part of the chapter we focus on some classical scheduling problems and summarize the state of the art. The second part of the chapter is devoted to energy-efficient scheduling, a topic that has received quite some research interest recently and promises to be an active theme for investigation in the future. Wherever possible results and theorems are accompanied by proofs. However, for many results, the corresponding analyses are quite involved and detailed proofs are beyond the scope of the chapter.

3.2 Classical Scheduling Problems

In the first part of this section we study online algorithms for makespan minimization, which represents one of the most basic problems in scheduling theory. The second part addresses flow time based objectives. Finally we consider load balancing problems where jobs have a temporary duration. Such jobs arise, e.g., in the context of telephone calls or network routing requests.

3.2.1 Makespan Minimization

Consider a basic scenario where we are given m identical machines working in parallel. As input we receive a sequence of jobs $I = J_1, J_2, \ldots, J_n$ with individual processing times, i.e., J_i has a processing time of p_i time units. The jobs arrive incrementally, one by one. Whenever a new job arrives, its processing time is known. The job has to be assigned immediately and irrevocably to one of the machines without knowledge of any future jobs. Preemption of jobs is not allowed. It is assumed that the processing of jobs may begin at a common starting point, say at time 0. The goal is to minimize the makespan, i.e., the completion time of the last job that finishes in the schedule.

This fundamental scenario was investigated by Graham [27] in 1966 who devised the famous *List* scheduling algorithm (see also Section 2.3.2). At any given time let the *load* of a machine be the sum of the processing times of the jobs currently assigned to it.

Algorithm List: Schedule any new job on the least loaded machine.

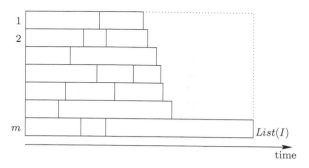

FIGURE 3.1: The schedule generated by *List*.

THEOREM 3.1 [27]
List is $(2 - \frac{1}{m})$*-competitive.*

PROOF   For an arbitrary job sequence $I = J_1, \ldots, J_n$, consider the schedule constructed by *List* and let $List(I)$ be the resulting makespan. Without loss of generality we number the machines in order of non-decreasing final loads in *List*'s schedule. Then the load on machine m defines the makespan. Consider a time interval of length $List(I)$ on any of the m machines, cf. Figure 3.1. The last machine m processes jobs continuously without interruption. The first $m-1$ machines each process a subset of the jobs and then experience an (possibly empty) idle period. During their active periods the m machines finish a total processing volume of $\sum_{i=1}^{n} p_i$. Consider the assignment of the last job on machine m. By the *List* scheduling rule, this machine had the smallest load at the time of the assignment. Hence, any idle period on the first $m-1$ machines cannot be longer than the processing time of the last job placed on machine m and hence cannot exceed the maximum processing time $\max_{1 \leqslant i \leqslant n} p_i$. We conclude

$$m\, List(I) \leqslant \sum_{i=1}^{n} p_i + (m-1) \max_{1 \leqslant i \leqslant n} p_i,$$

which is equivalent to

$$List(I) \leqslant \frac{1}{m} \sum_{i=1}^{n} p_i + \left(1 - \frac{1}{m}\right) \max_{1 \leqslant i \leqslant n} p_i.$$

Note that $\frac{1}{m} \sum_{i=1}^{n} p_i$ is a lower bound on $OPT(I)$ because the optimum makespan cannot be smaller than the average load on all the machines. Furthermore, $OPT(I) \geqslant \max_{1 \leqslant i \leqslant n} p_i$ because the largest job must be processed on some machine. We conclude $List(I) \leqslant (2 - \frac{1}{m})OPT(I)$. ▯

Graham also showed a matching lower bound on the performance of *List*.

FIGURE 3.2: Schedules generated by *List* and *OPT*.

THEOREM 3.2
List does not achieve a competitive ratio smaller than $2 - \frac{1}{m}$.

PROOF   Consider a job sequence I consisting of (a) $m(m-1)$ jobs, each having a processing time of 1, followed by (b) one job having a processing time of m time units. The resulting schedules generated by *List* and *OPT* are depicted in Figure 3.2. The *List* algorithm assigns the first $m(m-1)$ jobs in a *Round-Robin* fashion to machines so that a load of $m-1$ is generated. The final job of processing time m then causes a makespan of $2m-1$. On the other hand *OPT* schedules the initial small jobs on $m-1$ machines only, reserving the last machine for the final job. This gives a makespan of m, and the desired performance ratio follows.[1]

□

   Faigle, Kern, and Turan [22] showed that no deterministic online algorithm can have a competitive ratio smaller than $2 - \frac{1}{m}$ for $m = 2$ and $m = 3$. Thus, for these values of m, *List* is optimal. In the 1990s research focused on finding improved online algorithms for a general number m of machines. Galambos and Woeginger [24] presented an algorithm that is $(2 - \frac{1}{m} - \epsilon_m)$-competitive, where $\epsilon_m > 0$, but ϵ_m tends to 0 as m goes to infinity. The first online algorithm that achieved a competitive ratio asymptotically smaller than 2 was given by Bartal et al., [16]. Their algorithm is 1.986-competitive. The strategy was generalized by Karger, Phillips, and Torng [33] who proved an upper bound of 1.945. Later, Albers presented a new algorithm that is 1.923-competitive [1]. The strategy was modified by Fleischer and Wahl [23] who showed a bound of 1.9201, the best performance ratio known to date. We briefly describe the algorithm.
   The goal of all improved algorithms, beating the bound of $2 - \frac{1}{m}$, is to maintain an *imbalanced* schedule in which some machines are lightly loaded and some are heavily loaded. In case a large job arrives, it can be assigned to

---

[1]Note that we have independent jobs in this example, contrary to the more complicated offline example of Section 2.3.2.

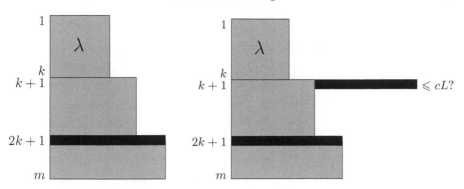

FIGURE 3.3: Schedules maintained by *Imbal*. Left part: A sample profile. Right part: The scheduling step.

a lightly loaded machine so that a makespan of $2 - \frac{1}{m}$ times the optimum value is prevented. Formally, let $c = 1 + \sqrt{(1 + \ln 2)/2}$ and, using this definition,

$$k = \left\lfloor \frac{2(c-1)^2 - 1}{c} m \right\rfloor + 1 \quad \text{and} \quad \alpha = \frac{2c - 3}{2(c - 1)}.$$

We note that $k \approx \lfloor 0.36m \rfloor + 1$ and $\alpha \approx 0.46$. The algorithm by Fleischer and Wahl, called *Imbal*, tries to maintain a schedule in which k machines are lightly loaded and $m - k$ are heavily loaded. At any time we number the machines in order of non-decreasing current load, where the load of a machine is again the total processing time of jobs presently assigned to that machine. Let l_i denote the load on the i-th smallest machine, $1 \leqslant i \leqslant m$. The left part of Figure 3.3 shows a sample schedule. Moreover, let $\lambda = \frac{1}{k} \sum_{i=1}^{k} l_i$ be the average load on the k smallest machines. This average load is always compared to the load on machine $2k + 1$. The goal is to maintain an *imbalanced* schedule in which $\lambda \leqslant \alpha l_{2k+1}$.

Each new job J_t is scheduled either on the machine with the smallest load or on the machine with the $(k+1)$-st smallest load. The decision which machine to choose depends on the current schedule. If the schedule is imbalaced, job J_t is placed on the machine with the smallest load. On the other hand, if the desired invariant $\lambda \leqslant \alpha l_{2k+1}$ does not hold, the algorithm considers scheduling J_t on the machine with the $(k + 1)$-st smallest load, cf. the right part of Figure 3.3. The algorithm computes the resulting load $l_{k+1} + p_t$ on that machine and compares it to the average load $L = \frac{1}{m} \sum_{j=1}^{t} p_j$ on the machines after J_t is assigned. Note that L is a lower bound on the optimum makespan. If $l_{k+1} + p_t \leqslant cL$, then J_t is placed on machine $k + 1$. Otherwise this assignment is risky and J_t is processed on the machine with the smallest load.

Algorithm Imbal: Schedule a new job J_t on the machine with the $(k+1)$st smallest load if $\lambda > \alpha l_{2k+1}$ and $l_{k+1} + p_t \leqslant cL$. Otherwise schedule J_t on the

FIGURE 3.4: An ideal schedule produced by *Rand-2*.

machine having the smallest load.

THEOREM 3.3 [23]
Imbal achieves a competitive ratio of $1 + \sqrt{(1 + \ln 2)/2} < 1.9201$ *as* $m \to \infty$.

We next turn to lower bounds for deterministic online algorithms. Recall that a result by Faigle, Kern, and Turan [22] states that no deterministic strategy can be better than $(2 - \frac{1}{m})$-competitive, for $m = 2$ and $m = 3$. Lower bounds for a general number m of machines were developed, for instance, in [1, 16, 22]. The best lower bound currently known is due to Rudin [45].

THEOREM 3.4 [45]
No deterministic online algorithm can achieve a competitive ratio smaller than 1.88.

An interesting open problem is to determine the exact competitiveness achievable by deterministic online strategies.

Since the publication of the paper by Bartal et al. [16], there has also been research interest in developing randomized online algorithms for makespan minimization. Bartal et al. gave a randomized algorithm for two machines that achieves an optimal competitive ratio of $4/3$. This algorithm, called *Rand-2* operates as follows. For a given two-machine schedule let the *discrepancy* be the load difference on the two machines. The algorithm tries to always maintain a schedule in which the expected discrepancy is $\frac{1}{3}L$, where $L = \sum_{j=1}^{t} p_j$ is the total processing time of jobs that have arrived so far. Figure 3.4 shows a sample schedule.

Algorithm Rand-2: Maintain a set of all schedules generated so far together with their probabilities. When a new job J_t arrives, compute the overall expected discrepancy E_1 that results if J_t were placed on the least loaded machine in each schedule. Similarly, compute the expected discrepancy E_2 if J_t were assigned to the most loaded machine in each schedule. Determine a p, $0 \leqslant p \leqslant 1$, such that $pE_1 + (1 - p)E_2 \leqslant \frac{1}{3}L$. If such a p exists, with probability p schedule J_t on the least loaded machine and with probability $1 - p$ assign it to the most loaded machine in each schedule. If such p does not exist, assign J_t to the least loaded machine.

THEOREM 3.5 [16]
Rand-2 achieves a competitive ratio of 4/3. This is optimal.

Chen, van Vliet, and Woeginger [21] and Sgall [47] proved that no randomized online algorithm can have a competitiveness smaller than $1/(1 - (1 - 1/m)^m)$. This expression tends to $e/(e - 1) \approx 1.58$ as $m \to \infty$. Seiden [46] presented a randomized algorithm whose competitive ratio is smaller than the best known deterministic ratio for $m \in \{3, \ldots, 7\}$. The competitiveness is also smaller than the deterministic lower bound for $m = 3, 4, 5$.

Recently, Albers [2] developed a randomized online algorithm that is 1.916-competitive, for all m, and hence gave the first algorithm that performs better than known deterministic algorithms for general m. She also showed that a performance guarantee of 1.916 cannot be proven for a deterministic online algorithm based on analysis techniques that have been used in the literature so far. An interesting feature of the new randomized algorithm, called *Rand*, is that at most two schedules have to be maintained at any time. In contrast, the algorithms by Bartal et al. [16] and by Seiden [46] have to maintain t schedules when t jobs have arrived. The *Rand* algorithm is a combination of two deterministic algorithms A_1 and A_2. Initially, when starting the scheduling process, *Rand* chooses A_i, $i \in \{1, 2\}$, with probability $\frac{1}{2}$ and then serves the entire job sequence using the chosen algorithm. Algorithm A_1 is a conservative strategy that tries to maintain schedules with a low makespan. On the other hand, A_2 is an aggressive strategy that aims at generating schedules with a high makespan. A challenging open problem is to design randomized online algorithms that beat the deterministic lower bound, for all m.

3.2.2 Flow Time Objectives

Minimizing the flow time of jobs is another classical objective in scheduling. In an online setting we receive a sequence $I = J_1, \ldots, J_n$ of jobs, where each job J_i is specified by an arrival time r_i and a processing time p_i. Clearly the arrival times satisfy $r_i \leqslant r_{i+1}$, for $1 \leqslant i \leqslant n - 1$. Preemption of jobs is allowed, i.e., the processing of a job may be stopped and resumed later. We are interested in minimizing the *flow time* of jobs. The flow time of a job is the length of the time period between arrival time and completion of the job. Formally, suppose that a job released at time r_i is completed at time c_i. Then the flow time is $f_i = c_i - r_i$.

Two scenarios are of interest. In *clairvoyant scheduling*, when J_i arrives, its processing time p_i is known. This assumption is realistic in classical manufacturing or, w.r.t. new application areas, in the context of a web server delivering static web pages. In *non-clairvoyant scheduling*, when J_i arrives, p_i is unknown and becomes known only when the job finishes. This assumption is realistic in operating systems.

We first study clairvoyant scheduling and focus on the objective of minimizing the sum of flow times $\sum_{i=1}^{n} f_i$. The most classical scheduling algorithm is

Shortest Remaining Processing Time.

Algorithm Shortest Remaining Processing Time (SRPT): At any time execute the job with the least remaining work.

It is well known and easy to verify that *SRPT* constructs optimal schedules on one machine, see e.g. [12].

THEOREM 3.6
For one machine, SRPT is 1-competitive.

For m machines, *SRPT* also achieves the best possible performance ratio, but the analysis is considerably more involved. A first, very sophisticated analysis was given by Leonardi and Raz [37]. A simplified proof was later presented by Leonardi [36]. In the following let $p_{\min} = \min_{1 \leqslant i \leqslant n} p_i$ and $p_{\max} = \max_{1 \leqslant i \leqslant n} p_i$ be the smallest and largest processing times, respectively. Moreover, $P = p_{\max}/p_{\min}$.

THEOREM 3.7 [38]
For m machines, SRPT has a competitive ratio of $O(\min\{\log P, \log(n/m)\})$.

The above competitiveness is best possible. Even randomized strategies cannot achieve better asymptotic bounds.

THEOREM 3.8 [38]
For m machines, any randomized online algorithm has a competitive ratio of $\Omega(\log(n/m))$ and $\Omega(\log P)$.

While *SRPT* is a classical algorithm and achieves an optimal performance ratio, it uses *migration*, i.e., a job when being preempted may be moved to another machine. In many practical settings, this is undesirable as the incurred overhead is large. Awerbuch et al. [7] gave a refined algorithm that does not use migration and is $O(\min\{\log P, \log n\})$-competitive. Chekuri, Khanna, and Zhu [20] gave a non-preemptive algorithm achieving an optimal performance of $O(\min\{\log P, \log(n/m)\})$.

The algorithms mentioned above have optimal competitive ratios; however the performance guarantees are not constant. Interestingly, it is possible to improve the bounds using *resource augmentation*, i.e., an algorithm is given processors of higher speed. Here it is assumed that an optimal offline algorithms operates with machines of speed 1, while an online strategy may use machines running at speed $s \geqslant 1$. In this context, the best result is due to McCullough and Torng [40] who showed that, using speed $s \geqslant 2 - 1/m$, *SRPT* is $(1/s)$-competitive.

An interesting, relatively new performance measure is the *stretch* of a job, which is defined as the flow time divided by the processing time, i.e. $st_i = f_i/p_i$. The motivation for this metric is that a user is willing to wait longer for the completion of long jobs; a quick response is expected for short jobs. Consider the objective of minimizing the total stretch $\sum_{i=1}^{n} f_i/p_i$ of jobs. Muthukrishan et al. [42] showed that *SRPT* performs very well.

THEOREM 3.9 [42]
On one machine SRPT is 2-competitive.

Muthukrishan et al. [42] established an almost matching lower bound on the performance of *SRPT*. Legrand, Su, and Vivien [35] proved that no online algorithm can be better than 1.1948-competitive. We next turn to parallel machines.

THEOREM 3.10 [42]
For m machines, SRPT is 14-competitive.

Chekuri, Khanna, and Zhu [20] developed an improved 9.82-competitive algorithm and presented a 17.32-competitive strategy not using any migration.

We next address non-clairvoyant scheduling where the processing time of an incoming job is not known in advance. We focus again on the objective of minimizing the total flow time $\sum_{i=1}^{n} f_i$ of jobs and concentrate on one machine. In this setting a natural algorithm is *Round Robin*, which always assigns an equal amoung of processing resources to all the jobs. Kalyanasundaram and Pruhs [31] showed that *Round Robin* does not perform well relative to the optimum; its competitiveness is at least $\Omega(n/\log n)$. In fact any deterministic algorithm has a high competitive ratio.

THEOREM 3.11 [41]
Any deterministic online algorithm has a competitive ratio of $\Omega(n^{1/3})$.

Again, resource augmentation proves to be a very powerful tool in this context. Kalyanasundaram and Pruhs [31] analyzed the following algorithm.

Algorithm Shortest Elapsed Time First (SETF): At any time execute the job that has been processed the least.

THEOREM 3.12 [31]
For any $\epsilon > 0$, using a speed of $1 + \epsilon$, SETF achieves a competitive ratio of $1 + 1/\epsilon$.

Finally, using randomization it is possible to get down to a logarithmic bound of $\Theta(\log n)$ without resource augmentation, see [17] and [32].

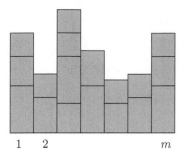

FIGURE 3.5: A load profile with m machines.

3.2.3 Load Balancing

In this section we study load balancing problems that arise in new applications. Consider, for instance, a set of satellite links on which phone calls have to be scheduled or, alternatively, a set of network links on which data transmission requests have to be served. In these scenarios the requests have an unknown duration and incur a certain load or congestion on the chosen link. The goal is to minimize the total load that ever occurs on any of the links. These problem settings can be formalized as follows.

We are given a set of m machines (representing the available links). A sequence of jobs $I = J_1, J_2, \ldots, J_n$ arrives online. Let r_i denote the arrival time of J_i. Job J_i has an unknown duration and incurs a load of l_i when assigned to a machine. For any time t, let $L_j(t)$ denote the load of machine j, $1 \leqslant i \leqslant m$, which is the sum of the loads of the jobs present on machine j at time t. The goal is to minimize the maximum load that occurs during the processing of I. Figure 3.5 depicts a sample schedule with m machines. The rectangles represent jobs, where the height of a rectangle corresponds to the load of the job.

We concentrate on settings with m identical machines. A natural online strategy is *Greedy* which assigns an incoming job to a machine currently having the smallest load. Azar and Epstein [10] observed that *Greedy* is $(2 - \frac{1}{m})$-competitive. They also proved that this is the best competitiveness achievable by a deterministic strategy. In the following we will study the scenario with identical machines and *restricted assignment*, i.e., each job J_i can only be assigned to a subset M_i of admissible machines. Azar, Broder, and Karlin [8] showed that *Greedy* is $\Theta(m^{2/3})$-competitive. They also proved that no online algorithm can achieve a competitiveness smaller than $\Omega(\sqrt{m})$. Azar et al. [11] gave a matching upper bound. The corresponding algorithm is called *Robin Hood*. Since both the algorithm and its analysis are simple and elegant, we give the details.

At any time *Robin Hood* maintains a lower bound B on the optimum load $OPT(I)$ incurred on the given job sequence I. Initially, let $B(0) = 0$. At any time t, this bound is updated as follows. If no new job arrives at time t, then $B(t) = B(t-1)$. If a new job J_i arrives at time $t = r_i$, then the update is as follows. Again, let $L_j(t-1)$ denote the load on machine j at the end of time step $t-1$.

$$B(t) := \max \left\{ B(t-1), \quad l_i, \quad \frac{1}{m}\left(l_i + \sum_{j=1}^{m} L_j(t-1)\right) \right\}.$$

Clearly, $B(t) \leqslant OPT(I)$ as we are interested in the maximum load that ever occurs on the machines. At any time t a machine is called *rich* if its current load is at least $\sqrt{m}B(t)$; otherwise the machine is called *poor*.

Algorithm Robin Hood: When a new job J_i arrives, if possible assign it to a poor machine. Otherwise assign it to the rich machine that became rich most recently.

THEOREM 3.13 [11]
Robin Hood is $O(\sqrt{m})$-competitive.

PROOF   We start with a simple lemma estimating the number of rich machines.

LEMMA 3.1
There always exist at most $\sqrt{m}$ machines that are rich.

PROOF   The number of rich machines can only increase when new jobs are assigned to machines. So consider any time $t = r_i$. If more than $\sqrt{m}$ machines were rich after the assignment of J_i, then the aggregate load on the m machines would be strictly greater than $\sqrt{m}\sqrt{m}B(t) = mB(t)$. However, by the definition of $B(t)$, $B(t) \geqslant \frac{1}{m}(l_i + \sum_{j=1}^{m} L_j(t-1))$, i.e. $mB(t)$ is an upper bound on the aggregate load. ☐

In order to establish the theorem we prove that whenever *Robin Hood* assigns a job J_i to a machine j, the resulting load on the machine is upper bounded by $(2\sqrt{m} + 1)OPT(I)$.

First suppose that j is poor. Then the new load on the machine is $L_j(r_i - 1) + l_i < \sqrt{m}B(r_i) + B(r_i) \leqslant (\sqrt{m} + 1)OPT(I)$.

Next suppose that machine j is rich when J_i is assigned. Let $r_{t(i)}$ be the most recent point in time when machine j became rich. Furthermore, let S be the set of jobs that are assigned to machine j in the interval $(r_{t(i)}, r_i]$. Any job $J_k \in S$ could only be assigned to machines that were rich at time $r_{t(i)}$ because *Robin Hood* places a job on the machine that became rich most recently if no poor machines are available. Let

$$h = \left| \bigcup_{J_k \in S} M_k \right|.$$

Recall that M_k is the set of machines to which J_k can be assigned. Then, by the Lemma 3.1 $h \leqslant \sqrt{m}$. Since only $\sqrt{m}$ machines are available for jobs in S we have $OPT(I) \geqslant \frac{1}{\sqrt{m}} \sum_{J_k \in S} l_k$ and hence

$$\sum_{J_k \in S} l_k \leqslant \sqrt{m} \, OPT(I).$$

Hence the resulting load on machine j after the assignment of J_i is at most

$$L_j(r_{t(i)} - 1) + l_{t(i)} + \sum_{J_k \in S} l_k < \sqrt{m} B(r_{t(i)}) + l_{t(i)} + \sqrt{m} \, OPT(I)$$

$$\leqslant (2\sqrt{m} + 1) OPT(I).$$

This concludes the proof. $\qquad\qquad\qquad\qquad\qquad\qquad\qquad\qquad\quad$ ⬜

Further work on online load balancing can be found, e.g., in [5, 9].

3.3 Energy-Efficient Scheduling

In many computational environments energy has become a scare and/or expensive resource. Consider, for instance, battery-operated devices such as laptops or mobile phones. Here the amount of available energy is severely limited. By performing tasks with low total energy, one can considerably extend the lifetime of a given device. Generally speaking, the energy consumption in computer systems has grown exponentially over the past years. This increase is strongly related to Moore's law which states that the number of transistors that can be placed on an integrated circuit doubles approximately every two years. Since transistors consume energy, increased transistor density leads to increased energy consumption. Moreover, electricity costs impose a substantial strain on the budget of data and computing centers, where servers and, in particular, CPUs account for 50–60% of the energy consumption. In fact, Google engineers, maintaining thousands of servers, recently warned that if power consumption continues to grow, power costs can easily overtake hardware costs by a large margin [15]. Finally, a high energy consumption is critical because most of the consumed energy is eventually converted into heat which can harm the electronic components of a system.

In this section we study algorithmic techniques to save energy. It turns out that these techniques are actually scheduling strategies. There exist basically two approaches.

- **Power-down mechanisms:** When a system is idle, move it into lower power stand-by or sleep modes. We are interested in scheduling algorithms that perform the transitions at the "right" points in time.

- **Dynamic speed scaling:** Modern microprocessors, such as Intel XScale, Intel Speed Step, or AMD Power Now, can run at variable speed. The higher the speed, the higher the power consumption. We seek algorithms that use the speed spectrum in an optimal way.

Obviously, the goal of both the above techniques is to minimize the consumed energy. However, this has to be done subject to certain constraints, i.e., we have to design feasible schedules or must provide a certain quality of service to a user.

Over the last years there has been considerable research interest in scheduling strategies saving energy. In the following we first address power-down mechanisms and then study dynamic speed scaling.

3.3.1   Power-Down Mechanisms

We start by analyzing a simple two-state system. Consider a processor or machine that can reside in one of two possible states. Firstly, there is an *active state* that consumes one energy unit per time unit. Secondly, there exists a *sleep state* consuming zero energy units per time unit. Transitioning the machine from the active to the sleep state and, at some later point, back to the active state requires D energy units, where $D \geqslant 1$. As input we receive an alternating sequence of active and idle time periods. During each active period the machine has to be in the active mode. During any idle period the machine has the option of powering down to the sleep state. The goal is to find a schedule, specifying the (possible) power-down time for each idle interval, that minimizes the total energy consumption.

Since the energy consumption is fixed in the active periods, optimization strategies focus on the idle periods. In the offline scenario, the length of any idle period is known in advance. In the online setting the length of an idle period is not known in advance and becomes known only when the period ends. We remark that we ignore the latency incurred by a power-up operation to the active mode; we focus on the mere energy consumption. In the following we present online algorithms that, for any idle time period I, incur an energy consumption that is at most c times the optimum consumption in I, for some $c \geqslant 1$. Obviously, such an algorithm is c-competitive, for any alternating sequence of active and idle time periods. We note that the problem of minimizing energy consumption in any idle period I is equivalent to the famous ski rental problem [29].

In the following we focus on one particular idle time period I. An optimal offline algorithm is easy to state. If the length of I is larger D time units, power down immediately at the beginning of I. Otherwise reside in the active mode throughout I. Next we present an optimal online algorithm.

Algorithm Alg-2: Power down to the sleep mode after D time units if the idle period I has not ended yet.

THEOREM 3.14
Alg-2 is 2-competitive.

PROOF Let T denote the length of I. If $T < D$, then *Alg-2* does not power down in I and incurs an energy consumption of T, which is equal to the consumption of an optimal offline algorithm OPT. On the other hand, if $T \geqslant D$, *Alg-2* powers down after D time units and the total energy consumption is $D + D = 2D$. In this case OPT pays a cost of D, and the desired performance ratio follows. ▯

THEOREM 3.15
No deterministic online algorithm A can achieve a competitive ratio smaller than 2.

PROOF An adversary observes the behavior of A during an idle time period I. As soon as A powers down to the sleep state, the adversary terminates I. The energy consumption of A is $T + D$, where $T = |I|$ is the length of the idle period. The theorem now follows because an optimal offline algorithm pays $\min\{T, D\}$. ▯

Using randomization one can improve the competitive ratio.

Algorithm RAlg-2: In an idle period power down to the sleep mode after t time units according to the probability density function p_t, where

$$p_t = \begin{cases} \frac{1}{(e-1)D}e^{t/D} & 0 \leqslant t \leqslant D \\ 0 & \text{otherwise} \end{cases}$$

THEOREM 3.16 [34]
RAlg-2 achieves a competitive ratio of $e/(e-1) \approx 1.58$.

The above performance guarantee is best possible.

THEOREM 3.17 [34]
No randomized online algorithm achieves a competitive ratio smaller than $e/(e-1) \approx 1.58$.

In practice, rather than in worst-case analysis, one might be interested in a probabilistic setting where the length of idle time periods is governed by a probabiliy distribution. Let $Q = (q_T)_{0 \leqslant T < \infty}$ be a probability distribution on the length of idle periods. Let A_t be the deterministic algorithm that powers down after t time units. The expected energy consumption of A_t on idle periods whose length is generated according to Q is

$$E[A_t(I_Q)] = \int_0^t T q_T dT + (t + D) \int_t^\infty q_T dT.$$

Let A_Q^* be the algorithm A_t that minimizes the above expression. This algorithm performs well relative to the expected optimum cost $E[OPT(I_Q)]$.

THEOREM 3.18 [34]
For any probability distribution Q, the best corresponding algorithm A_Q^* satisfies

$$E[A_Q^*(I_Q)] \leqslant \frac{e}{(e-1)} E[OPT(I_Q)].$$

Irani, Gupta, and Shukla [28] and Augustine, Irani, and Swamy [6] extended many of the above results to systems/machines that can reside in several states. A specification of such systems is given, for instance, in the Advanced Configuration and Power Interface (ACPI), which establishes industry-standard interfaces enabling power management and thermal management of mobile, desktop and server platforms.

In general, consider a system/machine consisting of $l + 1$ states $s_0, \ldots, s_l$, where s_0 is the active mode. Let e_i denote the energy consumption per time unit in state s_i, $0 \leqslant i \leqslant l$. These rates satisfy $e_i > e_{i+1}$, for $0 \leqslant i \leqslant l - 1$. Furthermore, let δ_{ij} denote the cost of transitioning from state s_i to s_j. We assume that the triangle inequality $\delta_{ij} \leqslant \delta_{ik} + \delta_{kj}$ holds for any i, j and k.

Augustine, Irani, and Swamy [6] first argue that if the machine powers up, then it powers up to the active mode. Furthermore, we can assume without loss of generality that the power-up cost to the active state is zero, i.e., $\delta_{i0} = 0$, for any $1 \leqslant i \leqslant l$. If $\delta_{i0} > 0$, for some i, then we can define a new system with $\delta'_{ij} = \delta_{ij} + \delta_{j0} - \delta_{i0}$ for $i < j$ and $\delta'_{ij} = 0$, for $j < i$. The total transition cost in the new system is exactly equal to the cost in the original one.

Let $D(i) = \delta_{0i}$ be the power-down cost into state i. Then the energy consumption of an optimal offline algorithm OPT in an idle period of length t is

$$OPT(t) = \min_{0 \leqslant i \leqslant l} \{D(i) + e_i t\}.$$

Interestingly, the optimal cost has a simple graphical representation, cf. Figure 3.6. If we consider all linear functions $f_i(t) = D(i) + e_i t$, then the optimum energy consumption is given by the lower envelope of the arrangement of lines. We can use this lower envelope to guide an online algorithm which state to use at any time. Let $S(t)$ denote the state used by OPT in an idle period

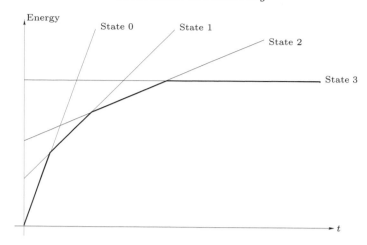

FIGURE 3.6: The optimum solution in a multi-state system.

of length t, i.e. $S(t)$ is the state $\arg\min_{0\leqslant i\leqslant l}\{D(i) + e_i t\}$. The following algorithm traverses the state sequence as suggested by the optimum offline algorithm.

Algorithm Lower Envelope (LE): In an idle period, at any time t, use state $S(t)$.

Irani, Gupta, and Shukla [28] analyzed the above algorithm in *additive systems*, where for any states $i < k < j$ we have $\delta_{ij} = \delta_{ik} + \delta_{kj}$.

THEOREM 3.19 [28]
Algorithm LE is 2-competitive.

Algorithm *LE* can be extended to work in non-additive systems where, in general, the triangle inequality $\delta_{ij} \leqslant \delta_{ik} + \delta_{kj}$ holds for any $i < k < j$. Let $\mathcal{S} = \{s_0, \ldots, s_l\}$ be the original set of states. We first construct a state set $\mathcal{S}' \subseteq \mathcal{S}$ such that for any states $s_i, s_j \in \mathcal{S}'$, with $i < j$, relation $\gamma D(i) \leqslant D(j)$ is satisfied, where $\gamma = 1 + 1/\sqrt{2}$. Such a set is easy to construct. Initially, let $\mathcal{S}' = \{s_l\}$. We then traverse the original states in order of decreasing index. Let s_j be the last state added to $\mathcal{S}'$. We now determine the largest i, where $i < j$, such that $\gamma D(i) \leqslant D(j)$. This state s_i is next added to $\mathcal{S}'$. At the end of the construction we have $s_0 \in \mathcal{S}'$ because $D(0) = 0$. Let OPT' denote the optimum offline algorithm using state set $\mathcal{S}'$ and let $S'(t)$ be the state used by OPT' in an idle period of length t.

Algorithm LE': In an idle period, at any time t, use state $S'(t)$.

THEOREM 3.20 [6]
Algorithm LE' achieves a competitive ratio of $3 + 2\sqrt{2} \approx 5.8$.

The above theorem ensures a competitiveness of $3 + 2\sqrt{2}$ in *any* multi-state system satisfying the triangle inequality. Better performance ratios are possible for specific systems. Augustine, Irani, and Swamy [6] developed an algorithm that achieves a competitive ratio of $c^* + \epsilon$, for any $\epsilon > 0$, where c^* is the best competitive ratio achievable for a given system. The main idea of the solution is to construct an efficient strategy A that decides, for a given c, if a c-competitive algorithm exists for the given architecture. The best value of c can then be determined using binary search in the interval $[1, 3 + 2\sqrt{2}]$.

THEOREM 3.21 [6]
A $(c^* + \epsilon)$-competitive online algorithm can be constructed in $O(l^2 \log l \log(\frac{1}{\epsilon}))$ time.

Irani, Gupta, and Shukla [28] and Augustine, Irani, and Swamy [6] also present various results for scenarios where the length of idle periods is governed by probability distributions. Furthermore, the first paper [28] contains an interesting experimental study on an IBM Hard Drive with four states.

3.3.2 Dynamic Speed Scaling

In this section we study the problem of dynamically adjusting the speed of a variable-speed processor/machine so as to minimize the total energy consumption. Consider a machine that can run at variable speed s. The higher the speed, the higher the energy consumption is. More formally, at speed s, the energy consumption is $E(s) = s^\alpha$ per time unit, where $\alpha > 1$ is a constant. In practical applications, α is typically in the range $[2, 3]$.

Over the past years, dynamic speed scaling has received considerable research interest and several scheduling problems have been investigated. However, most of the previous work focuses on deadline-based scheduling, a scenario considered in a seminal paper by Yao, Demers, and Shenker [49]. In this setting we are given a sequence $I = J_1, \ldots, J_n$ of jobs. Job J_i is released at time r_i and must be finished by a deadline d_i. We assume $r_i \leq r_{i+1}$, for $1 \leq i < n$. To finish J_i a processing volume of p_i must be completed. This processing volume, intuitively, can be viewed as the number of CPU cycles necessary to complete the job. The time it takes to finish the job depends on the processor speed. Using, for instance, a constant speed s, the execution time is p_i/s. Of course, over time, a variable speed may be used. Preemption of jobs is allowed. The goal is to construct a feasible schedule, observing the release times and deadlines, that minimize the total energy consumption.

The paper by Yao, Demers, and Shenker [49] assumes that (a) the machine can run at a continuous spectrum of speeds and (b) there is no upper

bound on the maximum speed. Condition (b) ensures that there is always
a feasible schedule. Later, we will discuss how to remove these constraints.
For the above deadline-based scheduling problem, again, two scenarios are of
interest. In the offline setting, all jobs of I along with their characteristics are
completely known in advance. In the online variant of the problem, the jobs
arrive over time. At any time future jobs are unknown. It turns out that the
offline problem is interesting in itself and can be used to design good online
strategies. For this reason, we first address the offline problem and present
an algorithm proposed by Yao, Demers, and Shenker [49].

The strategy is known as the *YDS* algorithm, referring to the initials of the
inventors, and computes the *density* of time intervals. Given a time interval
$I = [t, t']$, the density is defined as

$$\Delta_I = \frac{1}{|I|} \sum_{[r_i, d_i] \subseteq I} p_i.$$

Intuitively, Δ_I is the minimum average speed necessary to complete all jobs
that must be scheduled in I. Let S_I be the set of jobs J_i that must be processed
in I, i.e., that satisfy $[r_i, d_i] \subseteq I$. Algorithm *YDS* repeatedly determines the
interval I of maximum density. In I it schedules the jobs of S_I at speed Δ_I
using the *Earliest Deadline First* policy. Then set S_I as well as time interval
I are removed from the problem instance.

Algorithm YDS: Initially, let $\mathcal{J} = \{J_1, \ldots, J_n\}$. While $\mathcal{J}$ is not empty,
execute the following two steps. (1) Determine the time interval $I = [t, t']$ of
maximum density Δ_I along with the job set S_I. In I process the jobs of S_I
at speed Δ_I according to the *Earliest Deadline First* policy. (2) Reduce the
problem instance by I. More specifically, set $\mathcal{J} := \mathcal{J} \setminus S_I$. For any $J_i \in \mathcal{J}$
with $r_i \in I$, set $r_i := t'$. For any $J_i \in \mathcal{J}$ with $d_i \in I$, set $d_i := t$. Remove I
from the time horizon.

Obviously, when identifying intervals of maximum density, it suffices to
consider $I = [t, t']$ for which the interval boundaries are equal to the release
times r_i and deadlines d_i of the jobs.

THEOREM 3.22 [49]
*Algorithm YDS constructs a feasible schedule that minimizes the total energy
consumption.*

Feasibility of the constructed schedule follows from the fact that, for each
interval I of maximum density identified in the various iterations of *YDS*,
the algorithm constructs a feasible schedule in I. Optimality follows from
the convexity of the energy consumption function $E(s) = s^\alpha$: Suppose that
the machine runs at speed s_1 for ϵ time units and at speed s_2 for another ϵ
time units. Assume $s_1 < s_2$. Then a schedule with a strictly smaller energy
consumption can be achieved by using speed $(s_1 + s_2)/2$ in both periods

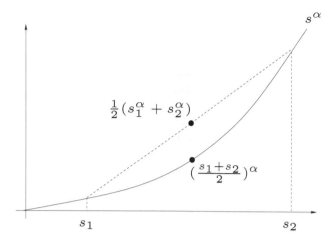

FIGURE 3.7: The convexity of the energy consumption function.

of length ϵ. This holds because $\epsilon s_1^\alpha + \epsilon s_2^\alpha > 2\epsilon(\frac{s_1+s_2}{2})^\alpha$ is equivalent to $\frac{1}{2}(s_1^\alpha + s_2^\alpha) > (\frac{s_1+s_2}{2})^\alpha$, and the latter inequality can easily be verified using Figure 3.7. This convexity argument implies that it is not reasonable to vary the speed in an interval I of maximum density. Moreover, it is not reasonable to increase the speed in I while reducing the speed outside I. Optimality of the resulting schedule then follows.

A straightforward implementation of *YDS* runs in time $O(n^3)$. Gaujal, Navet, and Walsh [26] and Gaujal and Navet [25] gave algorithms achieving improved running times for some specific classes of input instances.

Algorithm *YDS* assumes a continuous spectrum of speeds. In practice only a finite set of speed levels $s_1 < s_2 < \ldots < s_d$ is available. *YDS* can be adapted easily for feasible job instances, i.e., a feasible schedule exists for the available set of speeds. Obviously, feasibility can be checked easily by always using the maximum speed s_d and scheduling the available jobs according to the *Earliest Deadline First* policy. Given a feasible job instance, we first construct the schedule according to *YDS*. For each identified interval I of maximum density we approximate the desired speed Δ_I by the two adjacent speed levels $s_{k+1} > \Delta_I > s_k$. Speed s_{k+1} is used for the first δ time units and s_k is used for the last $|I| - \delta$ in I time units, where δ is chosen such that $\delta s_{k+1}+(|I|-\delta)s_k = |I|\Delta_I$. Here $|I|$ denotes the length of I. An algorithm with an improved running time of $O(dn \log n)$ was presented by Li and Yao [39].

We next turn to online algorithms and consider again a continuous unbounded spectrum of speeds. We assume that whenever a new job J_i arrives at time r_i, its deadline d_i and processing volume p_i are known. Yao, Demers, and Shenker [49] devised two elegant online strategies called *Average Rate* and *Optimal Available*. For any incoming job J_i, *Average Rate* considers the *density* $\delta_i = p_i/(d_i - r_i)$, which is the minimum average speed necessary to complete the job in time if no other jobs were present. At any time t the speed $s(t)$ is set to the accumulated density of unfinished jobs present at time t.

Algorithm Average Rate: At any time t use a speed $s(t) = \sum_{J_i : t \in [r_i, d_i]} \delta_i$. Among the available unfinished jobs, process the one whose deadline is earliest in the future.

THEOREM 3.23 [49]
For any $\alpha \geqslant 2$, the competitive ratio c of Average Rate satisfies $\alpha^\alpha \leqslant c \leqslant 2^{\alpha-1}\alpha^\alpha$.

The strategy *Optimal Available* is computationally more expensive in that it always computes an optimal schedule for the currently available work load. This can be done using algorithm *YDS*.

Algorithm Optimal Available: Whenever a new job arrives, compute an optimal schedule for the currently available, unfinished jobs.

Bansal, Kimbrel, and Pruhs [13] gave a comprehensive analysis of the above algorithm. It shows that *Optimal Available* is at least as good as *Average Rate*.

THEOREM 3.24 [13]
Optimal Available achieves a competitive ratio of exactly α^α.

Bansal, Kimbrel, and Pruhs [13] also presented a new online algorithm that tries to approximate the optimal speeds of *YDS* more closely. For times t, t_1 and t_2, let $w(t, t_1, t_2)$ be the total processing volume of jobs that have arrived by time t have a release time of at least t_1 and a deadline of at most t_2.

Algorithm BKP: At any time t use a speed of

$$s(t) = \max_{t' > t} \frac{w(t, et - (e-1)t', t')}{t' - t}.$$

Always process the available unfinished job whose deadline is earliest in the future.

THEOREM 3.25 [13]
Algorithm BKP achieves a competitive ratio of $2(\frac{\alpha}{\alpha-1})^\alpha e^\alpha$.

The competitiveness of all online algorithms presented so far depends exponentially on α. Bansal, Kimbrel, and Pruhs [13] demonstrated that this exponential dependence is inherent to the problem.

THEOREM 3.26 [13]
The competitive ratio of any randomized online algorithm is at least $\Omega((\frac{4}{3})^\alpha)$.

An interesting open problem is to settle the exact competitiveness that can be achieved by online strategies.

Chan et al. [19] present an online algorithm for the scenario that there exists a maximum speed $s_{\max}$; below this upper bound the spectrum of speeds is still continuous. The algorithm has the interesting feature that it can handle infeasible job instances by discarding jobs if the work load becomes too high. The proposed strategy achieves a constant factor in terms of throughput and in terms of energy. The throughput is the total processing volume of successfully completed jobs. Since we aim at throughput maximization, a strategy A is c-competitive w.r.t. this performance measure if the throughput A is at least $1/c$ times the optimum throughput.

Whenever a new job J_i arrives, the algorithm proposed by Chan et al. [19] checks if the job should be admitted. In this context, a set S of jobs is called *full-speed admissible* if all jobs in S can be completed by their deadline using a maximum speed of $s_{\max}$. Speed values are used according to *Optimal Available*.

Algorithm FSA(OAT): At any time maintain a set S of admitted jobs. Number the jobs $J_{i_1}, \ldots, J_{i_k}$ in S in order of non-decreasing deadlines. A new job J_i is admitted to S if (a) $S \cup \{J_i\}$ is full-speed admissible or if (b) $p_i > 2(p_{i_1} + \ldots + p_{i_l})$ and $\{J_i, J_{i_{l+1}}, \ldots, J_{i_k}\}$ is full-speed admissible, where $l \in [1, k]$ is the smallest integer satisfying this constraint. In the latter case $\{J_{i_1}, \ldots, J_{i_l}\}$ are removed from S. Whenever a job is finished or its deadline has passed, it is discarded from S. At any time t use the speed that *Optimal Available* would use for the workload in S, provided that this speed is not larger then $s_{\max}$. Otherwise use $s_{\max}$.

THEOREM 3.27 [19]
FSA(OAT) achieves a competitive ratio of 14 in terms of throughput and a competitive ratio of $\alpha^\alpha + \alpha^2 4^\alpha$ in terms of energy.

If there is no upper bound on the speed, *FSA(OAT)* mimicks *Optimal Available*. Hence it achieves an optimal throughput and a competitive ratio of α^α in terms of energy consumption.

So far in this section we have studied single-machine architectures. However, power consumption is also a major concern in multi-processor environments. Modern server systems are usually equipped with several CPUs. Furthermore, many laptops today feature a dual-processor architecture and some already include a quad-core chip. In the following we investigate the problem of minimizing energy consumption in parallel machine environments. We assume that we are given m identical variable-speed machines working in parallel. As before, we consider deadline-based scheduling where a sequence

$I = J_1, \ldots, J_n$, each specified by a release time, a deadline and a processing volume must be scheduled. Preemption of jobs is allowed. However migration of job is disallowed, i.e., whenever a job is preempted, it may not be moved to another machine as such an operation incurs considerable overhead in practice. The goal is to minimize the total energy consumed on all the m machines.

Albers, Müller, and Schmelzer [4] present a comprehensive study of the problem. They first consider the offline scenario and settle the complexity of settings with unit-size jobs, i.e. $p_i = 1$ for all $1 \leqslant i \leqslant n$. They show that problem instances with *agreeable deadlines* are polynomially solvable while instances with arbitrary deadlines are NP-hard. In practice, problem instances with agreeable deadlines form a natural input class where, intuitively, jobs arriving at later times may be finished later. Formally, deadlines are agreeable if, for any two jobs J_i and $J_{i'}$, relation $r_i < r_{i'}$ implies $d_i \leqslant d_{i'}$. We briefly describe the proposed algorithm as it is an interesting application of the *Round-Robin* strategy.

Algorithm RR: Given a sequence of jobs with agreeable deadlines, execute the following two steps. (1) Number the jobs in order of non-decreasing release dates. Jobs having the same release date are numbered in order of non-decreasing deadlines. Ties may be broken arbitrarily. (2) Given the sorted list of jobs computed in step (1), assign the jobs to machines using the *Round-Robin* policy. For each machine, given the jobs assigned to it, compute an optimal service schedule using, e.g., *YDS*.

THEOREM 3.28 [4]
For a set of unit size jobs with agreeable deadlines, algorithm RR computes an optimal schedule.

Paper [4] also develops various polynomial time approximation algorithms for both unit-size and arbitrary-size jobs. Again, let $\delta_i = p_i/(d_i - r_i)$ be the density of job J_i. The next algorithm partitions jobs into classes such that, within each class, job densities differ by a factor of at most 2. Formally, let $\Delta = \max_{1 \leqslant i \leqslant n} \delta_i$ be the maximum job density. Partition jobs $J_1, \ldots, J_n$ into classes C_k, $k \geqslant 0$, such that class C_0 contains all jobs of density Δ and C_k, $k \geqslant 1$, contains all jobs i with density $\delta_i \in [\Delta 2^{-k}, \Delta 2^{-(k-1)})$.

Algorithm Classified Round Robin (CRR): Execute the following two steps. (1) For each class C_k, first sort the jobs in non-decreasing order of release dates. Jobs having the same release date are sorted in order of non-decreasing deadline. Then assign the jobs of C_k to processors according to the *Round-Robin* policy, ignoring job assignments done for other classes. (2) For each processor, given the jobs assigned to it, compute an optimal service schedule.

THEOREM 3.29 [4]
CRR achieves an approximation factor of $\alpha^\alpha 2^{4\alpha}$ for problem instances consisting of (a) unit size jobs with arbitrary deadlines or (b) arbitrary size jobs with agreeable deadlines.

Albers, Müller, and Schmelzer [4] show that improved approximation guarantees can be achieved for problem settings where all jobs have a common release time or, symmetrically, have a common deadline. Furthermore, the authors give various competitive online algorithms.

In this section we have focused on speed scaling algorithms for deadline based scheduling problems. The literature also contains results on other objectives. References [3, 14, 44] consider the minimization of flow time objectives while keeping energy consumption low. Bunde additionally investigates makespan minimization [18].

3.4 Conclusion

In this chapter, we have surveyed important results in the area of online scheduling, addressing both classical results and contributions that were developed in the past few years. The survey is by no means exhaustive. As for results on classical scheduling problems, an extensive survey article was written by Pruhs, Sgall, and Torng [43] and is part of a comprehensive handbook on scheduling. The most promising and fruitful direction for future research is the field of energy-efficient scheduling. In this chapter we have presented some basic results. Another survey summarizing the state of the art was presented by Irani and Pruhs [30] in 2005. The field of energy-efficient scheduling is extremely active and new results are being published in the ongoing conferences. There is a host of open problems that deserves investigation.

References

[1] S. Albers. Better bounds for online scheduling. *SIAM Journal on Computing*, 29:459–473, 1999.

[2] S. Albers. On randomized online scheduling. In *34th ACM Symposium on Theory of Computing*, pages 134–143, 2002.

[3] S. Albers and H. Fujiwara. Energy-efficient algorithms for flow time

minimization. In *23rd International Symposium on Theoretical Aspects of Computer Science (STACS)*, volume 3884 of *LNCS*, pages 621–633. Springer, 2006.

[4] S. Albers, F. Müller, and S. Schmelzer. Speed scaling on parallel processors. In *19th ACM Symposium on Parallelism in Algorithms and Architectures*, pages 289–298, 2007.

[5] A. Armon, Y. Azar, and L. Epstein. Temporary tasks assignment resolved. *Algorithmica*, 36(3):295–314, 2003.

[6] J. Augustine, S. Irani, and C. Swamy. Optimal power-down strategies. *SIAM Journal on Computing*, 37(5):1499–1516, 2008.

[7] B. Awerbuch, Y. Azar, S. Leonardi, and O. Regev. Minimizing the flow time without migration. *SIAM Journal on Computing*, 31(5):1370–1382, 2002.

[8] Y. Azar, A. Broder, and A. Karlin. On-line load balancing. *Theoretical Computer Science*, 130(1):73–84, 1994.

[9] Y. Azar, A. Epstein, and L. Epstein. Load balancing of temporary tasks in the l_p norm. *Theoretical Computer Science*, 361(2-3):314–328, 2006.

[10] Y. Azar and L. Epstein. On-line load balancing of temporary tasks on identical machines. *SIAM Journal on Discrete Mathematics*, 18(2):347–352, 2004.

[11] Y. Azar, B. Kalyanasundaram, S. Plotkin, K. Pruhs, and O. Waarts. On-line load balancing of temporary tasks. *Journal of Algorithms*, 22(1):93–110, 1997.

[12] K. Baker. *Introduction to Sequencing and Scheduling*. John Wiley & Sons, New York, 1974.

[13] N. Bansal, T. Kimbrel, and K. Pruhs. Dynamic speed scaling to manage energy and temperature. *Journal of the ACM*, 54(1), 2007.

[14] N. Bansal, K. Pruhs, and C. Stein. Speed scaling for weighted flow time. In *18th Annual ACM-SIAM Symposium on Discrete Algorithms*, pages 805–813, 2007.

[15] L. Barroso. The price of performance. *ACM Queue*, 3(7):48–53, Sept. 2005.

[16] Y. Bartal, A. Fiat, H. Karloff, and R. Vohra. New algorithms for an ancient scheduling problem. *J. of Computer and System Sciences*, 51:359–366, 1995.

[17] L. Becchetti and S. Leonardi. Nonclairvoyant scheduling to minimize the total flow time on single and parallel machines. *Journal of the ACM*, 51(4):517–539, 2004.

[18] D. Bunde. Power-aware scheduling for makespan and flow. In *18th Annual ACM Symposium on Parallel Algorithms and Architectures*, pages 190–196, 2006.

[19] H.-L. Chan, W.-T. Chan, T.-W. Lam, L.-K. Lee, K.-S. Mak, and P. Wong. Energy efficient online deadline scheduling. In *18th Annual ACM-SIAM Symposium on Discrete Algorithms*, pages 795–804, 2007.

[20] C. Chekuri, S. Khanna, and A. Zhu. Algorithms for minimizing weighted flow time. In *33rd Annual ACM Symposium on Theory of Computing*, pages 84–93, 2001.

[21] B. Chen, A. van Vliet, and G. Woeginger. A lower bound for randomized on-line scheduling algorithms. *Information Processing Letters*, 51:219–222, 1994.

[22] U. Faigle, W. Kern, and G. Turan. On the performance of on-line algorithms for particular problems. *Acta Cybernetica*, 9:107–119, 1989.

[23] R. Fleischer and M. Wahl. Online scheduling revisited. *Journal of Scheduling*, 3:343–353, 2000.

[24] G. Galambos and G. Woeginger. An on-line scheduling heuristic with better worst case ratio than Graham's list scheduling. *SIAM Journal on Computing*, 22:349–355, 1993.

[25] B. Gaujal and N. Navet. Dynamic voltage scaling under edf revisited. *Real-Time Systems*, 37(1):77–97, 2007.

[26] B. Gaujal, N. Navet, and C. Walsh. Shortest path algorithms for real-time scheduling of fifo tasks with optimal energy use. *ACM Transactions on Embedded Computing Systems*, 4(4):907–933, 2005.

[27] R. L. Graham. Bounds for certain multiprocessor anomalies. *Bell System Technical Journal*, 45:1563–1581, 1966.

[28] S. Irani, R. Gupta, and S. Shukla. Competitive analysis of dynamic power management strategies for systems with multiple power savings states. In *Design, Automation and Test in Europe, Conference and Exposition*, pages 117–123, 2002.

[29] S. Irani and A. Karlin. Online computation. In D. Hochbaum, editor, *Approximation Algorithms for NP-Hard Problems*, pages 521–564. PWS Publishing Company, 1997.

[30] S. Irani and K. Pruhs. Algorithmic problems in power management. *ACM SIGACT News*, 36(2):63–76, 2005.

[31] B. Kalyanasundaram and K. Pruhs. Speed is as powerful as clairvoyance. *Journal of the ACM*, 47(4):617–643, 2000.

[32] B. Kalyanasundaram and K. Pruhs. Minimizing flow time nonclairvoy-
antly. *Journal of the ACM*, 50(4):551–567, 2003.

[33] D. Karger, S. Phillips, and E. Torng. A better algorithm for an ancient
scheduling problem. *Journal of Algorithms*, 20:400–430, 1996.

[34] A. Karlin, M. Manasse, L. McGeoch, and S. Owicki. Competitive randomized algorithms for nonuniform problems. *Algorithmica*, 11(6):542–
571, 1994.

[35] A. Legrand, A. Su, and F. Vivien. Minimizing the stretch when scheduling flows of divisible requests. *Journal of Scheduling*, 11(5):381–404,
2008.

[36] S. Leonardi. A simpler proof of preemptive total flow time approximation
on parallel machines. In *Efficient Approximation and Online Algorithms*,
volume 3484 of *LNCS*, pages 203–212. Springer, 2003.

[37] S. Leonardi and D. Raz. Approximating total flow time on parallel
machines. In *29th ACM Symposium on Theory of Computing*, pages
110–119, 1997.

[38] S. Leonardi and D. Raz. Approximating total flow time on parallel
machines. *J. of Computer and System Sciences*, 73(6):875–891, 2007.

[39] M. Li and F. Yao. An efficient algorithm for computing optimal discrete
voltage schedules. *SIAM Journal on Computing*, 35(3):658–671, 2005.

[40] J. McCullough and E. Torng. Srpt optimally utilizes faster machines to
minimize flow time. In *15th Annual ACM-SIAM Symposium on Discrete
Algorithms*, pages 350–358, 2004.

[41] R. Motwani, S. Phillips, and E. Torng. Non-clairvoyant scheduling. *Theoretical Computer Science*, 130:17–47, 1994.

[42] S. Muthukrishnan, R. Rajaraman, A. Shaheen, and J. Gehrke. Online
scheduling to minimize average stretch. *SIAM Journal on Computing*,
34(2):433–452, 2004.

[43] K. Pruhs, J. Sgall, and E. Torng. Online scheduling. In J. Leung, editor,
Handbook of Scheduling, chapter 15. Chapman & Hall/CRC, 2004.

[44] K. Pruhs, P. Uthaisombut, and G. Woeginger. Getting the best response for your erg. In *9th Scandinavian Workshop on Algorithm Theory
(SWAT)*, volume 3111 of *LNCS*, pages 14–25. Springer, 2004.

[45] J. Rudin III. *Improved bounds for the on-line scheduling problem*. PhD
thesis, The University of Texas at Dallas, May 2001.

[46] S. Seiden. Online randomized multiprocessor scheduling. *Algorithmica*,
28:73–216, 2000.

[47] J. Sgall. A lower bound for randomized on-line multiprocessor scheduling. *Information Processing Letters*, 63:51–55, 1997.

[48] D. Sleator and R. Tarjan. Amortized efficiency of list update and paging rules. *Communications of the ACM*, 28:202–208, 1985.

[49] F. Yao, A. Demers, and S. Shenker. A scheduling model for reduced cpu energy. In *36th Annual Symposium on Foundations of Computer Science*, pages 374–382, 1995.

Chapter 4

Job Scheduling

Uwe Schwiegelshohn

Technische Universität Dortmund

Abstract   In this chapter, we address scheduling problems with independent jobs on a single machine or on parallel identical machines. The jobs are either sequential or parallel, that is, they may require concurrent access to several machines at the same time. Our objectives, like the makespan or the total weighted completion time, are commonly used in scheduling literature. We consider deterministic and online problems. Our focus is on simple algorithms, like various forms of List Scheduling, that provide optimal or good solutions based on an evaluation with approximation or competitive factors. With the help of some example schedules and exemplary proofs, we explain common approaches to address this kind of problems.

4.1   Introduction

In a scheduling problem, several jobs are executed on one or more machines to satisfy a given objective. Based on this definition, every scheduling problem is a job scheduling problem. But, commonly, job scheduling problems assume independent jobs, that is, there are no precedence constraints between different jobs while scheduling problems with precedence constraints are often referred to as task scheduling problems. In contrast to job shop scheduling problems, a job in a job scheduling problem is atomic and does not require subsequent processing on different machines. Finally, jobs in job scheduling problems usually do not have deadlines or due dates. But unfortunately it is not always possible to use this classification for real scheduling problems:

For instance, although Grid Scheduling is usually considered to be part of job scheduling, Grid jobs often have several stages, like data transmission, job execution, and result transmission. Similarly, advance reservation or backfill algorithms may induce due dates for some Grid jobs.

Many people think of multiprocessor scheduling when talking about job scheduling, although this type of problem also arises in bandwidth scheduling, airport gate scheduling, and repair crew scheduling to mention just a few different areas. Nevertheless in this chapter, we will use the multiprocessor environment when connecting abstract problems to the real world.

As already addressed in Section 1.3, the notation $\alpha|\beta|\gamma$ is commonly used to characterize a scheduling problem with α, β, and γ denoting the machine model, the scheduling restrictions, and the scheduling objective, respectively. In this chapter, we first consider a single machine ($\alpha = 1$) and then focus on m parallel identical machines ($\alpha = P_m$). In the scheduling literature, see, for instance, the textbook by Pinedo [22], we also find machines with different speed ($\alpha = Q_m$) or unrelated machines ($\alpha = R_m$). While in deterministic scheduling many results using the P_m model can be transferred to the Q_m model, this is not true in online scheduling. Moreover, the practical applicability of the Q_m model is somewhat limited as processor performance evaluation has shown that the speedup on different processors usually depends on the application. Therefore, the R_m model is more appropriate but it requires the knowledge of the performance data of an application on each processor which is often not available in practice. As most processor nodes of a multiprocessor system are similarly equipped the P_m model is used frequently although there are some studies explicitly addressing other models. Here, we only discuss the single machine and the P_m models.

Furthermore, we restrict ourselves to few different constraints. p_j denotes the processing time of job J_j for the 1 and P_m models, that is, p_j is the difference between the completion time C_j and the start time of job J_j if the job is not interrupted during its execution. If the scheduling algorithm cannot use p_j to determine a good schedule the problem is called *nonclairvoyant* ($\beta = ncv$). Also we discuss parallel *rigid* jobs ($\beta = size_j$). A rigid job J_j requires the concurrent availability of $size_j$ (identical) machines during its whole processing. In addition to the *run-to-completion* scenario, we also consider preemption ($\beta = pmtn$), that is, the execution of a job can be interrupted and later be resumed. In the case of parallel jobs, we only look at gang scheduling, that is, the execution of a parallel job must be interrupted (and resumed) on all allocated machines at the same time. While the number of allocated machines remains invariant during different execution phases the set of allocated machines may change, that is, we allow preemption with migration. There is a large variety of different models for parallel jobs, like divisible load scheduling, malleable jobs, and moldable jobs. Here, we restrict ourselves to the rigid model, which we consider to be the basic model of a parallel job.

Jobs are either available at time 0 or each job J_j may have an individual release date r_j ($\beta = r_j$), that is, job J_j cannot start its execution before time r_j. In deterministic scheduling problems, the scheduler already knows at time 0 when each job is ready to be scheduled and can plan its schedule accordingly. However, in most real job scheduling problems, jobs are submitted over time and the scheduler only learns about the job after it has been submitted. In these cases, the submission time and the release time of a job typically are identical. Advance reservation in Grid scheduling is an exception to this rule as the desired start time of a job is provided at its submission.

$C_j(S)$ denotes the completion time of job J_j in schedule S. We omit the schedule identifier if the schedule is non-ambiguous. Generally, we can distinguish between machine and job related objectives. The simplest machine related objective is *utilization* $U_t(S)$, that is the ratio of the sum of the time instances a machine is active to the elapsed time t. This objective can easily be extended to the P_m model. In most theoretical scheduling studies, the *makespan* objective $C_{\max}(S) = \max_j\{C_j(S)\}$ is used instead of utilization. Clearly, there is a close correspondence between utilization and makespan in a completed schedule. However, differences exist when evaluating intermediate results in online scheduling, see Section 4.3. In a job related objective, the completion time of each job contributes to the objective. In this case, different job priorities are often expressed with the help of a job weight $w_j \geqslant 0$. Then a weighted sum is often used to combine the job related values resulting, for instance, in the total weighted completion time $\sum_j w_j C_j$. There are also variants like the total weighted flow time $\sum_j w_j(C_j - r_j)$ or the total weighted waiting time $\sum_j w_j(C_j - r_j - p_j)$. Although these different objectives are closely related as $\sum_j w_j p_j$ and $\sum_j w_j r_j$ are constants, some differences become apparent when discussing approximation results, see Section 4.2. Almost all studies allow arbitrary positive weights as input values. But, in practice, a user typically cannot select any weight by himself: Either he can pick weights from a small set provided by the system or it is the machine owner who determines the weights. Otherwise small variations of a weight may significantly influence the start time of a job in an optimal schedule. Nevertheless in this chapter, we focus on the makespan and the total (weighted) completion time. These objectives are regular, that is, they are non-decreasing functions of the completion times of the individual jobs.

The various machine models, scheduling constraints, and scheduling objectives expose a complexity hierarchy. We also say that a problem $\alpha_1|\beta_1|\gamma_1$ reduces ($\propto$) to a problem $\alpha_2|\beta_2|\gamma_2$ if a general solution to the second problem leads to a general solution to the first problem [22]. For instance, we can form the following complexity chain:

$$1||\sum C_j \ \propto \ 1||\sum w_j C_j \ \propto \ P_m||\sum w_j C_j \ \propto \ P_m|size_j|\sum w_j C_j$$

In particular, the reductions of Table 4.1 hold for all scheduling problems. As already noted, a problem with the objective $\sum C_j$ is a special case of the same problem with the objective $\sum w_j C_j$ as we can simply set all weights to 1. Sim-

general case	P_m	$\sum w_j C_j$	r_j	w_j	$size_j$	ncv
specific case	1	$\sum C_j$	-	-	-	-

Table 4.1: Elementary reductions.

ilarly, a problem without release dates, that is, where all release dates are 0, is a special case of the same problem with release dates. Note that there is no general reduction between problems with and without preemption. In deterministic offline scheduling problems, all job information is available at time 0. We discuss optimal polynomial time algorithms and simple approximation algorithms but ignore polynomial time approximation schemes for the most parts as FPTAS and PTAS have already been addressed in Chapter 2. Often the results can be improved using randomization but this is also not the subject of this chapter. Further, we present some simple deterministic online algorithms with their competitive factors. But contrary to Chapter 3, we assume that the scheduler can wait with the allocation of a job until a machine is available. As already mentioned many of our algorithms are simple and produce so-called nondelay schedules [22], that is, no machine is kept idle while a job is waiting for processing. However, note that some problems have no optimal nondelay schedule, see Table 4.2 and Figure 4.1.

Jobs	p_j	r_j	w_j
J_1	1	1	2
J_2	3	0	1

Table 4.2: A $1|r_j|\sum w_j C_j$ problem without optimal nondelay schedules.

4.2   Single Machine Problems

Job scheduling problems on a single machine with the objective makespan (or utilization) are easy: Every nondelay schedule is an optimal schedule independent of the scheduling constraints. Therefore, we only consider the total completion time and the total weighted completion time for the single machine model. A nondelay schedule in descending job order of the *Smith ratio* $\frac{w_j}{p_j}$ [30] produces an optimal schedule for the basic problem $1||\sum w_j C_j$. This approach is called *weighted shortest processing time first* (WSPT) rule.

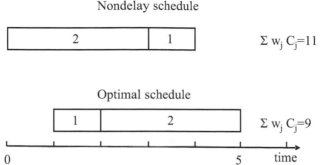

Nondelay schedule

Optimal schedule

FIGURE 4.1: A $1|r_j|\sum w_jC_j$ problem without optimal nondelay schedules.

The proof uses contradiction and is based on localization: Assume that two jobs in a schedule violate the WSPT rule. Then there are two neighboring jobs in the schedule that also violate the WSPT rule. A local exchange of these jobs does not influence the completion time of any other job and improves the total weighted completion time of the original schedule, see Figure 4.2.

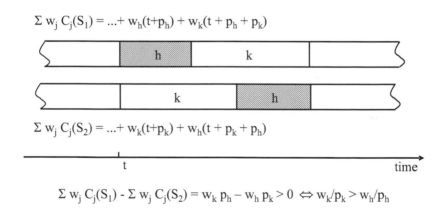

FIGURE 4.2: WSPT proof with local exchange of jobs.

The permission of preemption does not generate new optimal schedules, that is, the problems $1|pmtn|\sum C_j$ and $1|pmtn|\sum w_jC_j$ have only non-preemptive optimal schedules. However, preemption makes a difference if the jobs have release dates. Then the problem $1|r_j, pmtn|\sum C_j$ is still easy: We apply a nondelay schedule and determine the *shortest remaining processing time* (SRPT) of all available jobs when a job has completed its execution or a new release date is encountered. But the problem $1|r_j|\sum C_j$ is strongly NP-hard: Intuitively, it is better to leave the machine idle instead of occupying it with a long running job if many jobs with a short processing time will be released soon. For this problem, Chekuri et al. [3] presented an algorithm with the approximation factor $\frac{e}{e-1} \approx 1.58$.

This problem also offers an opportunity to demonstrate the difference between the total completion time $(\sum C_j)$ and the total flow time $(\sum(C_j - r_j))$ objectives. The problem $1|r_j, pmtn| \sum(C_j - r_j)$ is easy as every optimal schedule for the $1|r_j, pmtn| \sum C_j$ problem is optimal for $1|r_j, pmtn| \sum(C_j - r_j)$, as well. Remember that $\sum_j r_j$ is a constant. However, there is no algorithm with a constant approximation factor for the problem $1|r_j| \sum(C_j - r_j)$, see Kellerer, Tautenhahn, and Wöginger [15], although the corresponding problem $1|r_j| \sum C_j$ has an approximation factor of 1.58. Formally, these results can be explained by comparing the ratios of the achieved objectives to the optimal objectives and by noticing that the ratio $\frac{\sum r_j}{\sum(C_j(OPT) - r_j)}$ is not bounded:

$$\frac{\sum(C_j(S) - r_j)}{\sum(C_j(OPT) - r_j)} = \frac{\frac{\sum C_j(S)}{\sum C_j(OPT)} \sum(C_j(OPT) - r_j) + (\frac{\sum C_j(S)}{\sum C_j(OPT)} - 1)\sum r_j}{\sum(C_j(OPT) - r_j)}$$

$$= \frac{\sum C_j(S)}{\sum C_j(OPT)} + \left(\frac{\sum C_j(S)}{\sum C_j(OPT)} - 1\right)\frac{\sum r_j}{\sum(C_j(OPT) - r_j)}$$

If we allow weights then the problem with release dates is already strongly NP-hard even if preemption is allowed $(1|r_j, pmtn| \sum w_j C_j)$. This means that the *weighted shortest remaining processing time first* (WSRPT) rule is not necessarily optimal, see the problem in Table 4.3. The example shows that it is better to execute job J_2 first as this job will be completed before the high priority job J_3 is started. The WSRPT rule will start job J_1 first and delay its completion until job J_3 has completed.

Jobs	p_j	r_j	w_j
J_1	20	0	20
J_2	10	0	9
J_3	10	10	100

Table 4.3: Nonoptimal example for the WSRPT rule.

For total weighted completion time problems with release dates, there are algorithms with relatively small approximation factors: Goemans et al. [8] proved an approximation factor of 1.6853 for the $1|r_j| \sum w_j C_j$ problem while Schulz and Skutella [24] showed an approximation factor of $\frac{4}{3}$ for the $1|r_j, pmtn| \sum w_j C_j$ problem. The algorithm of Schulz and Skutella is based on WSRPT but transforms this preemptive schedule into another preemptive schedule using the order of non-decreasing α-points. An α-point in a schedule is the first point in time when an α-fraction of the job has completed.

Different values of α may lead to different job orders. First, these α values are determined randomly and later derandomization techniques are applied without reducing the performance bound. Random α-points and derandomization are also used in the algorithm of Goemans et al. [8]. Note that Afrati et al. [1] have shown the existence of polynomial time approximation schemes for the $1|r_j, pmtn| \sum w_j C_j$ problem. Therefore, better bounds are achievable with algorithms that have limited applicability in practice.

There are also several online results. Schulz and Skutella [24] showed that WSRPT will result in a competitive factor of 2 for the $1|r_j, pmtn| \sum w_j C_j$ problem. Anderson and Potts [2] demonstrated the same competitive factor for the $1|r_j| \sum w_j C_j$ problem. They use a delayed form of WSRPT, that is, in some circumstances, no job is scheduled although jobs are available. Moreover, they showed that a better competitive factor is not possible. Their algorithm had already been proposed by Hoogeveen and Vestjens [12] in 1996 who proved the same competitive factor of 2 for the simpler problem $1|r_j| \sum C_j$. All these algorithms use the processing time of the jobs. Therefore, they cannot be applied in a nonclairvoyant scenario. Clearly, without preemption $(1|ncv| \sum C_j)$, any algorithm will produce very bad results if the job that is started first has a sufficiently long processing time. In practice, this problem arises in multi-tasking operating systems and is addressed with preemption and a simple *Round-Robin* approach. This algorithm guarantees the competitive factor $2 - \frac{2}{n+1}$ for the $1|pmtn, ncv| \sum C_j$ problem with n being the number of jobs. Motwani, Phillips, and Torng [18] proved this competitive factor and showed that it is tight. Their algorithm can be extended to the weighted case $(1|pmtn, ncv| \sum w_j C_j)$:

PROOF Assume an indexing of the jobs in descending order of their Smith ratios, that is, $\frac{w_i}{p_i} \geq \frac{w_j}{p_j}$ for $i < j$. Therefore, a nondelay schedule using this order is optimal. In this optimal schedule, we have $C_j = \sum_{i=1}^{j} p_i$. This results in

$$\sum_{j=1}^{n} w_j C_j(OPT) = \sum_{j=1}^{n} w_j p_j + \sum_{j=2}^{n} w_j \sum_{i=1}^{j-1} p_i.$$

Our algorithm produces a schedule S by assigning time slices to the jobs in a *Round-Robin* fashion such that the ratio of time slices of jobs J_i and J_j equals the ratio of the corresponding weights. Further, let the time slices be sufficiently small in comparison to the processing times. Then the jobs complete in index order in S and job J_j delays the completion time of job J_i by $\frac{w_j \cdot p_i}{p_j}$ if $j > i$. This leads to

$$\sum_{j=1}^{n} w_j C_j(S) = \sum_{j=1}^{n} w_j p_j + \sum_{j=2}^{n} \left(w_j \sum_{i=1}^{j-1} p_i \right) + \sum_{j=1}^{n-1} \left(w_j \sum_{i=j+1}^{n} \frac{w_i p_j}{w_j} \right)$$

$$= \sum_{j=1}^{n} w_j p_j + \sum_{j=2}^{n} w_j \sum_{i=1}^{j-1} p_i + \sum_{j=1}^{n-1} p_j \sum_{i=j+1}^{n} w_i$$

$$= \sum_{j=1}^{n} w_j p_j + \sum_{j=2}^{n} w_j \sum_{i=1}^{j-1} p_i + \sum_{i=2}^{n} w_i \sum_{j=1}^{i-1} p_j$$

$$= \sum_{j=1}^{n} w_j p_j + 2 \sum_{j=2}^{n} w_j \sum_{i=1}^{j-1} p_i$$

$$< 2 \sum_{j=1}^{n} w_j C_j(OPT).$$

□

Motwani, Phillips, and Torng [18] also discussed the problem $1|r_j, pmtn, ncv|$ $\sum C_j - r_j$ and showed that there is no constant competitive factor for it in agreement with the results of Kellerer, Tautenhahn, and Wöginger [15]. Further, the problems $1|r_j, ncv| \sum C_j$ and $1|r_j, ncv| \sum w_j C_j$ have no constant approximation factor as the job that is scheduled first may have a very long processing time and therefore a small Smith ratio independent of its weight.

4.3   Makespan Problems on Parallel Machines

Next we consider m parallel identical machines with m being arbitrary and the makespan being the scheduling objective. Note that most associated offline problems are NP-hard and possess a PTAS, that is, we can come arbitrarily close to the optimum at the expense of a rather complex algorithm. In practice, simple algorithms are preferred as they often enable the addressing of additional constraints and as practical workloads almost never exhibit a worst case behavior. Therefore, we are looking at simple algorithms with good performance. Often these algorithms can also be adapted to the online case. First we assume that all jobs are sequential, that is, they only need a single machine for their execution. Independent of the scheduling objective, the scheduling process for sequential jobs on parallel identical machines can be divided into two steps:

1. Allocation of a job to a machine;

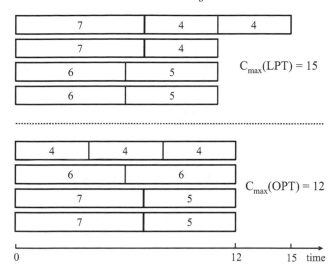

FIGURE 4.3: Worst case example for LPT with $m = 4$.

2. Generation of a sequence of jobs on each machine.

As already discussed in Section 4.2, the sequence generation is easy for the makespan objective and will lead to a nondelay schedule on each machine. Therefore, it is the main goal of a scheduling algorithm to obtain a good load balance among the machines. For the problem $P_m||C_{max}$, we have the lower bound

$$C_{max} \geqslant \max \left\{ \max_j \{p_j\}, \frac{1}{m} \sum_j p_j \right\}.$$

Graham [9] suggested the *longest processing time first* (LPT) rule to schedule all jobs in descending order of their processing times and proved a bound of $\frac{4}{3} - \frac{1}{3m}$ that is tight for this algorithm, see Table 4.4. Figure 4.3 shows for this example that we have $C_{max}(LPT) = 15$ and $C_{max}(OPT) = 12$, respectively.

Jobs	1	2	3	4	5	6	7	8	9
p_j	7	7	6	6	5	5	4	4	4

Table 4.4: Worst case example for LPT and $m = 4$.

If an arbitrary list is used (List Scheduling) the bound increases to $2 - \frac{1}{m}$, see Graham [10] (see also Sections 2.3.2 and 3.2.1). This bound is also tight for the algorithm, see Figure 4.4 depicting an example with five machines. In this figure, the number within a job denotes its processing time. In contrast to LPT, List Scheduling is also valid for nonclairvoyant scheduling problems ($P_m|ncv|C_{\max}$) as the processing times of the jobs are not considered in the algorithm.

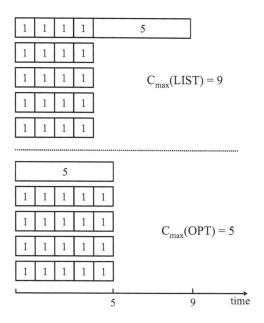

FIGURE 4.4: Worst case example for List Scheduling with $m = 5$.

As already mentioned there are few problems in job scheduling on parallel machines that are easy. $P_m|pmtn|C_{\max}$ is one of them with

$$C_{\max} = \max\left\{ \max_j\{p_j\}, \frac{1}{m}\sum_j p_j \right\}.$$

McNaughton [16] proposed a simple algorithm that first generates an arbitrary non-preemptive, nondelay schedule on a single machine. This schedule is then divided into m parts of length $C_{\max}$ that are afterward distributed to the machines. The resulting schedule has at most $m - 1$ preemptions. Alternatively, the preemptive version of LPT named *longest remaining processing time first* (LRPT) can be used. On the one hand, LRPT will produce an optimal schedule with a potentially large number of preemptions. On the other hand, LRPT optimally solves the $P_m|r_j, pmtn|C_{\max}$ problem even in the online clairvoyant case with jobs being submitted over time. In this scenario, a job is allocated to a machine as soon as this machine becomes idle and the processing time of a job becomes known at its submission time. In the non-preemptive online case $(P_m|r_j|C_{\max})$, LPT produces a competitive factor of $\frac{3}{2}$ which is slightly larger than the approximation factor of LPT if all jobs are released at time 0, see Chen and Vestjens [4]. In the nonclairvoyant case $P_m|r_j, ncv|C_{\max}$, the list scheduling bound $2 - \frac{1}{m}$ remains valid, see Hall and Shmoys [11]. This is also true if preemption is allowed $(P_m|r_j, pmtn, ncv|C_{\max})$, see Shmoys, Wein, and Williamson [29].

Shmoys, Wein, and Williamson [29] also introduced a general method to transform any offline scheduling algorithm A (all jobs are released at time 0) to an online algorithm A' (jobs are released over time) if the makespan objective is used. The online algorithm guarantees a competitive factor that is at most twice as large as the approximation factor of the original algorithm:

$$\frac{C_{\max}(A)}{C_{\max}(OPT)} \leqslant k \quad \rightarrow \quad \frac{C_{\max}(A')}{C_{\max}(OPT)} \leqslant 2k$$

PROOF   Before starting the proof, we introduce the notation $C_{\max}(A)|_{\tau_i}$ to describe the makespan of a schedule generated by algorithm A for a job subset τ_i. The proof is based on several epochs. In epoch i, the online algorithm A' uses the offline algorithm A to schedule all jobs that have been submitted during the previous epoch $i - 1$. These jobs constitute the job set τ_{i-1}. Epoch i ends with the completion of the last of these jobs, that is at time $T_i = C_{\max}(A')|_{\tau_{i-1}}$. Without loss of generality, we assume that all jobs from τ_{i-1} are released just after time T_{i-2}. Therefore, we have

$$C_{\max}(A)|_{\tau_{i-1}} = T_i - (T_{i-1} - T_{i-2}).$$

This does not change $C_{\max}(A')|_{\tau_{i-1}}$ while $C_{\max}(OPT)|_{\tau_{i-1}}$ cannot increase. Finally, we define $T_{-2} = T_{-1} = 0$. This leads to

$$T_i = C_{\max}(A')|_{\bigcup_{j=0}^{i-1} \tau_j} = C_{\max}(A)|_{\tau_{i-1}} + T_{i-1} - T_{i-2}$$
$$\leqslant T_{i-1} + k \cdot (C_{\max}(OPT)|_{\tau_{i-1}} - T_{i-2})$$
$$T_{i-2} - T_{i-1} + T_i \leqslant k \cdot C_{\max}(OPT)|_{\tau_{i-1}} \leqslant k \cdot C_{\max}(OPT)|_{\bigcup_{j=0}^{i-1} \tau_j}$$

and

$$T_{i-1} - T_{i-2} \leqslant T_{i-3} - T_{i-2} + T_{i-1} \leqslant k \cdot C_{\max}(OPT)|_{\bigcup_{j=0}^{i-2} \tau_j}$$
$$\leqslant k \cdot C_{\max}(OPT)|_{\bigcup_{j=0}^{i-1} \tau_j}.$$

The addition of both inequalities yields the final result

$$C_{\max}(A')|_{\bigcup_{j=0}^{i-1} \tau_j} \leqslant 2k \cdot C_{\max}(OPT)|_{\bigcup_{j=0}^{i-1} \tau_j}.$$

$\square$

Next we address rigid parallel jobs. Such a job J_j requires the concurrent availability of $size_j$ machines during its whole processing. As the allocated machines need not be contiguous there is a difference to bin packing. To support an easier intuition, we use a contiguous rectangle representation in our figures, nevertheless. Garey and Graham [7] showed already in 1975 that List Scheduling can handle rigid parallel jobs with the same performance guarantee $(2 - \frac{1}{m})$ as for sequential jobs. Therefore, List Scheduling can also be used for the problem $P_m|size_j, ncv|C_{\max}$. But it is worth to note that the proof technique is slightly different.

PROOF   In the case of sequential jobs, all machines are always busy until the last job starts its execution. For parallel jobs, some machines may be idle before the start time of the last job. We divide the schedule into intervals such that jobs can only start or complete at the beginning or the end of an interval. Due to the List Scheduling property, the sum of machines used in any two intervals is larger than m unless all jobs being processed in one interval are a subset of the jobs being processed in the other. Moreover, the $2 - \frac{1}{m}$ bound holds if during the whole schedule S there is no interval with at least $\frac{m}{2}$ idle machines:

$$C_{\max}(OPT) \geqslant \frac{1}{m} \sum_j size_j \cdot p_j \geqslant \frac{m+1}{2m} C_{\max}(S)$$

$$\geqslant \frac{m}{2m-1} C_{\max}(S) \geqslant \frac{1}{2 - \frac{1}{m}} C_{\max}(S)$$

This yields

$$C_{\max}(S) \leqslant \max\left\{ \left(2 - \frac{1}{m}\right) \max_j\{p_j\}, \left(2 - \frac{1}{m}\right) \sum_j \frac{size_j \cdot p_j}{m} \right\}.$$

☐

Finally, Naroska and Schwiegelshohn [20] showed that List Scheduling also guarantees the $2 - \frac{1}{m}$ bound for the online problem $P_m|size_j, r_j, ncv|C_{\max}$. While this bound is tight for this nonclairvoyant problem there are some better results for the clairvoyant offline problem $P_m|size_j, r_j|C_{\max}$ like the one of Mounié, Rapine, and Trystram [19] who proved a $1.5 + \epsilon$ bound for any $\epsilon > 0$. However, a small factor ϵ produces a large runtime of the algorithm.

In Section 4.1, we have already stated that the makespan represents machine utilization. This is true for any complete schedule using our models but gives misleading results for intermediate schedules: For sequential jobs and all release dates being 0, List Scheduling produces 100% utilization until the last job is started. But in online scheduling, there may be no last job. In this case, it is interesting to determine the utilization of the machines in the time interval from time 0 to the actual time. Hussein and Schwiegelshohn [13] proved that List Scheduling has a tight competitive factor of $\frac{4}{3}$ for the online problem $P_m|r_j, ncv|U_t$, see Table 4.5 for a worst case example with $m = 2$. The worst case occurs if job J_3 is scheduled last.

Jobs	p_j
J_1	1
J_2	1
J_3	2

Table 4.5: Worst case example for utilization with $m = 2$ and $t = 2$.

Note that the competitive factor of $\frac{4}{3}$ is different from the List Scheduling makespan bound.

Moreover, List Scheduling utilization may vary over time for the online problem $P_m|size_j, r_j, ncv|U_t$ while the intermediate makespan values of a List Schedule can not decrease. For instance, Table 4.6 and Figure 4.5 show that utilization at intermediate time instances may be very low although the final utilization is more than 50 percent. This underlines the differences between both types of online problems.

4.4 Completion Time Problems on Parallel Machines

In this section, we consider the total completion time and the total weighted completion time objectives for parallel identical machines. The $P_m||\sum C_j$ problem is easy and can be solved using the SPT rule, see Conway, Maxwell, and Miller [5].

PROOF From Section 4.2, we know that the SPT schedule is a nondelay schedule. Therefore, the completion time on a machine that executes k jobs

Jobs	p_j	r_j	$size_j$
J_1	$1+\epsilon$	0	1
J_2	$1+\epsilon$	1	1
J_3	$1+\epsilon$	2	1
J_4	$1+\epsilon$	3	1
J_5	1	4	1
J_6	5	0	5

Table 4.6: Example of a $P_m|size_j, r_j, ncv|U_t$ problem with $m = 5$.

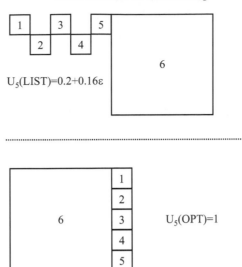

FIGURE 4.5: Example of a $P_m|size_j, r_j, ncv|U_t$ problem with $m = 5$.

is

$$\sum C_j = k \cdot p_{(1)} + (k-1) \cdot p_{(2)} + \ldots + p_{(k)}$$

with $p_{(i)}$ being the processing time of the ith job on this machine.

Without loss of generality we assume that the number of jobs n is a multiple of the number of machines m. Otherwise we simply add dummy jobs with processing time 0. If there is a machine P_1 executing $n_1 > \frac{n}{m}$ jobs then there must be at least one machine P_2 executing $n_2 < \frac{n}{m}$ jobs. Then we move the first job J_j of machine P_1 into the first position of machine P_2 increasing the positions of all other jobs on this machine. Then the total completion time is reduced by $n_1 \cdot p_j - n_2 \cdot p_j > 0$. Therefore, all machines execute the same number of jobs and each factor $1, \ldots, \frac{n}{m}$ occurs exactly m-times. In order to minimize the total completion time, we assign the jobs with the smallest processing times to the largest factors which is guaranteed by the SPT rule.
□

As the $P_m||\sum w_j C_j$ problem is strongly NP-hard, the WSPT rule cannot always achieve an optimal solution but it guarantees the small approximation factor $\frac{1+\sqrt{2}}{2}$, see Kawaguchi and Kyan [14]. Unfortunately, the proof presented in their publication is rather intricate. It consists of two parts: In one part, it is shown that the approximation factor cannot decrease by assuming that the Smith ratio is identical for all jobs. To this end, the authors iteratively reduce the number of different Smith ratio values of an arbitrary problem instance and its associated WSPT schedule without decreasing the approximation factor until there is only a single Smith ratio value left. For the second part, we, therefore, assume that $\frac{w_j}{p_j} = 1$ holds for all jobs J_j and provide an alternative proof:

PROOF Let S and OPT be an arbitrary and the optimal nondelay schedule of a problem instance with $\frac{w_j}{p_j} = 1$ for all jobs J_j, respectively.

First we introduce two already known corollaries, see Schwiegelshohn and Yahyapour [28] for the first one:

COROLLARY 4.1
Let S' and OPT' be a schedule and the optimal schedule of n jobs on m parallel identical machines, respectively. We split an arbitrary job J_j into two jobs J_{n+1} and J_{n+2} such that $0 < p_{n+1} < p_j$ and $p_{n+2} = p_j - p_{n+1}$ hold. Schedule S is derived from schedule S' by simply starting job J_{n+1} instead of job J_j and starting job J_{n+2} immediately after the completion of job J_{n+1}.

This results in the inequality $\dfrac{\sum_j p_j C_j(S')}{\sum_j p_j C_j(OPT')} \leqslant \dfrac{\sum_j p_j C_j(S)}{\sum_j p_j C_j(OPT)}$.

PROOF Note that the completion time of each job $J_{j'} \neq \{J_j, J_{n+1}, J_{n+2}\}$ is identical in both schedules S and S'.

Therefore, $C_{n+1}(S) = C_j(S') - p_{n+2}$ and $C_{n+2}(S) = C_j(S')$ hold, and S is a legal schedule as no machine is used to execute two jobs at the same time. Then we have

$$\sum_j p_j C_j(S') - \sum_j p_j C_j(S) = p_j C_j(S') - p_{n+1} C_{n+1}(S) - p_{n+2} C_{n+2}(S)$$
$$= (p_{n+1} + p_{n+2}) C_j(S') - p_{n+1}(C_j(S') - p_{n+2})$$
$$- p_{n+2} C_j(S')$$
$$= p_{n+1} p_{n+2}.$$

This result is independent of S'. Therefore, it also holds for the optimal schedule of the original problem. However, the schedule derived from the optimal schedule needs not be the new optimal schedule after the job split. With $\sum_j p_j C_j(OPT') - p_{n+1} p_{n+2} \geqslant \sum_j p_j C_j(OPT)$, this leads to

$$\frac{\sum_j p_j C_j(S)}{\sum_j p_j C_j(OPT)} \geqslant \frac{\sum_j p_j C_j(S') - p_{n+1} p_{n+2}}{\sum_j p_j C_j(OPT') - p_{n+1} p_{n+2}} \geqslant \frac{\sum_j p_j C_j(S')}{\sum_j p_j C_j(OPT')}.$$

$\square$

The second corollary has been presented by Queyranne [23].

COROLLARY 4.2
Let S be a nondelay schedule on a single machine. Then we have

$$\sum_j p_j C_j(S) = \frac{1}{2}\left(\left(\sum_j p_j\right)^2 + \sum_j p_j^2\right).$$

PROOF The proof is done by induction on the number of jobs in the schedule. It is clearly true for schedules with a single job. Let J_{n+1} be the last job in S. Then we have

$$\sum_{j=1}^{n+1} p_j C_j(S) = \frac{1}{2}\left(\left(\sum_{j=1}^{n} p_j\right)^2 + \sum_{j=1}^{n} p_j^2\right) + p_{n+1}\left(p_{n+1} + \sum_{j=1}^{n} p_j\right)$$

$$= \frac{1}{2}\left(\left(\sum_{j=1}^{n} p_j\right)^2 + 2p_{n+1}\sum_{j=1}^{n} p_j + p_{n+1}^2\right) + \frac{1}{2}\left(\sum_{j=1}^{n} p_j^2 + p_{n+1}^2\right)$$

$$= \frac{1}{2}\left(\left(\sum_{j=1}^{n+1} p_j\right)^2 + \sum_{j=1}^{n+1} p_j^2\right).$$

☐

For the general proof, we assume that time is continuous and allow fractional values of machines. Let $t_{idle}(S)$ be the time when the first machine in schedule S becomes idle. Then by repeated application of Corollary 4.1, each job J_j that starts in a worst case WSPT schedule before time $t_{idle}(S)$ has a very short processing time $p_j \to 0$. These jobs are called short jobs. In addition, there are less than m long jobs that start at time $t_{idle}(S)$.

As the order of jobs on a machine and the assignment of jobs that finish before time $t_{idle}(OPT)$ do not matter, we can arrange the jobs such that no machine executes two long jobs and all long jobs start at time 0 in an optimal schedule.

We can replace the set L of long jobs with another set L' of k long jobs with equal processing time $\bar{p}$ such that $\sum_{J_j \in L} p_j = k \cdot \bar{p}$ and $\sum_{J_j \in L} p_j^2 = k \cdot \bar{p}^2$. This results in $\bar{p} = \frac{\sum_{J_j \in L} p_j^2}{\sum_{J_j \in L} p_j}$ and $k = \frac{(\sum_{J_j \in L} p_j)^2}{\sum_{J_j \in L} p_j^2}$ and yields the next corollary.

COROLLARY 4.3
Using the above described transformation, we have $|L| \geqslant k$.

PROOF For the sake of a simpler notation let us temporarily assume that each job is executed on a single machine, that is, no fractional machines are used. With $\Delta_j = p_j - \frac{\sum_{J_j \in L} p_j}{|L|}$ and $\sum_{J_j \in L} \Delta_j = \sum_{J_j \in L} p_j - |L| \frac{\sum_{J_j \in L} p_j}{|L|} = 0$, we have

$$\frac{\left(\sum_{J_j \in L} p_j\right)^2}{k} = \sum_{J_j \in L} p_j^2$$

$$= \sum_{J_j \in L} \left(\frac{\sum_{J_j \in L} p_j}{|L|} + \Delta_j\right)^2$$

$$= |L| \frac{\left(\sum_{J_j \in L} p_j\right)^2}{|L|^2} + 2 \frac{\sum_{J_j \in L} p_j}{|L|} \sum_{J_j \in L} \Delta_j + \sum_{J_j \in L} \Delta_j^2$$

$$= \frac{\left(\sum_{J_j \in L} p_j\right)^2}{|L|} + \sum_{J_j \in L} \Delta_j^2$$

$$\geqslant \frac{\left(\sum_{J_j \in L} p_j\right)^2}{|L|}.$$

$\square$

Let L be the set of all long jobs. We must distinguish three cases:

1. If $\bar{p} \leqslant \frac{\sum_{J_j} p_j}{m}$ then we simply apply the long job transformation to all jobs in L. This does not decrease the worst case ratio for reasons of convexity and due to Corollary 4.2.

2. If no short job and no long job are executed together on the same machine then we also apply the long job transformation to all long jobs. This does not decrease the worst case ratio due to Corollary 4.3.

3. Otherwise we remove enough jobs from L and the machines occupied by them such that the first case applies with equality to the remaining long jobs on the reduced number of machines. Remember that we allow jobs occupying fractional machines. After this transformation the second case applies, see Figure 4.6.

In order to determine the approximation factor, we divide the number of machines by m such that our machine number is 1. Remember that fractional use of machines is permitted. Further, we say that the total resource consumption of small jobs is 1, that is, if all small jobs start as early as possible, the last small job completes at time 1. This approach produces a simple numerical optimization problem with the variables y being the (rational) number of long jobs and x being the processing time of the long jobs:

$$\max_{x \geqslant 0, 0 \leqslant y < 1} \left\{ \begin{array}{c} \frac{1}{2} + yx(x+1) \\ \frac{1}{2(1-y)} + yx^2 \end{array} \right\}$$

In the optimal case, all long jobs start at time 0 and contribute yx^2 to the total completion time while the contribution of the small jobs is $\frac{1}{2(1-y)}$ due Corollary 4.2. In the bad case, we start all small jobs as early as possible resulting in $\frac{1}{2}$ while the long jobs start at time 1 and contribute $yx(x+1)$ to the total completion time. Then $x = 1 + \sqrt{2}$ and $y = \frac{1}{2+\sqrt{2}}$ produce the maximum ratio $\frac{1+\sqrt{2}}{2}$. $\quad\Box$

Allowing preemption does not lead to better schedules, see McNaughton [16].

Recently, there have been several results regarding polynomial time approximation schemes for total completion time and total weighted completion time problems on identical parallel machines, see, for instance, Afrati et al [1]. As in the previous sections, we do not discuss these results in this chapter but focus on rather simple algorithms. Schulz and Skutella [25] gave algorithms with approximation factor 2 for the problems $P_m | r_j | \sum C_j$, $P_m | r_j, pmtn | \sum C_j$, $P_m | r_j | \sum w_j C_j$, and $P_m | r_j, pmtn | \sum w_j C_j$. Their algorithms are based on linear programs and use randomization. They can be derandomized at the expense of an increased running time but without increasing the performance guarantee. Megow and Schulz [17] gave deterministic online algorithms with competitive factors 3.28 and 2 for the problems $P_m | r_j | \sum w_j C_j$ and $P_m | r_j, pmtn | \sum w_j C_j$, respectively. Their algorithms are based on the WSPT rule. The competitive factor for the preemptive case is tight.

There have been few results considering parallel jobs. Schwiegelshohn [26] has shown that a preemptive algorithm using the WSPT rule has an approximation factor of 2.37 for the $P_m | size_j, pmtn | \sum w_j C_j$ problem. In case of parallel jobs, the Smith ratio is defined to be $\frac{w_j}{size_j \cdot p_j}$. Further for this result, preemption of a parallel job is executed via gang scheduling, that is, the execution of a job is interrupted and later resumed on all processors concurrently. Using the ideas of Philipps, Stein, and Wein [21], the preemptive schedule can be transformed into a non-preemptive one $(P_m | size_j | \sum w_j C_j)$ with a performance guarantee of 7.11. To the best of our knowledge, there are no simple deterministic algorithms yet with a constant performance guarantee for the problems $P_m | size_j, r_j | \sum w_j C_j$ and $P_m | size_j, r_j, pmtn | \sum w_j C_j$ although there is a polynomial time approximation scheme for the problem $P_m | size_j, r_j | \sum w_j C_j$, see Fishkin, Jansen, and Prokolab [6]. Similarly, there are no known results on the corresponding clairvoyant online problems. Regarding nonclairvoyant variants, we show that there is an online algorithm

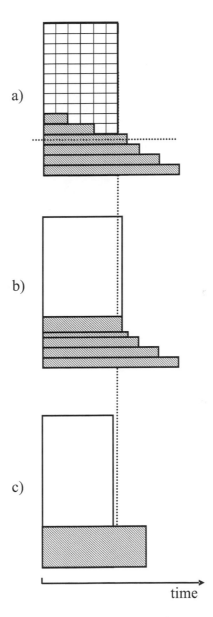

FIGURE 4.6: Application of the Long Job transformation in the WSPT proof
for P_m:
a) Optimal schedule of a problem instance.
b) Transformation of a subset of the long jobs.
c) Transformation of all long jobs.

for the problem $P_m|size_j, ncv, pmtn|\sum w_j C_j$ with a competitive factor of less than 4.

PROOF   As lower bound we use the squashed area lower bound A introduced by Schwiegelshohn et al. [27]: Every job J_j of a problem instance is assumed to be fully parallelized such that it uses all m machines and maintains its workload $size_j \cdot p_j$. Therefore, its fully parallel processing time is $\frac{size_j \cdot p_j}{m}$. For such a job set, the problem $P_m|size_j, ncv, pmtn|\sum w_j C_j$ is transformed into a single machine problem $(1|ncv, pmtn|\sum w_j C_j)$ with the completion times C'_j and the optimal total completion time $A = \sum_j w_j C'_j \leqslant \sum_j w_j C_j(OPT)$. For the second problem, the *Round-Robin* algorithm generates a schedule S with $\sum_j w_j C_j(S) < 2A \leqslant 2\sum_j w_j C_j(OPT)$, see Section 4.2.

Within one round, job J_j must receive $\frac{w_j \cdot size_i}{w_i \cdot size_j}$ times the processing time of job J_i. We denote the required processing times of jobs J_i and J_j within a round by π_i and π_j, respectively. Therefore, we are looking for a schedule that obeys the new processing times and the size restrictions of all jobs. List Scheduling generates such a schedule and guarantees $\frac{m \cdot C_{\max}(S)}{\sum_j size_j \cdot \pi_j} < 2$ if $\max_j\{\pi_j\} \leqslant \frac{1}{m}\sum_j size_j \cdot \pi_j$ holds. A new *Round-Robin* schedule based on the original data must be determined whenever a job completes. Let S' be such a schedule consisting of the rounds generated by List Scheduling. Then we have

$$\sum_j w_j C_j(S) < 2 \cdot 2A \leqslant 4 \sum_j w_j C_j(OPT).$$

The applied preemption uses gang scheduling and requires migration, that is, the execution may be resumed on a different set of machines as a new *Round-Robin* need not preserve the allocation of the previous one.

But there may be jobs J_i with $\pi_i > \frac{1}{m}\sum_j size_j \cdot \pi_j$ if they have a high weight and a small size. We appropriately reduce their processing times until for every such job J_i, we have $\pi_i = \frac{1}{m}\sum_j size_j \cdot \pi_j$. This corresponds to a problem instance with w_j being replaced by a smaller weight $\bar{w}_j$ for each reduced job J_j while $w_j = \bar{w}_j$ holds for all other jobs. In every round, a reduced job will execute during more than 50 percent of the length of the round due to List Scheduling. On the one hand, such a reduced job will delay the other jobs less than originally calculated. This will reduce the total weighted completion time of the other jobs. On the other hand, the completion time of this job is less than twice as large than its processing time and therefore less than twice as large than its completion time in the optimal schedule. Hence,

$$\sum_j w_j C_j(S) = \sum_j (w_j - \bar{w}_j) C_j(S) + \sum_j \bar{w}_j C_j(S)$$
$$< 2 \sum_j (w_j - \bar{w}_j) C_j(OPT) + 4 \sum_j \bar{w}_j C_j(OPT)$$
$$\leqslant 4 \sum_j w_j C_j(OPT).$$

□

Note that this bound is not tight.

It is obviously not possible to find a polynomial time algorithm with a constant approximation factor for the problem $P_m|size_j, ncv| \sum w_j C_j$, see Section 4.2.

4.5 Conclusion

In this chapter, we discussed and presented results for basic job scheduling problems. While there are still some open problems, particularly in the area of nonclairvoyant scheduling, good solutions exist for most of the basic problems, although some bounds are not tight. But it is most likely that new problems will arise particularly if parallel jobs are involved. Although there are already many additional results that address various forms of job parallelization and have not been mentioned in this chapter, the variety of the existing and foreseeable restrictions generate a large problem space with many yet unanswered questions.

But before exploring this space with the eyes of a theoretician looking for new challenges, it is appropriate to maintain the connection with problems in the real world. In general, it is not possible to find optimal algorithms and constant approximation or competitive factors for real-world job scheduling problems. Therefore, we need proved and simple methods like List Scheduling or the WSPT rule that can be extended to handle some additional constraints of the real world. Moreover, our simple scheduling objectives like total completion time or makespan may not be appropriate for many real world scheduling problems. But they can often be used as building blocks for more complex objectives. Finally, the real world deals with problems that often use job systems with a specific composition. Exotic worst case situations may be of little relevance. In these cases, it is necessary to prove the performance of an algorithm for specific workloads. While common theoretical methods may not be directly applicable to solve these problems it is our belief that the systematic approaches of scheduling theory may also be helpful in these cases.

References

[1] F. Afrati, E. Bampis, C. Chekuri, D. Karger, C. Kenyon, S. Khanna, I. Milis, M. Queyranne, M. Skutella, C. Stein, and M. Sviridenko. Approximation schemes for minimizing average weighted completion time with release dates. In *Proceedings of the 40th Annual Symposium on Foundations of Computer Science (FOCS)*, pages 32–44, 1999.

[2] E. Anderson and C. Potts. Online scheduling of a single machine to minimize total weighted completion time. *Mathematics of Operations Research*, 29(3):686–697, 2004.

[3] C. Chekuri, R. Motwani, B. Natarajan, and C. Stein. Approximation techniques for average completion time scheduling. *SIAM Journal on Computing*, 31(1):146–166, 2001.

[4] B. Chen and A. Vestjens. Scheduling on identical machines: How good is LPT in an on-line setting? *Operations Research Letters*, 21(4):165–169, 1997.

[5] R. Conway, W. Maxwell, and L. Miller. *Theory of Scheduling*. Addison-Wesley, 1967.

[6] A. Fishkin, K. Jansen, and L. Porkolab. On minimizing average weighted completion time of multiprocessor tasks with release dates. In *Proceedings of 28th International Colloquium on Automata, Languages and Programming ICALP 2001*, volume 2076 of *Lecture Notes in Computer Science*, pages 875–886. Springer-Verlag, 2001.

[7] M. Garey and R. Graham. Bounds for multiprocessor scheduling with resource constraints. *SIAM Journal on Computing*, 4(2):187–200, June 1975.

[8] M. Goemans, M. Queyranne, A. Schulz, M. Skutella, and Y. Wang. Single machine scheduling with release dates. *SIAM Journal on Discrete Mathematics*, 15:165–192, 2002.

[9] R. Graham. Bounds on multiprocessor timing anomalies. *SIAM Journal on Applied Mathematics*, 17:416–429, 1969.

[10] R. L. Graham. Bounds for certain multiprocessor anomalies. *Bell System Technical Journal*, 45:1563–1581, 1966.

[11] L. Hall and D. Shmoys. Approximation schemes for constrained scheduling problems. In *Proceedings of the 30th Annual Symposium on Foundations of Computer Science*, pages 134–139, 1989.

[12] H. Hoogeveen and A. Vestjens. Optimal on-line algorithms for single-machine scheduling. In *5th International Integer Programming and Combinatorial Optimization Conference*, volume 1084 of *Lecture Notes in Computer Science*, pages 404–414. Springer-Verlag, 1996.

[13] M. Hussein and U. Schwiegelshohn. Utilization of nonclairvoyant online schedules. *Theoretical Computer Science*, 362:238–247, 2006.

[14] T. Kawaguchi and S. Kyan. Worst case bound of an LRF schedule for the mean weighted flow-time problem. *SIAM Journal on Computing*, 15(4):1119–1129, Nov. 1986.

[15] H. Kellerer, T. Tautenhahn, and G. Wöginger. Approximability and nonapproximability results for minimizing total flow time on a single machine. *SIAM Journal on Computing*, 28(4):1155–1166, 1999.

[16] R. McNaughton. Scheduling with deadlines and loss functions. *Management Science*, 6(1):1–12, Oct. 1959.

[17] N. Megow and A. Schulz. On-line scheduling to minimize average completion time revisited. *Operations Research Letters*, 32(5):485–490, 2004.

[18] R. Motwani, S. Phillips, and E. Torng. Non-clairvoyant scheduling. *Theoretical Computer Science*, 130:17–47, 1994.

[19] G. Mounié, C. Rapine, and D. Trystram. A $\frac{3}{2}$ approximation algorithm for scheduling independent monotonic malleable tasks. *SIAM Journal on Computing*, 37(2):401–412, 2007.

[20] E. Naroska and U. Schwiegelshohn. On an online scheduling problem for parallel jobs. *Information Processing Letters*, 81(6):297–304, Mar. 2002.

[21] C. Phillips, C. Stein, and J. Wein. Minimizing average completion time in the presence of release dates. *Mathematical Programming*, 82:199–223, 1998.

[22] M. Pinedo. *Scheduling: Theory, Algorithms, and Systems*. Prentice-Hall, New Jersey, second edition, 2002.

[23] M. Queyranne. Structure of a simple scheduling polyhedron. *Mathematical Programming*, 58(2):263–285, 1993.

[24] A. S. Schulz and M. Skutella. The power of α-points in preemptive single machine scheduling. *Journal of Scheduling*, 5:121–133, 2002.

[25] A. S. Schulz and M. Skutella. Scheduling unrelated machines by randomized rounding. *SIAM Journal on Discrete Mathematics*, 15(4):450–469, 2002.

[26] U. Schwiegelshohn. Preemptive weighted completion time scheduling of parallel jobs. *SIAM Journal on Computing*, 33(6):1280–1308, 2004.

[27] U. Schwiegelshohn, W. Ludwig, J. Wolf, J. Turek, and P. Yu. Smart SMART bounds for weighted response time scheduling. *SIAM Journal on Computing*, 28(1):237–253, Jan. 1999.

[28] U. Schwiegelshohn and R. Yahyapour. Fairness in parallel job scheduling. *Journal of Scheduling*, 3(5):297–320, 2000.

[29] D. B. Shmoys, J. Wein, and D. P. Williamson. Scheduling parallel machines on-line. *SIAM Journal on Computing*, 24(6):1313–1331, Dec. 1995.

[30] W. Smith. Various optimizers for single-stage production. *Naval Research Logistics Quarterly*, 3:59–66, 1956.

Chapter 5

Cyclic Scheduling

Claire Hanen

Université Paris Ouest Nanterre-La Défense

Abstract   This chapter is an introduction to cyclic scheduling. We focus on a useful generalization of precedence constraints in a cyclic context, called uniform task systems. We show how it can be used to model problems from different applications fields, and introduce the main concepts and results that have been proposed to solve cycle time minimization problems, with and without resource constraints.

5.1   Introduction

Cyclic scheduling problems occur when a finite set of jobs is to be repeatedly executed a large (and assumed to be infinite) number of times. Such problems have several applications: compilation of loops for parallel computers, mass production in flexible manufacturing systems, hoist scheduling in electroplating facilities, design of embedded architectures, network scheduling. Several models have been proposed to handle such kind of problems: Petri Nets [35, 8], Max-plus algebra [11, 2], graphs [26, 22].

The wide range of applications and models induced many independent publications of the same results, and extensive bibliography is thus quite hard to make. However, we try in this chapter to give an insight of basic ideas that have been developed in different contexts, and we make the choice to use graph based models, which are interesting to derive efficient algorithms, although other models are equally suitable to get theoretical results. Most of the results mentioned here are recent generalizations of results or applications that have been referred in the survey [22]. Complementary surveys can be found in [20, 28, 16].

As ordinary scheduling problems, a cyclic problem can be described by three fields: the resource environment, the task system description, and the optimization criteria. In the next section we describe the specificities of the two last fields, and we focus on a particular model of cyclic precedence constraints called uniform constraints which is of interest in many applications. One of the main problems that arises in cyclic scheduling is that an infinite schedule must have a compact description in order to be computed. Much work has been devoted to periodic schedules, in which each job J_i is repeated every w_i time units. Periodic schedules of uniform task systems without resource constraints are thus the subject of Section 5.3. But another way to handle a compact definition of infinite schedule is to study schedules generated by a simple dynamic rule, for example without resource: start a job as soon as possible. In Section 5.4 we recall (without proofs) the main known results about the earliest schedule of a uniform task system. Then, we address problems with resource constraints. We show in Section 5.5 some theoretical tools that have been designed to build periodic schedules. Then in Section 5.6 we briefly describe the approach of defining dynamic rules to build an infinite schedule. At last, we list some open questions that are of interest from a practical or theoretical point of view.

5.2 Cyclic Scheduling and Uniform Constraints

We first introduce some notations about cyclic scheduling, and then we define the uniform constraints which will be used throughout all the chapter.

5.2.1 Common Features of Cyclic Scheduling Problems

A cyclic scheduling problem is defined by:

- A set of n jobs $\mathcal{T} = \{J_1, \ldots, J_n\}$ with processing times $\{p_1, \ldots, p_n\}$ to be repeated a large (and assumed to be infinite) number of times.

- For $J_i \in \mathcal{T}$, $< J_i, k >$ denotes the kth occurrence of J_i.

- Jobs are assumed to be non reentrant: for all $k > 0$, $< J_i, k + 1 >$ cannot start before the end of $< J_i, k >$. This hypothesis can be lightened by assuming only that $< J_i, k + 1 >$ starts at least one time unit after $< J_i, k >$ starts (partial reentrance).

- We call Iteration k the set composed of the kth occurrence of each job:

$$\{< J_1, k >, \ldots, < J_n, k >\}.$$

The jobs of $\mathcal{T}$ might be subject to constraints (task system constraints and resource constraints), which will be described later.

Thus, a solution of a cyclic scheduling problem is an infinite schedule σ which defines:

- $\forall k \geq 1, \quad S^\sigma_{<J_i,k>}$ the starting time of $< J_i, k >$;

- Resources for each job execution.

We define, for an infinite schedule σ:

DEFINITION 5.1

- *The* **average cycle time of job** J_i **in** σ *is the mean time interval between two executions of J_i in σ:*

$$W_i^\sigma = \lim_{k \to +\infty} \frac{S^\sigma_{<J_i,k>}}{k}$$

- *The* **throughput of job** J_i **in** σ *is* $\tau_i^\sigma = \frac{1}{W_i^\sigma}$

- *The* **average cycle time of** σ *is* $W^\sigma = \max_{J_i \in \mathcal{T}} W_i^\sigma$

- **The throughput of** σ *is* $\tau^\sigma = \frac{1}{W^\sigma}$

- *The* **maximum iteration completion time** *is the maximum duration of an iteration:*

$$T^\sigma = \lim_{k \to +\infty} \left(\max_{J_i \in \mathcal{T}} (S^\sigma_{<J_i,k>} + p_i) - \min_{J_i \in \mathcal{T}} S^\sigma_{<J_i,k>} \right).$$

Now, the most usual optimization criteria is to minimize the average cycle time of σ, or equivalently to maximize the throughput of σ.

But sometimes the maximum iteration completion time might be minimized for a given average cycle time bound. Those criteria are antagonist in practice. Other antagonist criteria are considered in [4] and [25, 35], which could be formulated as maximum weighted flow time of a cycle or work-in-progress (maximum number of iterations in progress at the same time). Chapter 6 focuses on similar criteria.

DEFINITION 5.2   *A schedule σ is said to be stable if its maximum iteration completion time T^σ is finite.*

Stable schedules are often sought since unstable schedules lead usually to unbounded storage capacities in practice. Indeed, if a uniform precedence constraint between job J_i and J_j models a material or data transfer betwen occurrence $< J_i, k >$ and $< J_j, k > \forall k > 0$, and if σ is a schedule such that $W_i^\sigma < W_j^\sigma$. Then σ is unstable, since for large enough k, $S^\sigma_{<J_j,k>} + p_j - S^\sigma_{<J_i,k>} \approx k(W_j^\sigma - W_i^\sigma) \to +\infty$. Now a data or material is produced by J_i every W_i^σ time units on average, and is consumed or used every W_j^σ on average. One can easily see that at time t the number of data or material already produced by occurrences J_i and not yet used by occurrences of J_j increases to infinity with t.

5.2.2   Uniform Task Systems

A particular attention is to be paid to the definition of precedence constraints in cyclic scheduling problems. Indeed, such constraints should concern the infinite set of job occurrences, and thus in order to be described should be expressed on the generic set of tasks $\mathcal{T}$.

Among different models, we choose here to present one of the most used ones, Chapter 6 will present another.

DEFINITION 5.3   *A uniform constraint between two jobs J_i and J_j is defined by two integers $l, h \in \mathbb{Z}$. It induces the following constraints on any infinite schedule σ:*

$$\forall k \geq \max(1, 1-h), \quad S^\sigma_{<J_i,k>} + l \leq S^\sigma_{<J_j,k+h>}$$

Figure 5.1 illustrates the constraints on the task occurrences induced by a uniform constraint.

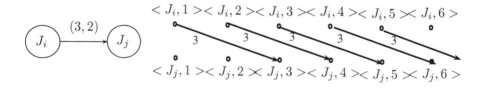

FIGURE 5.1: Constraints induced by a uniform constraint with $l = 3$, $h = 2$.

Note that non-reentrance of a job J_i (resp. partial reentrance) can be modeled by a uniform constraint between J_i and J_i with $l = p_i$ (resp. $l = 1$) and $h = 1$. Hence, from now on, we assume that such constraints are added to the set of uniform constraints.

DEFINITION 5.4   *A* **uniform task system** *is given by a multi-graph* $G = (\mathcal{T}, A)$, *and two valuations* $L, H : A \to \mathbb{Z}$ *also called* **uniform graph**. *Special cases are considered in this chapter:* **nonnegative uniform task systems**, *in which* L, H *have nonnegative values, and* **uniform precedence task systems** *in which for each arc* a, $L(a) = p_{b(a)}$, *and* $H(a) \geq 0$.

For any arc $a \in A$ we denote by $J_{b(a)}$ the input node of a, and by $J_{e(a)}$ the output node of a. Arc a models a uniform constraint between $J_{b(a)}$ and $J_{e(a)}$ with values $L(a), H(a)$. $L(a)$ is called the *length of the arc* a, and can also be called *delay* or *latency* in the literature. $H(a)$ is called the *height of the arc* a, and can also be called the *iteration distance* or the *dependence distance* in the literature. For any path μ of G we denote by $L(\mu), H(\mu)$ the sum of lengths and heights of the arcs of μ.

Uniform task systems are useful to model cyclic problems from computer science applications as well as manufacturing applications. In the two next subsections, we show two examples, one issued from a loop parallelization problem, the other from a manufacturing system.

5.2.2.1   Loop Example

Assume that arrays A, B, C, and D are stored in the memory of a computer which can compute several instructions in parallel, although each functional unit is a 7 stage pipeline, which means that if an instruction starts at time t, the result of its calculus is stored in the memory at time $t + 7$.

Moreover, we assume that a pipelined functional unit will be devoted to all occurrences of the same job, thus two successive occurrence of job J_i cannot start at the same time, which means that we have the partial reentrance assumption. Figure 5.2 shows the uniform constraints induced by the loop semantic.

When the number of iterations is very large, it can be assumed to be infinite, and the problem is then to find a schedule that minimizes the average cycle time, and that can be described in a few parallel instructions.

Many papers on so-called software pipelining are concerned with similar applications [18, 33, 17, 16, 5, 13]. Recent work like [15, 34] try to introduce memory requirements (registers) in the model as well as limited resource for task executions.

5.2.2.2   Manufacturing Example

We consider here a part, many copies of which must be produced. This part is manufactured successively in two shops. The first shop can handle at most two parts at the same time (a), and the second shop at most three parts (b). In the first shop, the part is manufactured successively by three machines, which can be modeled by 3 jobs J_1, J_2, J_3 with processing times $1, 2, 2$ (c). Then the part has to wait at least 6 time units (d) and at most 10 time units (e) before entering the second shop, in which it is processed by two machines

for $k = 2$ to N do		
$B(k) = A(k-1) + 1$	job J_1	$< J_4, k-1 >$ precedes $< J_1, k >$
$C(k) = B(k) + 5$	job J_2	$< J_1, k >$ precedes $< J_2, k >$
$D(k) = B(k-2) * D(k)$	job J_3	$< J_1, k-2 >$ precedes $< J_3, k >$
$A(k) = C(k-2) + B(k)$	job J_4	$< J_2, k-2 >, < J_1, k >$ precede $< J_4, k >$

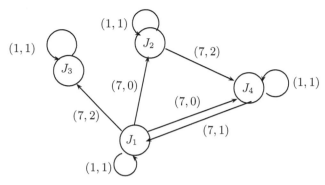

FIGURE 5.2: Uniform task system associated with a loop. Each arc a is labeled with $(L(a), H(a))$.

(jobs J_4, J_5, with processing times $2, 3$ (f)). Moreover, it is wished that the processing of one part does not take more than 30 time units (g). This can be summarized by the following system:

$$\forall k > 0, \begin{cases} (a) \ S^\sigma_{< J_3, k >} + 2 & \leq S^\sigma_{< J_1, k+2 >} \\ (b) \ S^\sigma_{< J_5, k >} + 3 & \leq S^\sigma_{< J_4, k+3 >} \\ (c) \ S^\sigma_{< J_1, k >} + 1 & \leq S^\sigma_{< J_2, k >}, \quad S^\sigma_{< J_2, k >} + 2 \leq S^\sigma_{< J_3, k >} \\ (d) \ S^\sigma_{< J_3, k >} + 2 + 6 \leq S^\sigma_{< J_4, k >} \\ (e) \ S^\sigma_{< J_4, k >} & \leq S^\sigma_{< J_3, k >} + 10 + 2 \\ (f) \ S^\sigma_{< J_4, k >} + 2 & \leq S^\sigma_{< J_5, k >} \\ (g) \ S^\sigma_{< J_5, k >} + 3 & \leq S^\sigma_{< J_1, k >} + 30 \end{cases}$$

Figure 5.3 shows how all these constraints are modeled by a uniform graph.

A lot of papers handle similar constraints. Uniform constraints and their properties studied below are in particular used to solve cyclic shop problems as well as Hoist scheduling problems in which the cyclic sequence of operations on any machine is fixed. Parametric path approach by Levner et al. [26, 27, 1, 28] is one of the main examples, but one can also mention earlier approaches by McCormick and Rao [29] or more recent ones by Caggiano and Jackson [4].

5.2.3   Questions

If we are given a uniform task system (G, L, H), several questions arise, independently of additional resource constraints.

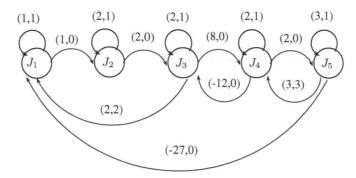

FIGURE 5.3: Uniform graph $G2$ for cyclic manufacturing.

- Does an infinite schedule exist?

- What is the minimum average cycle time of a schedule?

- Can we build optimal static schedules in polynomial time?

In the next section we consider a particular subclass of infinite schedules, called periodic schedules, and we show how the previous questions can be answered for this subclass.

5.3 Periodic Schedules of Uniform Task Systems

Infinite schedules might be difficult to describe in polynomial time. Many authors consider only periodic schedules, that have been shown dominant for the average cycle time when there are no resource constraints [8, 31]. In this section, we show the main theoretical tools that allow to compute optimal periodic schedules of uniform task systems. All missing proofs can be found in [31].

5.3.1 Properties of Periodic Schedules

DEFINITION 5.5 *A schedule σ is periodic if each job J_i has a period w_i^σ such that:*

$$\forall k \geq 1, \quad S_{<J_i,k>}^\sigma = s_i^\sigma + (k-1)w_i^\sigma$$

s_i^σ is the starting time of the first occurrence of J_i which is repeated every w_i^σ time units.

Note that the average cycle time of a periodic schedule equals the maximum period: $w_{max}^{\sigma} = \max\limits_{J_i \in \mathcal{T}} w_i^{\sigma}$. Moreover, a periodic schedule is stable if and only if all its periods are equal.

Let us now express an existence condition for a periodic schedule of a uniform task system (G, L, H). This condition states, for each arc, the relations between periods and first starting times of the tasks connected by this arc in order to meet the uniform precedence constraint. The proof of the lemma is left to the reader.

LEMMA 5.1
A periodic schedule σ is feasible if and only if: $\forall a \in A, \forall k \geq \min(1, 1 - H(a))$

$$s_{e(a)}^{\sigma} - s_{b(a)}^{\sigma} \geq L(a) - w_{e(a)}^{\sigma} H(a) + (k-1)(w_{b(a)}^{\sigma} - w_{e(a)}^{\sigma})$$

Note that as k can be chosen as large as possible, for any arc a, this Lemma implies that $w_{b(a)}^{\sigma} \leq w_{e(a)}^{\sigma}$. From this inequality we deduce easily that in any circuit μ of G, all periods of jobs are equal. Thus we can deduce that in any strong component of the graph all periods are equal. If there is an arc between two strong components, then the period of the first one is not greater than the period of the second.

Let us now consider a circuit $\mu = (J_{i_1}, \ldots, J_{i_u}, J_{i_1})$ of G with arcs $a_1, \ldots, a_u$. For the sake of simplicity, let us call w_{μ} the period of any job of μ. Using Lemma 5.1 we can see that

$$s_{i_2}^{\sigma} - s_{i_1}^{\sigma} \geq L(a_1) - w_{\mu} H(a_1)$$

$$\ldots$$

$$s_{i_u}^{\sigma} - s_{i_{u-1}}^{\sigma} \geq L(a_{u-1}) - w_{\mu} H(a_{u-1})$$
$$s_{i_1}^{\sigma} - s_{i_u}^{\sigma} \geq L(a_u) - w_{\mu} H(a_u)$$

Summing all these inequalities leads to the following one:

$$L(\mu) - w_{\mu} H(\mu) \leq 0.$$

Hence if $H(\mu) > 0$ then $w_{\mu} \geq \frac{L(\mu)}{H(\mu)}$, if $H(\mu) = 0$ then $L(\mu) \leq 0$ and if $H(\mu) < 0$ then $w_{\mu} \leq \frac{L(\mu)}{H(\mu)}$.

This result will lead to necessary existence conditions for periodic schedules.

For any circuit μ, we denote by $\alpha(\mu) = \frac{L(\mu)}{H(\mu)}$ the mean value of the circuit and by $\mathcal{C}_1, \ldots, \mathcal{C}_q$ the strong components of G. For any strong component $\mathcal{C}_s$, we denote by $\overline{w_s^{\sigma}}$ the common period of all jobs in $\mathcal{C}_s$.

Moreover, we say that $\mathcal{C}_{s'}$ precedes $\mathcal{C}_s$ if there is a path from a node of $\mathcal{C}_{s'}$ to a node of $\mathcal{C}_s$. Such precedence relations can be computed using the reduced graph of components.

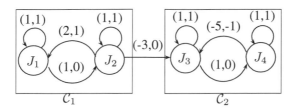

	α^+	α^-	α^*
C_1	3	$+\infty$	3
C_2	1	4	3

FIGURE 5.4: A graph $G3$ with two strong components.

Let us define:

$$\alpha^+(C_s) = \max_{\mu \ circuit \ \in C_s, H(\mu)>0} \alpha(\mu)$$

$$\alpha^-(C_s) = \min_{\mu \ circuit \ \in C_s, H(\mu)<0} \alpha(\mu) \ (+\infty \text{ if no such circuit exists})$$

$$\alpha^*(C_s) = \max \left(\alpha^+(C_s), \max_{C_{s'} \ precedes \ C_s} \alpha^+(C_{s'}) \right)$$

Figure 5.4 illustrates these values on a small example.

LEMMA 5.2
Any feasible periodic schedule σ meets the following conditions:

$$\forall s \in \{1, \ldots, q\}, \forall J_i, J_j \in C_s, w_i^\sigma = w_j^\sigma = \overline{w_s^\sigma} \quad and \quad \alpha^*(C_s) \leq \overline{w_s^\sigma} \leq \alpha^-(C_s)$$

Hence, circuits play a central role in the performance of uniform constrained cyclic schedules, as paths do for non-cyclic precedence constrained schedules.

DEFINITION 5.6 *We call* critical circuit *of G a circuit μ such that* $\alpha(\mu) = \alpha^+(G)$.

In order to prove that the conditions expressed in Lemma 5.2 are in fact necessary and sufficient, and to derive a polynomial algorithm to check feasibility and compute the optimal periodic schedule, we need to recall some basic graph results.

5.3.2   Critical Circuit of a Strongly Connected Graph

All computations in the rest of the paper are based on the following basic result. Let $G = (X, A)$ be an oriented multi-graph, with valuation $v : A \to \mathbb{R}$. For any path μ of G we denote by $v(\mu)$ the sum of its arc values.

LEMMA 5.3

The following potential equation system, or difference constraint system [12]:

$$\forall a = (i, j) \in A, \quad u_j - u_i \geq v(a)$$

has a solution if and only if for all circuits μ of G, $v(\mu) \leq 0$. A particular solution of the system is: $\forall i, \quad u_i = \max\limits_{\mu \ path \ to \ node \ i} v(\mu)$. It can be computed in polynomial time using the Bellman-Ford algorithm [12].

The proof is left to the reader. Now, consider a uniform task system and assume that G is strongly connected. According to Lemma 5.1 and 5.2, any periodic schedule σ must satisfy:

$$\forall a \in A, \quad s^\sigma_{e(a)} - s^\sigma_{b(a)} \geq L(a) - w^\sigma H(a)$$

Note that for a given value of w^σ this is a potential system with $u_i = s^\sigma_i$ and $\forall a, \quad v(a) = L(a) - w^\sigma H(a)$.

Assume now that for a given value of w^σ, the condition stated in Lemma 5.3 is met. Then any circuit μ has a negative value, which implies that:

$$\begin{cases} w^\sigma \geq \alpha(\mu) \geq \alpha^+(G) \ if \ H(\mu) > 0 \\ L(\mu) \leq 0 \qquad\qquad\quad if \ H(\mu) = 0 \\ w^\sigma \leq \alpha(\mu) \leq \alpha^-(G) \ if \ H(\mu) < 0 \end{cases}$$

Moreover, the particular solution given by Lemma 5.3 gives a feasible periodic schedule.

Otherwise, one can find a circuit (and Bellman-Ford algorithm can be used to produce such a circuit [12]) μ of G such that $L(\mu) - w^\sigma H(\mu) > 0$. From that we can deduce that:

$$\begin{cases} \alpha(\mu) \leq \alpha^+(G) \ if \ H(\mu) > 0 \\ \alpha(\mu) \geq \alpha^-(G) \ if \ H(\mu) < 0 \end{cases}$$

The properties stated previously are used in all algorithms that have been designed to compute the value of the critical circuit of a graph when it exists.

One can use a binary search for which Gondran and Minoux [19] proved that the number of iterations is polynomial in the input size, so that the overall complexity could be $O(n|A|(\log(n) + \log(\max\limits_{a \in A} |\max(L(a), H(a))|)))$. Algorithm 5.1 summarizes this approach.

However, according to [14] in which several algorithms are experimented with and compared, the most efficient algorithm in practice, although pseudo-polynomial in the worst case, is Howard's algorithm, which increases a lower

---

Algorithm 5.1: Critical circuit of a strongly connected graph

---

Compute a lower bound b on $\alpha^+(G)$ and an upper bound B (for example $\sum_{a \in A} L(a)$)

while $b < B$ **do**

$\quad$ set $w = \frac{b+B}{2}$

$\quad$ (*) check existence of a positive circuit in $(G, L - wH)$

$\quad$ **if** *no such circuit* **then**

$\quad\quad |$ set $b = w$

$\quad$ **else**

$\quad\quad$ **if** *(*) outputs a circuit μ with positive height* **then**

$\quad\quad\quad |$ set $b = \alpha(\mu)$

$\quad\quad$ **else**

$\quad\quad\quad$ **if** *(*) outputs a circuit with null height* **then**

$\quad\quad\quad\quad |$ infeasibility, exit

$\quad\quad\quad$ **else**

$\quad\quad\quad\quad |$ (*) outputs a circuit μ with negative height thus set

$\quad\quad\quad\quad$ $B = \alpha(\mu)$

if $b > B$ **then**

$\quad |$ the task system is infeasible

else

$\quad |$ $\alpha^+(G) = b$

---

bound b of $\alpha^+(G)$ until the critical circuit is reached or infeasibility is detected. Recent papers mention also an efficient parametric path algorithm with complexity $O(n^4)$ [1].

5.3.3 Computation of an Optimal Periodic Schedule

The results of the previous subsection can be extended when the graph is not strongly connected, and lead to an algorithm that computes an optimal periodic schedule.

THEOREM 5.1
A periodic schedule exists if and only if for all component C_s, $\alpha^*(C_s) \leq \alpha^-(C_s)$. An optimal periodic schedule such that $\forall s$, $\overline{w_s^\sigma} = \alpha(C_s)$ can be built in polynomial time.

Algorithm 5.2 outputs infeasibility or an optimal periodic schedule.

Figure 5.5 illustrates the construction of an optimal periodic schedule for the graph $G3$ of Figure 5.4.

Algorithm 5.2: Optimal periodic schedule

Compute the strong components of G

Sort components according to topological order of the reduced graph: $\mathcal{C}_1, \ldots, \mathcal{C}_q$

for $s = 1$ *to* q **do**

    Use algorithm 5.1 to compute $\alpha^+(\mathcal{C}_s)$ and check feasibility of $\mathcal{C}_s$;

    if $\mathcal{C}_s$ *is not feasible* **then**

        Output infeasibility

    Compute $\alpha^*(\mathcal{C}_s) = \max(\alpha^+(\mathcal{C}_s), \displaystyle\max_{\mathcal{C}_{s'} \text{ precedes } \mathcal{C}_s} \alpha^*(\mathcal{C}_{s'}))$

    Set $\overline{w_s^\sigma} = \alpha^*(\mathcal{C}_s)$

    For each arc a such that $e(a) \in \mathcal{C}_s$ set $v(a) = L(a) - \alpha^*(\mathcal{C}_s)H(a)$

    if $\mathcal{C}_s$ *has negative height values* **then**

        Use Bellman-Ford algorithm to check if $(\mathcal{C}_s, v)$ has a positive circuit

        if *positive circuit* **then**

            Output infeasibility

Add a dummy node J_0 to G, with for each job J_i an arc (J_0, J_i) with $v((J_0, J_i)) = 0$

for *each job* J_i **do**

    Compute s_i^{opt} the longest path from J_0 to J_i in (G, v)

    Output σ_{opt}, an optimal periodic schedule with cycle time

$$W^{opt} = \max_{s \in \{1, \ldots, q\}} \alpha^*(\mathcal{C}_s).$$

5.3.3.1 Stable Periodic Schedules

Note that in the schedule produced by Algorithm 5.2, the periods of two jobs might be different. Hence this schedule might not be stable. The existence of a stable schedule is not guaranteed, even when the task system is feasible. For example, if in graph $G3$ of Figure 5.4 we remove the arc between the two components and if we replace the arc values $(-5, -1)$ by $(-3, -1)$ then $\alpha^-(\mathcal{C}_2) = 2$, although $\alpha^+(\mathcal{C}_1) = 3$ so that no stable periodic schedule exists.

LEMMA 5.4

There exists a stable periodic schedule with period w if and only if

$$\max_{s \in \{1, \ldots, q\}} \alpha^+(\mathcal{C}_s) \leq w \leq \min_{s \in \{1, \ldots, q\}} (\alpha^-(\mathcal{C}_s)).$$

Such a value can be computed on the whole graph in polynomial time according to Algorithm 5.1.

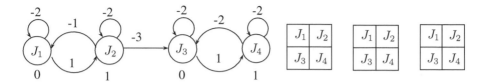

FIGURE 5.5: Computation of an optimal periodic schedule for $G3$ with valuation $v = L - 3.H$. The starting times s^σ are typed below the nodes.

Early results on cyclic scheduling [8, 6, 26, 25] were handling only nonnegative uniform task systems. This restriction makes the feasibility and stability questions easier.

LEMMA 5.5
Let (G, L, H) be a nonnegative uniform task system. It is feasible iff G does not have any circuit μ with $H(\mu) = 0$. Moreover, if the system is feasible, then there exists a stable optimal periodic schedule σ with a unique period $\forall J_i \in T, \quad w_i^\sigma = \alpha^+(G).$

5.3.3.2 Maximum Iteration Completion Time

Note that the tools presented in this chapter can also be used to find periodic schedules with a given period and minimum value of the maximum iteration completion time T^σ. Only a stable schedule can have a finite value of T^σ. So let us assume that there exist a stable periodic schedule with period w.

According to Lemma 5.1, we have to solve the following optimization problem:

$$\begin{cases} \min T \\ \forall a \in A \ \ s_{e(a)} - s_{b(a)} \geq v(a) = L(a) - wH(a) \\ \forall J_i \in T \ \ 0 \leq s_i \leq T \end{cases}$$

The last group of constraints can be expressed as: $\forall J_i \in T, \quad T - s_i \geq 0$. Thus adding a dummy node J_{n+1} and $\forall J_i \in T$, an arc (J_i, J_{n+1}) with $v(J_i, J_{n+1}) = 0$, will lead to a potential equation system on the modified graph G. Hence we can add, as in Algorithm 5.2, another dummy node J_0 with, $\forall J_i \in T$, an arc (J_0, J_i) with $v(J_0, J_i) = 0$, and we get the following result:

LEMMA 5.6
For a given period w, the minimum value of the maximum iteration completion time is the longest path from J_0 to J_{n+1} in $(G, v = L - wH)$.

Such a schedule can thus be computed in polynomial time. Note that for a fixed period value, other criteria could be considered with graph techniques. For example, minimizing the weighted flow time of a periodic schedule, i.e., minimizing $\sum_{J_i \in \mathcal{T}} u_i s_i$, induces a linear program with unimodular matrix, which can be solved using network flows algorithms

5.4 Earliest Schedule of Uniform Task Systems

But are periodic schedules optimal among all schedules? This question is answered "yes" by the results presented in this section. As in a uniform task system no resource constraint occurs, several authors studied the existence and properties of the earliest schedule, firstly with uniform systems with non-negative length and height functions [6, 8], and more recently for any integer length and height [31].

The authors carefully study the paths of G. Indeed, it can be proven quite easily that the earliest starting time of an occurrence of a job $< J_i, k >$ is the maximum length of a path in G to J_i with height k.

The authors first get the following results:

THEOREM 5.2
A uniform task system is feasible if and only if a periodic schedule exists: there is no circuit μ in G with $H(\mu) = 0$ and $L(\mu) > 0$ and

$$\forall s \in \{1, \ldots, q\} \quad \alpha^*(\mathcal{C}_s) \leq \alpha^-(\mathcal{C}_s).$$

Then, they studied the structure of the earliest schedule $\overline{\sigma}$. It happens that after a while, the starting times become K-periodic.

DEFINITION 5.7 *A schedule σ is asymptotically K-periodic if to each job J_i is associated two values K_i^σ and w_i^σ, and an integer k_0 so that:*

$$\forall k \geq k_0, \quad S_{<J_i,k>}^\sigma = w_i^\sigma + S_{<J_i,k-K_i^\sigma>}^\sigma$$

Note that periodic schedules defined in the previous section are 1-periodic schedules.

THEOREM 5.3
If the uniform task system is feasible, the earliest schedule $\overline{\sigma}$ is asymptotically K-periodic and the cycle time of any job $J_i \in \mathcal{C}_s$ can be computed in polynomial time:

$$W_i^{\overline{\sigma}} = \frac{w_i^{\overline{\sigma}}}{K_i^{\overline{\sigma}}} = \alpha^*(\mathcal{C}_s)$$

Note that the schedule description relies on two major elements:

1. The length of the transistory part of the schedule, before it becomes K-periodic;

2. The periodicity factor $K_i^{\overline{\sigma}}$ for each job J_i.

Bounds on the first item are provided in [24] for strong connected graphs. The length of the transitory part is strongly linked to the ratio between the critical circuit value $\alpha^*(G)$ and the value of the second critical circuit of the graph. The smaller this ratio is, the longer the transitory part might be. But this bound might not be tight. Bounds on the second item are described in [8, 31]: it is proven that the least common multiple of heights of critical circuits is a valid periodicity factor for all jobs, and thus is an upper bound of the minimal periodicity factor of the schedule. However, finding this minimal periodicity factor is an open problem.

5.5 Periodic Schedules of Uniform Task Systems with Resource Constraints

In this section we address problems given by a uniform task system (G, L, H) and resource constraints. Many resource frameworks have been considered in the literature, as for non cyclic scheduling problems: cyclic job-shop of flexible manufacturing with or without set-ups, pipelined and parallel machines in a computer science context, host scheduling where one or several robots move on a line to drag and drop parts in tanks.

In order to illustrate some of the concepts that can be used to solve such problems, let us consider a quite general resource setting, mentioned as *typed task systems* in the literature [16] in which there are u different resource (kind of machines), resource $R_j, j \in \{1, \ldots, u\}$ having m_j units. Each job J_i can be processed only by one resource R_{ρ_i}, and uses one unit of the resource during its processing.

In this case a schedule σ will assign not only a starting time to each occurrence of a job $< J_i, k >$ but also a unit $1 \leq m(< J_i, k >) \leq m_{\rho_i}$ of resource R_{ρ_i}.

5.5.1 Which Periodicity?

Most of the papers deal with the problem of finding an optimal stable periodic schedule, so that all job periods are equal.

But some questions about periodicity and resource arise.

DEFINITION 5.8 *A schedule σ is said periodic with constant assignment* *if it is periodic as defined previously, and if for any job J_i and any integer $k \geq 1$, $m(< J_i, k >) = m(< J_i, k + 1 >)$. Otherwise, a periodic schedule is assumed to be with variable assignment: two occurrences of a job J_i might be performed by different units of resource R_{ρ_i}.*

Unlike the previous section, such schedules are not dominating for the cycle time objective. As shown in [31] there are examples for which the optimal cycle time is not achieved by a periodic schedule (whatever the assignment is) even if the task system is nonnegative uniform. But in [7] it is proven that for uniform precedence task systems, the optimal schedule can be approximated by K-periodic schedules. Moreover periodic schedules with constant assignment don't dominate periodic schedules [22, 30].

5.5.2   Complexity and MIP Models

Most of the scheduling problems with resource constraints are NP-hard. As shown in this section, in terms of complexity a cyclic problem can be usefully linked to its underlying noncyclic problem. However, in the next section we prove that for acyclic uniform graph, the problem becomes solvable in polynomial time, even for complex resource settings.

THEOREM 5.4
An instance of a noncyclic scheduling problem

$$\alpha|prec|C_{max} \leq B$$

can be polynomially reduced to $\alpha|uniform\ prec|W \leq B + 2$.

PROOF   Indeed, if $\overline{G}$ is an acyclic precedence graph for the noncyclic scheduling problem, we can add one source node J_0 which precedes any job and one sink node J_{n+1} which is a successor of any job, with unit processing time to $\overline{G}$. Setting, for any arc a, $L(a) = p_{b(a)}, H(a) = 0$, and adding an arc from J_{n+1} to J_0 with length and height 1 will complete the input (G, L, H) of the cyclic problem.

Now, as $< J_{n+1}, k >$ precedes $< J_0, k + 1 >$, in any schedule of (G, L, H) iteration k completes before iteration $k + 1$ starts. Hence a schedule of the noncyclic problem with makespan B induces a periodic schedule of the cyclic problem of makespan B+2. Conversely a periodic schedule of (G, L, H) with period $B + 2$ induces a schedule of $\overline{G}$ with makespan B.

One of the corollary of this theorem is that cyclic dedicated or parallel machine problems with uniform task systems are NP-hard.   ▯

Mixed integer programming formulations have been proposed to model the problem of finding a stable periodic schedule of a given period for problems with dedicated processors, i.e., when each $m_j = 1$ in our model. According to

Roundy [32] and then Hanen [21], a disjunctive constraint between two jobs J_i and J_j, which are assigned to a same processor, can be modeled as follows.

Let σ be a stable periodic schedule with period w^σ that fulfills the resource constraints. Let $< J_i, k >$ be any occurrence of J_i. In the time interval $[S^\sigma_{<J_i,k>} + p_i, S^\sigma_{<J_i,k>} + w^\sigma]$, only one occurrence of J_j will start, let us call it $< J_j, k' >$. Due to the structure of the schedule, $k' - k$ is a constant h^σ_{ij} that does not depend on k. Moreover, as in a time interval $[S^\sigma_{<J_i,k>} - w^\sigma, S^\sigma_{<J_i,k>}]$ the occurrence $< J_j, k' - 1 >$ is performed, we get $h^\sigma_{ji} = k - k' + 1 = 1 - h^\sigma_{ij}$.

This results in three constraints depending on starting times of the two jobs and two integer variables h_{ij}, h_{ji}:

LEMMA 5.7
A periodic stable schedule σ meets the disjunctive constraints between two jobs J_i and J_j if and only if there exists two integers h_{ij}, h_{ji} such that:

$$\begin{cases} s^\sigma_j - s^\sigma_i \geq p_i - w^\sigma h_{ij} \\ s^\sigma_i - s^\sigma_j \geq p_j - w^\sigma h_{ji} \\ h_{ij} + h_{ji} = 1 \end{cases}$$

Note that once the new variables h_{ij} are given, the above constraints are equivalent to two uniform arcs (J_i, J_j) with length p_i and height h_{ij}, and (J_j, J_i) with length p_j and height h_{ji}. If w^σ is given, disjunctive constraints and uniform constraints lead to a mixed integer program.

This property have been used by Brucker and Kampmeyer to derive a Tabu search algorithm for the cyclic job-shop problem [3].

Time-indexed formulations have also been investigated [16, 17]. But no recent exact and specific branch and bound method has been developed for such problems.

Let us now introduce a theoretical tool that can be used to link a cyclic problem with a "classical" underlying noncyclic makespan minimization problem.

5.5.3 Patterns and Iteration Vectors

Let σ be a stable periodic schedule with period w^σ for a uniform task system submitted to resource constraints. The pattern of the schedule describes the starting times of jobs in any interval $[t, t + w^\sigma]$ in the stationary part of the schedule (for large enough t). The iteration vector of the schedule describes the relative iterations of tasks performed during such an interval:

DEFINITION 5.9 *The **pattern** π^σ and **iteration vector** η^σ of σ assign, respectively, to each job J_i a starting time $\pi^\sigma_i \in [0, w^\sigma)$ and an integer $\eta^\sigma_i \geq 0$ so that $s^\sigma_i = \pi^\sigma_i + \eta^\sigma_i w^\sigma$.*

Note that a stable periodic schedule is completely defined by its pattern and its iteration vector.

LEMMA 5.8
For any value $t \geq 0$, $t = kw^{\sigma} + r$ with $0 \leq r < w^{\sigma}$, let us consider the interval $I = [t, t + w^{\sigma})$:

- *if $\eta_i^{\sigma} > k$, no occurrence of J_i starts in I;*
- *if $\eta_i^{\sigma} \leq k$, then the unique occurrence of J_i which starts in I is:*

$$< J_i, k - \eta_i^{\sigma} + 1 > \text{ starting at time } t - r + \pi_i^{\sigma} \text{ if } \pi_i^{\sigma} \geq r$$
$$< J_i, k - \eta_i^{\sigma} + 2 > \text{ starting at time } t + w^{\sigma} - r + \pi_i^{\sigma} \text{ if } \pi_i^{\sigma} < r$$

Let us define for any $t \in [0, w^{\sigma})$ and for each resource R_j the number of jobs asking for resource R_j at time t in the pattern π^{σ}:

$$NP_t(R_j) = |\{J_i \in \mathcal{T}, \rho_i = j \quad \text{and} \quad \pi_i^{\sigma} \leq t < \pi_i^{\sigma} + p_i\}|$$
$$+ |\{J_i \in \mathcal{T}, \rho_i = j \quad \text{and} \quad \pi_i^{\sigma} - w^{\sigma} \leq t < \pi_i^{\sigma} + p_i - w^{\sigma}\}|$$

DEFINITION 5.10   *A pattern π^{σ} meets the resource constraints if*

$$\forall R_j, \forall t \in [0, w^{\sigma}), \quad NP_t(R_j) \leq m_j$$

One of the key points is the following Lemma:

LEMMA 5.9
If we are given a periodic schedule σ, then σ meets the resource constraints if and only if its pattern meets the resource constraints.

Indeed, it is possible to define from the pattern an allocation function of the resources so that at each time in the periodic schedule for each resource R_j no more than m_j resource units are needed. The shape of the allocation function has been studied in the special case of one resource (parallel processors), and uniform precedence task systems [23], and some of them (circular allocation shape) dominate all other shapes.

Now, let us express the uniform constraints using patterns and iteration vectors, using Lemma 5.1 (proof is left to the reader and can be found in [23]):

LEMMA 5.10
A stable periodic schedule σ satisfies a uniform constraint a if and only if:

$$\eta_{e(a)}^{\sigma} - \eta_{b(a)}^{\sigma} \geq \left\lceil \frac{L(a) + \pi_{b(a)}^{\sigma} - \pi_{e(a)}^{\sigma}}{w^{\sigma}} \right\rceil - H(a) = f_{\pi}^{\sigma}(a)$$

5.5.4 Decomposed Software Pipelining: A Generic Approach

The two Lemmas 5.9 and 5.10 can be used to decompose the solution of a problem with resource constraints into two successive sub-problems. Either compute first a pattern that meets the resource constraints and then try to build an iteration vector which satisfies the inequalities of Lemma 5.10, or fix first an iteration vector, and then try to compute a pattern which meets all the constraints. Such approaches have been investigated in [18, 5, 13] and by some other authors mentioned in [22].

But let us first show how the two Lemmas can be used to derive polynomial special cases.

5.5.4.1 Polynomial Special Cases

Note that, once a period w^σ and a pattern π^σ are fixed, then the inequality on the iteration vector induced by Lemma 5.10 is a potential inequality as presented in Section 5.3.

Hence we can state that:

LEMMA 5.11
There exists a feasible iteration vector associated with a pattern π^σ and a period w^σ if and only if (G, f_π^σ) has no circuit with positive value.

Now, if G is acyclic (except for the non-reentrance loops), this condition is met by any pattern, as soon as $w^\sigma \geq \max_{J_i \in \mathcal{T}} p_i$ (resp $w^\sigma \geq 1$ if partial reentrance is assumed).

Hence for an acyclic graph G, once we are able to build a pattern that meets the resource constraints, we can define an iteration vector and thus a periodic schedule based on the pattern. This property has been used in [22, 23], in results mentioned in Chapter 7 and [4] for different criteria (weighted flow time).

For the particular resource setting considered here, building a pattern is quite easy, as mentioned for parallel processors by [30]. Indeed, we can first set w^σ to be the lower bound due to resource utilization and non-reentrance:

$$w^\sigma = \max\left(\max_{J_i \in \mathcal{T}} p_i, \max_{R_j} \frac{\sum_{J_i \in \mathcal{T}, \rho_i = j} p_i}{m_j}\right).$$

Then we can use the classical MacNaughton algorithm (see for example [6]) to schedule the jobs preemptively on each resource as if they were independent. The nice property of this algorithm is that each job is preempted at most once, and in this case its first part starts at time 0, as its last part ends at time w^σ. Thus if a job J_i is not preempted, then we set π_i^σ to be its starting time in the MacNaughton schedule, while if it is preempted, π_i^σ is the starting time of its last part. This is summarized in the following theorem:

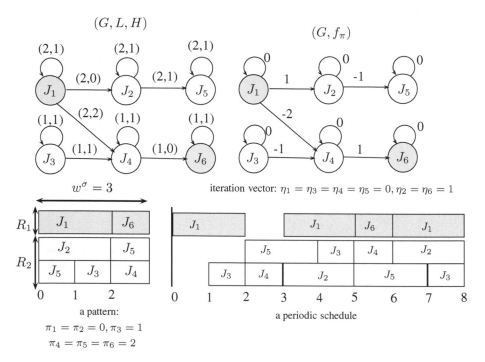

FIGURE 5.6: A periodic schedule for an acyclic graph (G, L, H) (upper left). J_1 and J_6 share resource R_1 as other jobs share resource R_2. A pattern is computed for $w^\sigma = 3$, the lower bound, according to the resource constraints (bottom left). Then an iteration vector according to (G, f_π) is produced (upper right), and the resulting optimal periodic schedule is shown (bottom right).

THEOREM 5.5

If G is acyclic then a periodic schedule with minimum cycle time can be found in polynomial time when the resource constraints can be formulated as "typed task systems".

Note that this result applies to job-shop or flow-shop like constraints as well as parallel processors. Figure 5.6 illustrates the construction of an optimal periodic schedule.

Assume now that we are given a uniform task system, and that the resources are disjunctive (for all R_j, $m_j = 1$). Some authors [28] observed that within a pattern, on each resource R_j, the jobs assigned to this resource follow a sequence.

They remarked that once both the iteration vector and the sequence are given, then the resource constraint can be expressed as a circuit of uniform constraints, with height 1. Indeed, if job J_i and $J_{i'}$ are processed successively in any pattern, this will induce a constraint $(J_i, J_{i'})$ with length p_i (or more if setups are considered) and height $\eta_{i'} - \eta_i$. If now J_i is the last job of the sequence and $J_{i'}$ the first one, then we know by Lemma 5.8 that in any schedule $< J_i, k >$ will precede $< J_{i'}, k + 1 + \eta_{i'} - \eta_i >$. This result in a uniform constraint $(J_i, J_{i'})$ with length p_i and height $1 + \eta_{i'} - \eta_i$. This lead to the following result:

THEOREM 5.6
For disjunctive resource settings, if the iteration vector as well as the sequence of jobs in the pattern are known, the problem of finding an optimal stable periodic schedule can be solved in polynomial time.

5.5.4.2 The Retiming Approach

Literature on parallel computing [5, 13, 18] use the idea of decomposition as follows, in order to solve a cyclic problem of a uniform precedence task system with parallel processors.

The first step is to compute an iteration vector η.

Then, according to Lemma 5.1, the pattern should meet the following constraint associated with each arc a:

$$\pi^\sigma_{e(a)} - \pi^\sigma_{b(a)} \geq p_{b(a)} - w^\sigma(H(a) + \eta_{e(a)} - \eta_{b(a)}). \qquad (*)$$

Note that as $\pi^\sigma_{e(a)} - \pi^\sigma_{b(a)} \in [-w^\sigma, w^\sigma]$, any feasible iteration vector satisifes $H(a) + \eta_{e(a)} - \eta_{b(a)} \geq 0$.

For a given feasible iteration vector (also called retiming by some authors [5]), note that if $H(a) + \eta_{e(a)} - \eta_{b(a)} \geq 1$ then $(*)$ is met by any pattern in which $w^\sigma \geq \pi_i + p_i$ for any job i.

Thus, the only constraints that are of interest for such patterns are those associated with arcs of G for which $H(a) + \eta_{e(a)} - \eta_{b(a)} = 0$. One can easily check that the subgraph G_η of G induced by these arcs is acyclic.

This leads to a generic algorithm given below.

Algorithm 5.3: The retiming approach

Step 1: Compute an iteration vector η s.t. $\forall a \in A, \eta_{e(a)} - \eta_{b(a)} \geq -H(a)$.
Step 2: Compute a schedule π of the acyclic graph G_η with initial resource constraints, subject to makespan minimization.
Step 3: Combine η and π to get a schedule with period C^π_{max}.

The authors who propose this general approach show several interesting properties. First at Step 1, one can derive in polynomial time (again with path techniques, since η is a potential on (G, H)) an iteration vector which minimizes the longest path of the generated G_η, and among such vectors, one which minimizes the number of arcs of G_η. This might be interesting since the length of path is one of the lower bounds of the minimum makespan.

Secondly, they prove that according to G_η with minimum longest path, if at Step 2 a makespan minimization algorithm with usual Graham's worst case performance ratio $2 - \frac{1}{m}$ is used, then the performance guarantee still holds for w^σ with a small additive term.

But note that any relaxation of the problem, in particular any feasible schedule that satisfies the uniform constraints leads to a feasible η. Moreover at Step 2, exact approaches as well as approximation algorithms that have been produced for noncyclic versions of the problems could be investigated to analyze the overall performance of this approach in different resource settings.

5.6 Dynamic Schedules

Periodic schedules are not the only way to produce efficient schedules that can be described shortly. Indeed, if we are able to describe a simple priority rule (such as: the larger the number of immediate successors of a job, the more urgent), we can use it to perform usual list scheduling (see Chapter 2) on any resource settings and any uniform graph. The question is then to analyze the performance of the resulting schedule with respect to the optimal one.

This approach has not been much investigated until now. Chrétienne [10, 9] gives a first insight and proves some basic promising results recalled here, that might be further investigated.

This author addresses the problem of scheduling a strong connected uniform precedence task system on m identical parallel machines. He shows in [10] that even if the non-reentrance hypothesis is not assumed, in any regular schedule of G, there are at least $H_{min} = \min_{\mu \ circuit \ of \ G} H(\mu)$ ready or active jobs at each time unit.

He uses this property to extend Graham's worst case performance ratio:

THEOREM 5.7
Let σ be a schedule generated by any list scheduling algorithm. Then

$$W^\sigma \leq \left(2 - \frac{\min(H_{min}, m)}{m} \right) W^{opt}.$$

This proves, in particular, that for any graph for which $H_{min} \geq m$, any list scheduling algorithm outputs an optimal schedule.

In another work he introduces some particular priority lists, called K-periodic linear orders. He shows that such priority lists used by a list scheduling algorithm induce K-periodic schedules.

Then, he considers the noncyclic version of the problem $P|prec|C_{max}$, and any algorithm $\mathcal{A}$ that computes a list for this problem so that the associated list schedule has worst case performance ratio γ.

Let G^k be the graph of the precedence relations induced by G on the k first iterations of jobs. He shows that if $\mathcal{A}$ computes for G^k a K-periodic linear order for large enough k, then it will induce a K-periodic schedule with the same performance ratio γ on the cycle time.

This result is then applied to the Coffman-Graham list, which is known to be optimal for the makespan problem minimization with unit processing times and 2 processors, and has a performance ratio $2 - \frac{2}{m}$ for $m \geq 3$.

But some questions still arise from this original approach: the computation of the list needs the expansion of G into G^k for large enough k. But is k polynomial? What is the maximum periodicity factor K of a given list algorithm?

5.7 Conclusion

Before concluding, let us just consider several open questions that could be of interest in the future. Surprisingly, the complexity of the problem $P2|uniform\ prec, p_j = 1|W$ is still unknown, although its noncyclic counterpart is solvable in polynomial time as well as some cyclic special cases mentioned in [22]. Secondly, convergence study of earliest schedule or similarly of list schedules with resource constraints has not been very much investigated.

Periodic schedules are not dominating schedules with resource constraints, but K-periodic schedules are. One of the question that arises, is to link the periodicity factor K with the resource settings. For example, in the problem for m parallel processor, are the m-periodic schedules dominating?

As mentioned in Section 5.5, the retiming approach could have many applications in exact algorithms, as well as in approximation for various resource settings, not only in the software pipelining application field. More generally, efficient branch and bound algorithms should be sought in this field, and thus new lower bounds based on both uniform and resource constraints.

And, finally, the definition of list schedules that could be computed in polynomial time with interesting performance guarantee and control on their structure (periodicity factor and transistory state) are very important in practice as well as for the theory.

In this chapter, we tried to give an insight of the most basic and useful concepts to solve cyclic scheduling problems. It is far from being exhaustive, and readers are kindly invited to read the next chapters, and some of the cited papers, which will give them many more references than those mentioned here.

References

[1] D. Alacaide, C. Chu, V. Kats, E. Levner, and G. Sierksma. Cyclic multiple robot scheduling with time-window constraints using a critical path approach. *European Journal of Operational Research*, 177:147–162, 2007.

[2] F. Bacelli, G. Cohen, G. J. Olsder, and J.-P. Quadrat. *Synchronisation and Linearity: An Algebra for Discrete Event Systems*. John Wiley & Sons, New York, 1992.

[3] P. Brucker and T. Kampmeyer. Tabu search algorithms for cyclic machine scheduling problems. *Journal of Scheduling*, 8(4):303–322, 2005.

[4] K. Caggiano and P. Jackson. Finding minimum flow time cyclic schedules for non identical, multistage jobs. *IIE Transactions*, 40:45–65, 2008.

[5] P.-Y. Calland, A. Darte, and Y. Robert. Circuit retiming applied to decomposed software pipelining. *IEEE Transactions on Parallel and Distributed Systems*, 9(1):24–35, 1998.

[6] J. Carlier and P. Chrétienne. *Problèmes D'ordonnancement: Modèlisation, Complexité, Algorithmes*. Masson, Paris, 1988.

[7] J. Carlier and P. Chrétienne. Timed Petri nets schedules. In *Advances in Petri nets*, volume 340 of *Lectures Notes in Computer Sciences*, pages 62–84. Springer-Verlag, 1988.

[8] P. Chrétienne. Transient and limiting behavior of timed event graphs. *RAIRO Techniques et Sciences Informatiques*, 4:127–192, 1985.

[9] P. Chrétienne. List schedules for cyclic scheduling. *Discrete Applied Mathematics*, 94:141–159, 1999.

[10] P. Chrétienne. On Graham's bound for cyclic scheduling. *Parallel Computing*, 26(9):1163–1174, 2000.

[11] G. Cohen, P. Moller, J.-P. Quadrat, and M. Viot. Algebraic tools for the performance evaluation of discrete event systems. *IEEE Proceeding: Special Issue on Discrete Event Dynamics Systems*, 77(1):39–58, 1989.

[12] T. Cormen, C. Leiserson, and R. Rivest. *Introduction to Algorithms*. MIT Press, 1990.

[13] A. Darte and G. Huard. Loop shifting for loop compaction. *International Journal of Parallel Programming*, 28(5):499–534, 2000.

[14] A. Dasdan, S. Irani, and R. K. Gupta. Efficient algorithms for optimum cycle mean and optimum cost to time ratio problems. In *Design Automation Conference*, pages 37–42, 1999.

[15] D. de Werra, C. Eisenbeis, S. Lelait, and B. Marmol. On a graph-theoretical model for cyclic register allocation. *Discrete Applied Mathematics*, 93:191–203, 1999.

[16] B. Dupont de Dinechin, C. Artiques, and S. Azem. Resource constrained modulo scheduling. In C. Artigues, S. Demassey, and E. Neron, editors, *Resource-Constrained Project Scheduling*, pages 267–277. ISTE and John Wiley, London, 2008.

[17] A. Eichenberger and E. Davidson. Efficient formulation for optimal modulo schedulers. In *ACM SIGPLAN Conference on Programming Language Design and Implementation*, pages 194–205, Las Vegas, Nevada, 1997.

[18] F. Gasperoni and U. Schwiegelshohn. Generating close to optimum loop schedules on parallel processors. *Parallel Processing Letters*, 4:391–403, 1994.

[19] M. Gondran and M. Minoux. *Graphs and Algorithms*. John Wiley & Sons, first edition, 1984.

[20] N. G. Hall, T.-E. Lee, and T.-E. Posner. The complexity of cyclic scheduling problems. *Journal of Scheduling*, 5:307–327, 2002.

[21] C. Hanen. Study of an NP-hard cyclic scheduling problem: the recurrent job-shop. *European Journal of Operational Research*, 72(1):82–101, 1994.

[22] C. Hanen and A. Munier. Cyclic scheduling on parallel processors: An overview. In P. Chrétienne, E. G. Coffman, J. K. Lenstra, and Z. Liu, editors, *Scheduling Theory and Its Applications*. John Wiley & Sons, 1994.

[23] C. Hanen and A. Munier. A study of the cyclic scheduling problem on parallel processors. *Discrete Applied Mathematics*, 57(2–3):167–192, 1995.

[24] M. Hartmann and C. Arguelles. Transience bounds for long walks. *Mathematics of Operation Research*, 24(2), 1999.

[25] H. Hillion and J.-M. Proth. Performance evaluation of job-shop systems using timed event graph. *IEEE Transactions on Automatic Control*, 34(1):3–9, 1989.

[26] E. Levner and V. Kats. A parametric critical path problem and an application for cyclic scheduling. *Discrete Applied Mathematics*, 87:149–158, 1998.

[27] E. Levner and V. Kats. Polynomial algorithms for periodic scheduling of tasks on parallel processors. In L. Yang and M. Paprzycki, editors, *Practical Applications of Parallel Computing: Advances in Computation Theory and Practice*, volume 12, pages 363–370. Nova Science Publishers, Canada, 2003.

[28] E. Levner, V. Kats, and D. Alcaide. Cyclic scheduling in robotic cells: An extension of basic models in machine scheduling theory. In E. Levner, editor, *Multiprocessor Scheduling: Theory and Applications*, pages 1–20. I-Tech Education and Publishing, Vienna, Austria, 2007.

[29] T. McCormick and U. S. Rao. Some complexity results in cyclic scheduling. *Mathematical and Computer Modelling*, 20:107–122, 1994.

[30] A. Munier. Contribution à l'étude des ordonnancements cycliques. Ph.D. thesis, 1991.

[31] A. Munier-Kordon. A generalization of the basic cyclic scheduling problem. *Journal of Scheduling*, 2009.

[32] R. Roundy. Cyclic schedules for job-shop with identical jobs. *Mathematics of Operation Research*, 17(4):842–865, 1992.

[33] P. Sucha, Z. Pohl, and Z. Hanzalek. Scheduling of iterative algorithms on FPGA with pipelined arithmetic unit. In *10th IEEE Real-Time and Embedded Technology and Applications Symposium, RTAS*, 2004.

[34] S. Touati. On the periodic register need in software pipelining. *IEEE Transactions on Computers*, 56(11):1493–1504, 2007.

[35] B. Trouillet, O. Korbaa, and J.-C. Gentina. Formal approach of FMS cyclic scheduling. *IEEE Transactions on Systems, Man and Cybernetics. Part C: Applications and Reviews*, 37(1):126–137, 2007.

Chapter 6

Cyclic Scheduling for the Synthesis of Embedded Systems

Olivier Marchetti

Université Pierre et Marie Curie Paris 6

Alix Munier-Kordon

Université Pierre et Marie Curie Paris 6

Abstract   This chapter is devoted to the study of timed weighted event graphs, which constitute a subclass of Petri nets often considered for modeling embedded applications such as video encoders. Some basic recent mathematical properties are presented, leading to algorithms checking the liveness and computing the optimum throughput of these systems.

6.1   Introduction

The design of embedded multi-media applications is a central industrial problem nowadays. These systems may often be modeled using Synchronous Dataflow Graphs (in short SDFs) introduced by Lee and Messerschmitt [20, 21]. The vertices of these graphs correspond to programs. Each arc models a buffer used for communications by the adjacent programs. SDFs are briefly recalled in Section 6.2.1. SDFs are totally equivalent to Weighted Event Graphs (in short, WEGs), which are a subclass of Petri nets [29]. In this chapter, we prefer WEG instead of SDF because of their importance in the computer science community.

There is important literature on timed (non weighted) event graphs. Indeed, as recalled in Section 6.3.2, they can be viewed as a subclass of uniform constraints as defined in Chapter 5. As recalled in that chapter, for this class of graphs, the liveness (i.e., the existence of a schedule) and the computation of the optimum throughput are polynomial problems. The consequence is that most of the optimization problems on these structures are in $\mathcal{NP}$. For example, many authors developed efficient algorithms to solve the minimization of the number of tokens of an event graph for a given throughput [10, 14, 18] or some interesting variants [13].

For timed WEG, there is, to our knowledge, no polynomial algorithm for the liveness or the computation of the maximum throughput despite original attempts to solve these problems [6]. Note that, for a slightly different formalism called computation graph [17], Karp and Miller have shown that the deadlock existence problem is in $\mathcal{NP}$.

From a practical point of view, the optimization problems are solved using pseudo-polynomial algorithms to evaluate the liveness and the maximal throughput, which limits dramatically the size of tractable instances. In [30, 32], the liveness problem of a WEG is solved by using pseudo-polynomial algorithms mainly based upon a transformation of the WEG into an event graph, called expansion, introduced in [25, 26]. In [11], the authors consider model checking methods to evaluate the liveness of a marked WEG. In the same way, they concluded in [12] that the computation of a state graph is not possible for large instances and that efficient methods are required to compute the optimal throughput. In [5, 8, 27], the authors considered the computation of a periodic schedule for a timed WEG with multi-objective functions such as the minimal schedule length and the minimal amount of memory. Several authors also consider the minimization of place capacities of a WEG for a given initial marking. This problem is NP-complete even for event graphs as proved in [28] and several heuristics were developed to solve it [1, 2]. Surprisingly, some polynomial algorithms exist for optimization problems on timed WEGs with additional assumptions on the structure of the graph or the values of the throughput: a polynomial algorithm is developed in [23] to compute an initial live marking of a WEG that minimizes the place capacities. An approximation algorithm was also developed in [22] which maximizes the throughput for a place capacity at most twice the minimum.

The aim of this chapter is to present some basic mathematical properties on timed WEGs leading to polynomial algorithms to evaluate a sufficient condition of liveness and lower bounds of the optimum throughput. SDFs and WEGs are briefly presented in Section 6.2. Section 6.3 is devoted to the characterization of the precedence relations associated with a WEG and some basic technical important lemmas. Section 6.4 is dedicated to unitary graphs, which constitute an important subclass of WEGs for the computation of the optimal throughput. Two important transformations of unitary WEGs, namely the normalization and the expansion, are detailed along some other important properties. We present in Section 6.5 a polynomial time algorithm to compute an optimal periodic schedule for a unitary WEG and leading to a sufficient condition of liveness and a lower bound to the optimum throughput.

Section 6.6 is our conclusion.

6.2   Problem Formulation and Basic Notations

This section is devoted to the presentation of the problem. Synchronous dataflow graphs and timed weighted event graphs are briefly introduced. The two problems tackled here, namely the liveness and the determination of the maximal throughput of a timed marked weighted event graph, are then recalled.

6.2.1   Synchronous Dataflow Graphs

Synchronous Dataflow Graphs (SDFs) are a formalism introduced and studied by Lee and Messerschmitt [20, 21] to model embedded applications defined by a set of programs $Pg_1, \ldots, Pg_n$ exchanging data using FIFO (First-In First-Out) queues. Each FIFO queue has exactly one input program Pg_i and one output program Pg_j and is modeled by an arc $e = (Pg_i, Pg_j)$ bi-valued by strictly positive integers val_i and val_j such that:

1. at the completion of one execution of Pg_i, $val_i(e)$ data are stored in the queue to be sent to Pg_j, and

2. at the beginning of one execution of Pg_j, $val_j(e)$ data are removed from the queue; if there are not enough data, Pg_j stops and waits for them.

This formalism suits particularly well for streaming applications such as video encoders and decoders which are nowadays crucial for economical reasons. Many real life examples of modeling such systems using SDF may be found in [4, 15]. Several complex environments for modeling and simulating these systems were also developed recently (see examples [19, 33]). Figure 6.1 presents the modeling of a H263 decoder using an SDF presented in [31].

FIGURE 6.1: Modeling the H263 decoder using an SDF [31].

6.2.2   Timed Weighted Event Graphs

A Weighted Event Graph $\mathcal{G} = (T, P)$ (WEG) is given by a set of transitions $T = \{t_1, \ldots, t_n\}$ and a set of places $P = \{p_1, \ldots, p_m\}$. Every place $p \in P$ is defined between two transitions t_i and t_j and is denoted by $p = (t_i, t_j)$. Arcs (t_i, p) and (p, t_j) are valued by strictly positive integers denoted respectively by $u(p)$ and $v(p)$. At each firing of the transition t_i (*resp.* t_j), $u(p)$ (*resp.* $v(p)$) tokens are added to (*resp.* removed from) place p. If $u(p) = v(p) = 1$ for every place $p \in P$, $\mathcal{G}$ is a (non weighted) event graph.

For every transition $t \in T$, $\mathcal{P}^+(t)$ (*resp.* $\mathcal{P}^-(t)$) denotes the set of places that are successors (*resp.* predecessors) of t in $\mathcal{G}$. More formally,

$$\mathcal{P}^+(t) = \{p \in P, \exists t' \in T/p = (t, t') \in P\} \text{ and}$$

$$\mathcal{P}^-(t) = \{p \in P, \exists t' \in T/p = (t', t) \in P\}.$$

For any integer $\nu > 0$ and any transition $t_i \in T$, $< t_i, \nu >$ denotes the νth firing of t_i.

An initial marking of the place $p \in P$ is usually denoted as $M_0(p)$ and corresponds to the initial number of tokens of p. A marked WEG $G = (T, P, M_0)$ is a WEG with an initial marking. Figure 6.2 presents a marked place $p = (t_i, t_j)$.

FIGURE 6.2: A marked place $p = (t_i, t_j)$.

A timed WEG $\mathcal{G} = (T, P, M_0, \ell)$ is a marked WEG such that each transition t has a processing time $\ell(t) \in \mathbb{N}^*$. Preemption is not allowed. The firing of a transition t at time μ then requires three steps:

1. if every place $p \in \mathcal{P}^-(t)$ has at least $v(p)$ tokens, then exactly $v(p)$ tokens are removed from p at time μ,

2. t is fired and is completed at time $\mu + \ell(t)$,

3. lastly, $u(p)$ tokens are placed in every place $p \in \mathcal{P}^+(t)$ at time $\mu + \ell(t)$.

A schedule σ is a function $S^\sigma : T \times \mathbb{N}^* \to \mathbb{Q}^+$ which associates, with any tuple $(t_i, k) \in T \times \mathbb{N}^*$, the starting time of the kth firing of t_i denoted by $S^\sigma_{<t_i, k>}$.

There is a strong relationship between a schedule σ and its corresponding instantaneous marking. Let $p = (t_i, t_j)$ be a place of P. For any value

$\mu \in \mathbb{Q}^{+*}$, let us denote by $E(\mu, t_i)$ the number of firings of t_i completed at time μ. More formally,

$$E(\mu, t_i) = \max\{q \in \mathbb{N}, S^{\sigma}_{<t_i, q>} + \ell(t_i) \leqslant \mu\}.$$

In the same way, $B(\mu, t_j)$ denotes the number of firings of t_j started up to time μ and

$$B(\mu, t_j) = \max\{q \in \mathbb{N}, S^{\sigma}_{<t_j, q>} \leqslant \mu\}.$$

Clearly,

$$M(\mu, p) = M(0, p) + u(p) \cdot E(\mu, t_i) - v(p) \cdot B(\mu, t_j).$$

The initial marking of a place $p \in P$ is usually denoted as $M_0(p)$ (*i.e.*, $M_0(p) = M(0, p)$).

A schedule (and its corresponding marking) is feasible if $M(\mu, p) \geqslant 0$ for every tuple $(\mu, p) \in \mathbb{Q}^{+*} \times P$. The throughput of a transition t_i for a schedule σ is defined by

$$\tau_i^{\sigma} = \lim_{q \to +\infty} \frac{q}{S^{\sigma}_{<t_i, q>}}.$$

The throughput of σ is the smallest throughput of a transition:

$$\tau^{\sigma} = \min_{t_i \in T}\{\tau_i^{\sigma}\}.$$

Throughout this chapter, we also assume that transitions are non-reentrant, *i.e.*, two successive firings of a same transition cannot overlap. This corresponds to

$$\forall t \in T, \forall q > 0, S^{\sigma}_{<t, q>} + \ell(t) \leqslant S^{\sigma}_{<t, q+1>}.$$

Non-reentrance of a transition $t \in T$ can be modeled by a place $p = (t, t)$ with $u(p) = v(p) = 1$ and $M_0(p) = 1$. In order to simplify the figures, they are not pictured. However, most of the results presented here may be easily extended if some transitions are reentrant.

SDFs and timed WEGs are equivalent formalisms: transitions may be associated to programs and places to FIFO queues. However, timed marked WEGs simply model the data exchanged using tokens and they define a subclass of Petri nets. We have selected this latter formalism in the rest of the chapter.

6.2.3 Problem Formulation

Let us consider a given timed marked WEG $\mathcal{G} = (T, P, M_0, \ell)$. The two basic problems considered here are formally defined as follows:

Liveness: May every transition be fired infinitely often?

$\mathcal{G}$ must be live since an embedded code has to be performed without interruption.

Maximal throughput: what is the maximal throughput of a feasible schedule?

Remark that the earliest schedule (that consists in firing the transitions as soon as possible) always exists for live timed marked WEGs and has a maximum throughput.

As recalled in Section 6.1, these two problems correspond to the checking stages [9] of most optimization problems on timed marked WEGs. They must be efficiently solved to compute good solutions to most optimization problems on timed WEGs.

6.3   Precedence Relations Induced by a Timed Marked WEG

This section is dedicated to several basic technical properties on timed marked WEGs. The precedence relations between the successive firings of two transitions adjacent to a place p are characterized first. Then, it is observed that for event graphs (*i.e.*, $u(p) = v(p) = 1$ for every $p \in P$), these relations are uniform precedence constraints as defined in Chapter 5. Finally, some additional technical lemmas on precedence relations are considered.

6.3.1   Characterization of Precedence Relations

As defined in Section 6.2.2, a schedule σ is feasible if the number of tokens remains positive in every place. This constraint generates precedence relations between the firings of every couple of transitions adjacent to a place. Strict precedence relations between two firings is defined and characterized in the following.

DEFINITION 6.1   *Let $p = (t_i, t_j)$ be a marked place and let (ν_i, ν_j) be a couple of strictly positive integers. There exists a (strict) precedence relation from $< t_i, \nu_i >$ to $< t_j, \nu_j >$ if*

Condition 1 $< t_j, \nu_j >$ *can be done after* $< t_i, \nu_i >$;

Condition 2 $< t_j, \nu_j - 1 >$ *can be done before* $< t_i, \nu_i >$ *while* $< t_j, \nu_j >$ *cannot.*

The following lemma characterizes the couples of firings constrained by precedence relations:

LEMMA 6.1

A place $p = (t_i, t_j) \in P$ with initially $M_0(p)$ tokens models a precedence relation between the ν_ith firing of t_i and the ν_jth firing of t_j iff

$$u(p) > M_0(p) + u(p) \cdot \nu_i - v(p) \cdot \nu_j \geqslant \max\{u(p) - v(p), 0\}.$$

PROOF   By Definition 6.1, a place $p = (t_i, t_j) \in P$ with initially $M_0(p)$ tokens models a precedence relation from $< t_i, \nu_i >$ to $< t_j, \nu_j >$ if Conditions 1 and 2 hold.

1. Condition 1 is equivalent to

$$M_0(p) + u(p) \cdot \nu_i - v(p) \cdot \nu_j \geqslant 0.$$

2. Condition 2 is equivalent to

$$v(p) > M_0(p) + u(p) \cdot (\nu_i - 1) - v(p) \cdot (\nu_j - 1) \geqslant 0.$$

Combining these two inequalities, we obtain the inequality required.   ◻

6.3.2   Timed Event Graphs

A timed event graph is a WEG such that $u(p) = v(p) = 1$ for every place $p \in P$. The set of precedence relations induced by a marked timed event graph $\mathcal{G} = (T, P, M_0, \ell)$ can be modeled using uniform constraints as defined previously in Chapter 5. Indeed, by Lemma 6.1, there exists a precedence relation between the ν_ith firing of t_i and the ν_jth firing of t_j induced by a place $p = (t_i, t_j)$ iff

$$1 > M_0(p) + \nu_i - \nu_j \geqslant 0,$$

which is equivalent to $\nu_j = \nu_i + M_0(p)$. A feasible schedule σ verifies then, for every place $p = (t_i, t_j) \in P$, the infinite set of precedence relations

$$\forall \nu > 0, S^{\sigma}_{<t_i, \nu>} + \ell(t_i) \leqslant S^{\sigma}_{<t_j, \nu + M_0(p)>}$$

which corresponds exactly to a uniform precedence constraint $a = (t_i, t_j)$ with length $L(a) = \ell(t_i)$ and height $H(a) = M_0(p)$. So, liveness and computation of the maximal throughput may be both polynomially computed for this subclass of timed marked WEGs using the algorithms recalled in Chapter 5.

6.3.3   Equivalent Places

For any place $p = (t_i, t_j) \in P$ with initially $M_0(p)$ tokens, $PR(p, M_0(p))$ denotes the infinite set of precedence relations between the firings of t_i and t_j induced by p.

DEFINITION 6.2   *Two marked places $p_1 = (t_i, t_j)$ and $p_2 = (t_i, t_j)$ are said equivalent if they induce the same set of precedence relations between the firings of t_i and t_j, i.e., $PR(p_1, M_0(p_1)) = PR(p_2, M_0(p_2))$.*

LEMMA 6.2
Two marked places $p_1 = (t_i, t_j)$ and $p_2 = (t_i, t_j)$ with $\frac{u(p_2)}{u(p_1)} = \frac{v(p_2)}{v(p_1)} = \frac{M_0(p_2)}{M_0(p_1)} = \Delta \in \mathbb{Q}^{+*}$ are equivalent.

PROOF   Let us assume the existence of a precedence relation induced by p_1 between $< t_i, \nu_i >$ and $< t_j, \nu_j >$. Then, by Lemma 6.1, we get

$$u(p_1) > M_0(p_1) + u(p_1) \cdot \nu_i - v(p_1) \cdot \nu_j \geqslant \max\{u(p_1) - v(p_1), 0\}$$

$$\Updownarrow \times \Delta$$

$$\Delta.u(p_1) > \Delta.(M_0(p_1) + u(p_1) \cdot \nu_i - v(p_1) \cdot \nu_j) \geqslant \Delta. \max\{u(p_1) - v(p_1), 0\}$$

$$\Updownarrow$$

$$u(p_2) > M_0(p_2) + u(p_2) \cdot \nu_i - v(p_2) \cdot \nu_j \geqslant \max\{u(p_2) - v(p_2), 0\}.$$

We conclude that p_2 induces a precedence relation between $< t_i, \nu_i >$ and $< t_j, \nu_j >$, which completes the proof.   □

For every place $p \in P$, the greatest common divisor of the integers $u(p)$ and $v(p)$ is denoted by gcd_p, i.e., $gcd_p = gcd(u(p), v(p))$. The following lemma limits the possible values of the initial markings of a place to the multiples of gcd_p:

LEMMA 6.3
The initial marking $M_0(p)$ of any place $p = (t_i, t_j)$ may be replaced by $M_0^\star(p) = \left\lfloor \frac{M_0(p)}{gcd_p} \right\rfloor \cdot gcd_p$ tokens without any influence on the precedence relations induced by p, i.e., $PR(p, M_0(p)) = PR(p, M_0^\star(p))$.

PROOF   Using the Euclidean division of $M_0(p)$ by gcd_p, we get

$$M_0(p) = M_0^\star(p) + R_{gcd}(M_0(p)),$$

with $R_{gcd}(M_0(p)) \in \{0, \ldots, gcd_p - 1\}$.

$PR(p, M_0(p)) \subseteq PR(p, M_0^\star(p))$. Let us suppose that there exists a precedence relation from $PR(p, M_0(p))$ between $< t_i, \nu_i >$ and $< t_j, \nu_j >$. By Lemma 6.1,

$$u(p) > M_0(p) + u(p) \cdot \nu_i - v(p) \cdot \nu_j \geqslant \{(u(p) - v(p), 0\}.$$

So, we get

$$u(p) > M_0^\star(p) + R_{gcd}(M_0(p)) + u(p) \cdot \nu_i - v(p) \cdot \nu_j \geqslant \{u(p) - v(p), 0\}.$$

Clearly,
$$u(p) > M_0^\star(p) + u(p) \cdot \nu_i - v(p) \cdot \nu_j.$$

Thus, since $M_0^\star(p) + u(p) \cdot \nu_i - v(p) \cdot \nu_j = 0 \mod (gcd_p)$, $\max(u(p) - v(p), 0) = 0 \mod (gcd_p)$, and $R_{gcd}(M_0(p)) \in \{0, \ldots, gcd_p - 1\}$, we get

$$M_0^\star(p) + u(p) \cdot \nu_i - v(p) \cdot \nu_j \geqslant \max\{u(p) - v(p), 0\}$$

and the precedence relation between $< t_i, \nu_i >$ and $< t_j, \nu_j >$ belongs also to $PR(p, M_0^\star(p))$.

$PR(p, M_0^\star(p)) \subseteq PR(p, M_0(p))$. Let us consider now a precedence relation from $PR(p, M_0^\star(p))$ between $< t_i, \nu_i >$ and $< t_j, \nu_j >$. By Lemma 6.1,

$$u(p) > M_0^\star(p) + u(p) \cdot \nu_i - v(p) \cdot \nu_j \geqslant \max\{u(p) - v(p), 0\}.$$

Clearly,

$$M_0^\star(p) + R_{gcd}(M_0(p)) + u(p) \cdot \nu_i - v(p) \cdot \nu_j \geqslant \max\{u(p) - v(p), 0\}.$$

Now, since $M_0^\star(p) + \nu_i \cdot u(p) - \nu_j \cdot v(p) = 0 \mod (gcd_p)$, we get

$$u(p) - gcd_p \geqslant M_0^\star(p) + u(p) \cdot \nu_i - v(p) \cdot \nu_j.$$

As $R_{gcd}(M_0(p)) < gcd_p$,

$$u(p) > M_0^\star(p) + R_{gcd}(M_0(p)) + u(p) \cdot \nu_i - v(p) \cdot \nu_j \geqslant \max\{u(p) - v(p), 0\}$$

and the precedence relation between $< t_i, \nu_i >$ and $< t_j, \nu_j >$ belongs also to $PR(p, M_0(p))$.

$\square$

In the rest of the chapter, it is assumed that the initial marking of any place p is a multiple of gcd_p.

6.4 Unitary WEGs

This section is dedicated to an important subclass of WEGs called unitary graphs, which we define first. We also recall briefly the interest of unitary graphs for checking the liveness or computing the optimal throughput of general timed marked WEGs. The normalization of a WEG is then presented: it is a transformation introduced in [24] which simplifies the values of the marking functions of a unitary WEG. As recalled in Section 6.5, this transformation is the first step for the computation of an optimal periodic schedule. The expansion, which is another transformation presented in [25], is also detailed and the relationship between expansion and normalization is investigated. Finally, we present a small example which illustrates the limit of the expansion for checking the liveness or computing the maximal throughput of a unitary WEG.

6.4.1 Definitions

A path μ of $\mathcal{G}$ is a sequence of k places such that $\mu = (p_1 = (t_1, t_2), p_2 = (t_2, t_3), \ldots, p_k = (t_k, t_{k+1}))$. If $t_{k+1} = t_1$ then μ is a circuit.

DEFINITION 6.3   *The weight (or gain) of a path μ of a WEG is the product* $\Gamma(\mu) = \displaystyle\prod_{p \in P \cap \mu} \frac{u(p)}{v(p)}$.

DEFINITION 6.4   *A strongly connected WEG $\mathcal{G}$ is unitary if every circuit c of $\mathcal{G}$ has a unit weight.*

Figure 6.3 presents a marked unitary WEG.

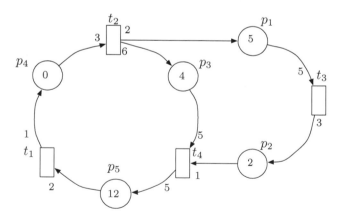

FIGURE 6.3: $\mathcal{G}$ is a marked unitary WEG.

Liveness and maximal throughput for general timed marked WEGs

Let us consider in this subsection that $\mathcal{G}$ is a general timed marked WEG. Then several authors [17, 25, 32] proved that if $\mathcal{G}$ is live, the weight of every circuit c of $\mathcal{G}$ is at least 1. This condition is clearly not sufficient for the liveness: indeed, this condition is fulfilled for any usual marked event graph with null markings, which is not live. However, this necessary condition of liveness allows to partition the transitions into unitary WEGs called unitary components of $\mathcal{G}$ [25].

1. There are then two kinds of deadlocks in a general timed marked WEG:

   - If the circuit c causing the deadlock has a unit weight, it is included in a unitary component. This deadlock can be detected by studying the liveness of the corresponding unitary components of $\mathcal{G}$.

   - Otherwise, $\Gamma(c) > 1$. The only known way to detect it is by computing the firings of the transitions. Since $\Gamma(c) > 1$, this deadlock might occur quite quickly. However, there is no bound for the maximum number of firings needed to ensure the liveness of this class of circuits.

2. It is proved in [25] that the maximum throughput can be computed in polynomial time from the maximum throughput of each unitary component. The problem is that the algorithms developed up to now to compute the maximum throughput of a unitary component are pseudo-polynomial.

So, the study of the unitary graphs is fundamental to obtain efficient algorithms for both problems. Moreover, they correspond to a wide class of interesting practical problems where the capacity of each place remains bounded [23].

6.4.2 Normalization of a Unitary WEG

We present here the normalization of a unitary WEG. This transformation was originally presented in [24] and simplifies the marking functions.

DEFINITION 6.5 *A transition t_i is called normalized if there exists $Z_i \in \mathbb{N}^*$ such that $\forall p \in \mathcal{P}^+(t_i)$, $u(p) = Z_i$ and $\forall p \in \mathcal{P}^-(t_i)$, $v(p) = Z_i$. A unitary WEG $\mathcal{G}$ is normalized if all its transitions are normalized.*

By Lemma 6.2, functions $u(p)$, $v(p)$, and $M_0(p)$ of a place $p \in P$ can be multiplied by any strictly positive integer without any influence on the precedence relations induced. Normalization of a unitary WEG consists then in finding a vector $\gamma = (\gamma_1, \dots, \gamma_m) \in \mathbb{N}^{*m}$ such that

$$\forall t_i \in T, \forall (p_a, p_b) \in \mathcal{P}^-(t_i) \times \mathcal{P}^+(t_i), \gamma_a \cdot v(p_a) = \gamma_b \cdot u(p_b) = Z_i.$$

THEOREM 6.1

Let $\mathcal{G}$ be a strongly connected WEG. $\mathcal{G}$ is normalizable iff $\mathcal{G}$ is unitary.

PROOF

$A \Rightarrow B$. Let us suppose that $\mathcal{G}$ is normalized. The weight of every circuit c is

$$\Gamma(c) = \prod_{p \in P \cap c} \frac{u(p)}{v(p)} = \left(\prod_{t_i \in T \cap c} Z_i \right) \cdot \left(\prod_{t_i \in T \cap c} \frac{1}{Z_i} \right) = 1$$

so $\mathcal{G}$ is unitary.

$B \Rightarrow A$. Let us suppose now that $\mathcal{G}$ is unitary. We must prove the existence of a vector $\gamma = (\gamma_1, \ldots, \gamma_m) \in \mathbb{N}^{*m}$ such that

$$\forall t_i \in T, \forall (p_a, p_b) \in \mathcal{P}^-(t_i) \times \mathcal{P}^+(t_i), \gamma_a \cdot v(p_a) = \gamma_b \cdot u(p_b) = Z_i.$$

Let us build a directed valued graph $G = (P, E)$ as follows:

1. the set of vertices is the set of places,

2. $\forall t \in T$ and for every couple of places $(p_a, p_b) \in \mathcal{P}^-(t) \times \mathcal{P}^+(t)$, the arc $e = (p_a, p_b)$ is built valued by $y(e) = \frac{v(p_a)}{u(p_b)}$.

The problem consists then in finding a vector $\gamma = (\gamma_1, \ldots, \gamma_m) \in \mathbb{N}^{*m}$ such that, for every arc $e = (p_a, p_b) \in E$, $\gamma_a \cdot y(e) \leqslant \gamma_b$.

From the proof of Bellman-Ford algorithm [7], γ exists iff every circuit c of G has a value $Y(c) = \prod_{e \in c} y(e) = 1$. Now, if C denotes the circuit of $\mathcal{G}$ associated with the circuit c of G, we get:

$$Y(c) = \prod_{e \in c} y(e) = \prod_{p \in c} v(p) \cdot \prod_{p \in c} \frac{1}{u(p)} = \frac{1}{\Gamma(C)}.$$

Since $\mathcal{G}$ is a unitary WEG, $\Gamma(C) = 1$ and thus $Y(c) = 1$. Now, since $\mathcal{G}$ is strongly connected, so is G and we get $\gamma_a \cdot y(e) = \gamma_b$ for every arc $e = (p_a, p_b) \in E$. An integer vector $\gamma^\star$ can be obtained from γ by setting $\gamma^\star = A \cdot \gamma$, where A is the least common multiple of the denominators of the components of γ^r. Thus, $\mathcal{G}$ is normalizable, which completes the proof.

□

The associated system of the example pictured by Figure 6.3 is:

$$\begin{aligned}
Z_1 &= 2\gamma_5 = \gamma_4 \\
Z_2 &= 3\gamma_4 = 2\gamma_1 = 6\gamma_3 \\
Z_3 &= 5\gamma_1 = 3\gamma_2 \\
Z_4 &= 5\gamma_3 = \gamma_2 = 5\gamma_5
\end{aligned}$$

A minimum integer solution is $\gamma = (3, 5, 1, 2, 1)$ with $Z_1 = 2$, $Z_2 = 6$, $Z_3 = 15$, and $Z_4 = 5$. Figure 6.4 presents the corresponding normalized marked WEG.

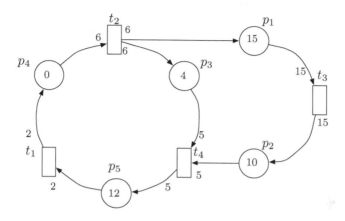

FIGURE 6.4: Normalized unitary graph of the marked WEG pictured by Figure 6.3.

6.4.3 Expansion of a Unitary Timed Marked WEG

Let us suppose that $\mathcal{G}$ is a unitary timed marked WEG. The main idea here is to prove that the sets of precedence relations induced by $\mathcal{G}$ can be modeled using a timed event graph.

6.4.3.1 Study of a Place

We consider here a place $p = (t_i, t_j)$ of a WEG. Each transition t_i may be replaced by N_i transitions denoted by $t_i^1, \ldots, t_i^{N_i}$ such that for any $k \in \{1, \ldots, N_i\}$ and $r > 0$, the rth firing of t_i^k corresponds to the $((r-1) \cdot N_i + k)$th firing of t_i. Transitions $t_i^1, \ldots, t_i^{N_i}$ are called the *duplicates* of t_i.

Since transitions are supposed to be non-reentrant, these duplicates are included in a circuit as pictured by Figure 6.5.

LEMMA 6.4

Let $p = (t_i, t_j)$ be a place from a timed marked WEG. Let us assume that the precedence relations induced by p are equivalent to those obtained from a finite set of non weighted places between the duplicates of t_i and t_j. Then, the number of duplicates N_i and N_j of t_i and t_j must verify $\dfrac{N_i}{v(p)} = \dfrac{N_j}{u(p)}$.

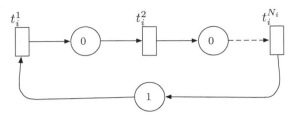

FIGURE 6.5: A circuit between N_i duplicates of t_i modeling the non-reentrant constraint.

PROOF Let us consider two positive integers ν_i and ν_j such that the inequality of Lemma 6.1 holds. It is assumed that the corresponding precedence relation is modeled by a place p_s between a duplicate of t_i and t_j. For any $r > 0$, the firings $\nu_i + r \cdot N_i$ and $\nu_j + r \cdot N_j$ are also constrained by a precedence relation induced by p_s, so

$$u(p) - M_0(p) > u(p) \cdot (\nu_i + r \cdot N_i) - v(p) \cdot (\nu_j + r \cdot N_j) \geqslant \max\{u(p) - v(p), 0\} - M_0(p).$$

These inequalities must be true for any value $r > 0$, so $N_i \cdot u(p) - N_j \cdot v(p) = 0$, which completes the proof. $\square$

Conversely, let us suppose now that $\dfrac{N_i}{v(p)} = \dfrac{N_j}{u(p)} \in \mathbb{N}^*$. Two subcases are considered:

LEMMA 6.5
Let us suppose that $u(p) > v(p)$. If $\dfrac{N_i}{v(p)} = \dfrac{N_j}{u(p)} \in \mathbb{N}^*$ then p may be modeled by N_i non weighted places between the N_i and N_j duplicates of transitions t_i and t_j.

PROOF If $u(p) > v(p)$, the inequality of Lemma 6.1 becomes

$$\frac{M_0(p) + u(p) \cdot (\nu_i - 1)}{v(p)} < \nu_j \leqslant \frac{M_0(p) + u(p) \cdot (\nu_i - 1)}{v(p)} + 1$$

and thus $\nu_j = \left\lfloor \dfrac{M_0(p) + u(p) \cdot (\nu_i - 1)}{v(p)} \right\rfloor + 1$.

For every integer $\nu_i > 0$, r and s are two integers with $\nu_i = (r - 1) \cdot N_i + s$, $r > 0$ and $s \in \{1, \ldots, N_i\}$. By definition of the duplicates, $< t_i, \nu_i > = < t_i^s, r >$.

We get $\nu_j = \left\lfloor \dfrac{M_0(p) + u(p) \cdot (s - 1)}{v(p)} \right\rfloor + 1 + (r - 1) \cdot N_j$. Let the sequences a_s and b_s be such that $\left\lfloor \dfrac{M_0(p) + u(p) \cdot (s - 1)}{v(p)} \right\rfloor + 1 = a_s \cdot N_j + b_s$ with

$b_s \in \{1, \ldots, N_j\}$ then $\nu_j = (r - 1 + a_s) \cdot N_j + b_s$. We deduce that $< t_j, \nu_j >=$ $< t_j^{b_s}, r + a_s >$.

Precedence relations between $< t_i^s, r >$ and $< t_j^{b_s}, r + a_r >$ with $s \in \{1, \ldots, N_i\}$ are modeled by a place $p'_s = (t_i^s, t_j^{b_s})$ with $M_0(p'_s) = a_s$ tokens. $\Box$

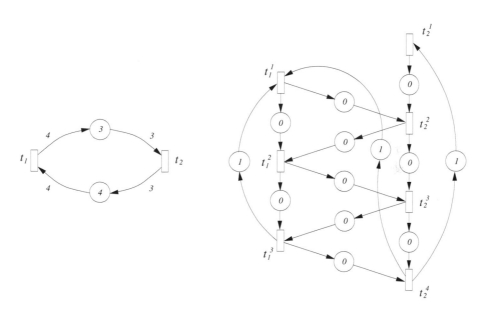

FIGURE 6.6: A unitary marked WEG (on the left) and its minimum expansion (on the right).

For example, let us consider the place $p = (t_1, t_2)$ of the marked unitary WEG pictured by Figure 6.6 and the number of duplicates $N_1 = 3$ and $N_2 = 4$. Sequences a_s and b_s, $s \in \{1, 2, 3\}$, must verify $\left\lfloor \dfrac{3 + 4 \cdot (s - 1)}{3} \right\rfloor + 1 = 4 \cdot a_s + b_s$ with $b_s \in \{1, \ldots, 4\}$. We obtain the couples $(a_1, b_1) = (0, 2)$, $(a_2, b_2) = (0, 3)$, and $(a_3, b_3) = (0, 4)$. Thus, p may be replaced by the places $p'_1 = (t_1^1, t_2^2)$ with $M_0(p'_1) = 0$, $p'_2 = (t_1^2, t_2^3)$ with $M_0(p'_2) = 0$, and $p'_3 = (t_1^3, t_2^4)$ with $M_0(p'_3) = 0$.

LEMMA 6.6

Let us suppose now that $u(p) \leqslant v(p)$. If $\dfrac{N_i}{v(p)} = \dfrac{N_j}{u(p)} \in \mathbb{N}^*$ then p may be modeled by N_j non weighted places between the N_i and N_j duplicates of transitions t_i and t_j.

PROOF   If $u(p) \leqslant v(p)$, the inequality of Lemma 6.1 becomes

$$\frac{v(p) \cdot \nu_j - M_0(p)}{u(p)} + 1 > \nu_i \geqslant \frac{v(p) \cdot \nu_j - M_0(p)}{u(p)}$$

and thus $\nu_i = \left\lceil \dfrac{v(p) \cdot \nu_j - M_0(p)}{u(p)} \right\rceil$ with $\nu_j > \dfrac{M_0(p)}{v(p)}$.

For every integer $\nu_j > 0$, r and s can be defined as $\nu_j = (r-1) \cdot N_j + s$ with $r > 0$ and $s \in \{1, \ldots, N_j\}$. By definition of the duplicates, $< t_j, \nu_j >= < t_j^s, r >$.

We get $\nu_i = \left\lceil \dfrac{v(p) \cdot \nu_j - M_0(p)}{u(p)} \right\rceil = (r-1) \cdot N_i + \left\lceil \dfrac{s \cdot v(p) - M_0(p)}{u(p)} \right\rceil$. Let the

sequences c_s and d_s be such that $\left\lceil \dfrac{s \cdot v(p) - M_0(p)}{u(p)} \right\rceil = c_s \cdot N_i + d_s$ with $d_s \in$

$\{1, \ldots, N_i\}$ then $\nu_i = (r - 1 + c_s) \cdot N_i + d_s$. We get $< t_i, \nu_i >=< t_i^{d_s}, r + c_s >$.

Remark that $\left\lceil \dfrac{s \cdot v(p) - M_0(p)}{u(p)} \right\rceil \leqslant \left\lceil \dfrac{N_j \cdot v(p) - M_0(p)}{u(p)} \right\rceil \leqslant N_i$. Thus $c_s \leqslant$

0 and the precedence relations between $< t_i^{d_s}, r + c_s >$ and $< t_j^s, r >$ for $k \in \{1, \ldots, N_j\}$ are modeled by a place $p_s =< t_i^{d_s}, t_j^s >$ with $M_0(p_s) = -c_k$ tokens.   ◻

For example, let us consider the place $p = (t_2, t_1)$ of the marked unitary WEG pictured by Figure 6.6 and the number of duplicates $N_1 = 3$ and $N_2 = 4$. Sequences c_s and d_s verify $s \in \{1, 2, 3\}$, $d_s \in \{1, \ldots, 4\}$, and $\left\lceil \dfrac{4 \cdot s - 4}{3} \right\rceil =$ $4 \cdot c_s + d_s$. We obtain the couples $(c_1, d_1) = (-1, 4)$, $(c_2, d_2) = (0, 2)$, and $(c_3, d_3) = (0, 3)$. Thus, p may be replaced by the places $p_1' = (t_2^4, t_1^1)$ with $M_0(p_1') = 1$, $p_2' = (t_2^2, t_1^1)$ with $M_0(p_2') = 0$, and $p_3' = (t_2^3, t_1^3)$ with $M_0(p_3') = 0$.

6.4.3.2   Minimum Expansion of a WEG

DEFINITION 6.6   *Let $\mathcal{G}$ be a strongly connected marked WEG. $\mathcal{G}$ is expansible if there exists a vector $(N_1, \ldots, N_n) \in \mathbb{N}^{*n}$ such that every place $p = (t_i, t_j)$ can be replaced by non weighted places between the N_i and N_j duplicates of transitions t_i and t_j following Lemmas 6.5 and 6.6, i.e., $\dfrac{N_i}{v(p)} = \dfrac{N_j}{u(p)}$.*

An expansion of $\mathcal{G}$ is then a timed event graph (*i.e.*, non weighted) which models exactly the same sets of precedence relations as $\mathcal{G}$. The number of duplicates of such a graph verifies the system $\Sigma(\mathcal{G})$ defined as:

$$\Sigma(\mathcal{G}): \quad \forall p = (t_i, t_j) \in P, \; \frac{N_i}{v(p)} = \frac{N_j}{u(p)} \in \mathbb{N}^*.$$

Let $\mathcal{S}$ be the set of solutions of $\Sigma(\mathcal{G})$.

Figure 6.6 presents a unitary WEG and an associated expansion. Since N_1 and N_2 are prime among them, the size of this expansion is minimum. Next theorem characterizes the expansion set associated with a strongly connected marked WEG:

THEOREM 6.2
Let $\mathcal{G}$ be a strongly connected marked WEG. If $\mathcal{G}$ is expansible, then there exists a minimum vector $N^* = (N_1^*, \ldots, N_n^*) \in \mathbb{N}^{*n}$ such that

$$\mathcal{S} = \{\lambda \cdot N^*, \lambda \in \mathbb{N}^*\}.$$

The marked event graph associated with N^* is then the minimum expansion of $\mathcal{G}$.

PROOF Let $N^* = (N_1^*, \ldots, N_n^*)$ be the element from $\mathcal{S}$ with N_1^* minimum and let $N = (N_1, \ldots, N_n) \in \mathcal{S}$. For every place $p = (t_i, t_j)$, $\dfrac{N_i}{N_i^*} = \dfrac{N_j}{N_j^*}$. As $\mathcal{G}$ is connected, there exists two coprime strictly positive integers a and b such that $\dfrac{N_1}{N_1^*} = \dfrac{N_2}{N_2^*} = \cdots = \dfrac{N_n}{N_n^*} = \dfrac{a}{b}$. Thus $N = \dfrac{a}{b} \cdot N^*$.

By contradiction, let suppose that $b > 1$ then $\forall l \in \{1, \ldots, n\}$, b is a divisor of N_l^*. Thus, there exists an integer vector $k = (k_1, \ldots, k_n)$ with $k_l = \dfrac{N_l^*}{b}$. Thus $k \in \mathcal{S}$ with $k_1 < N_1^*$, the contradiction.

Lastly, since elements from $\mathcal{S}$ are proportional, N^* has all its components minimum in $\mathcal{S}$. $\qquad\square$

The system $\Sigma(\mathcal{G})$ associated with the marked WEG pictured by Figure 6.3 is

$$\Sigma(\mathcal{G}) = \begin{cases} N_2/5 = N_3/2 \\ N_3/1 = N_4/3 \\ N_2/5 = N_4/6 \\ N_1/3 = N_2/1 \\ N_4/2 = N_1/5 \end{cases}$$

The minimum integer solution is then $N^* = (15, 5, 2, 6)$.

6.4.4 Relationship between Expansion and Normalization

THEOREM 6.3
Let $\mathcal{G}$ be a strongly connected WEG. $\mathcal{G}$ is expansible iff $\mathcal{G}$ is normalizable. Moreover, there exists $K \in \mathbb{N}^*$ such that, for any $t_i \in T$, $Z_i \cdot N_i = K$.

PROOF

$A \Rightarrow B$. If $\mathcal{G}$ is expansible then there exists a vector $N = (N_1, \ldots, N_n) \in \mathbb{N}^{*n}$ such that for any place $p = (t_i, t_j)$, $\dfrac{N_i}{v(p)} = \dfrac{N_j}{u(p)}$. Let us define M as the least common multiple of integers $N_1, \ldots, N_n$, $M = lcm_{t_i \in T} N_i$.

For every place $p = (t_i, t_j) \in P$, we set $\gamma_p = \dfrac{M}{N_i \cdot u(p)} = \dfrac{M}{N_j \cdot v(p)}$. Then, for any couple of places $(p_a, p_b) \in \mathcal{P}^+(t_i) \times \mathcal{P}^-(t_i)$, we have $\gamma_{p_a} = \dfrac{M}{N_i \cdot u(p_a)}$ and $\gamma_{p_b} = \dfrac{M}{N_i \cdot v(p_b)}$. So,

$$\gamma_{p_a} \cdot u(p_a) = \frac{M}{N_i} = \gamma_{p_b} \cdot v(p_b)$$

and, setting $Z_i = \dfrac{M}{N_i}$ for any $t_i \in T$, we get that $\mathcal{G}$ can be normalized.

$B \Rightarrow A$. Conversely, let us assume now that $\mathcal{G}$ is normalized. So, for every place $p = (t_i, t_j) \in P$, $v(p) = Z_j$ and $u(p) = Z_i$. Let us define $M = lcm_{t_i \in T} Z_i$ and $\forall t_i \in T$, $N_i = \dfrac{M}{Z_i}$. Then, for any place $p = (t_i, t_j) \in P$,

$$\frac{N_i}{v(p)} = \frac{N_i}{Z_j} = \frac{M}{Z_i \cdot Z_j} = \frac{N_j}{Z_i} = \frac{N_j}{u(p)},$$

and $\mathcal{G}$ is expansible.

Now, if $\mathcal{G}$ is normalized, we get $\dfrac{N_j}{Z_i} = \dfrac{N_i}{Z_j}$ for every place $p = (t_i, t_j) \in P$. Since $\mathcal{G}$ is strongly connected, there exists an integer $K > 0$ such that, for every $t_i \in T$, $Z_i \cdot N_i = K$. $\qquad\qquad\Box$

For the example pictured by Figure 6.3, we get

$$Z_1 \cdot N_1^\star = Z_2 \cdot N_2^\star = Z_3 \cdot N_3^\star = Z_4 \cdot N_4^\star = 30.$$

Liveness and maximal throughput of a unitary timed marked WEG using its minimum expansion

The main interest of the expansion is to get an algorithm to check the liveness and to compute the optimal throughput of a unitary timed marked WEG. Indeed, the expansion is a timed marked event graph. As noticed in Section 6.2.2, it corresponds then to uniform constraints, for which there exists polynomial algorithms for the these two problems (see Chapter 5).

The main drawback of this method is that the minimum number of vertices of an expansion may be exponential. Thus, computing the expansion may not be possible for a wide class of graphs.

For example, let us consider the circuit of n transitions pictured by Figure 6.7. The numbers of duplicates of an expansion verify $N_n = \dfrac{N_1}{2^{n-1}}$ and for every $i \in \{1, \ldots, n-1\}$, $N_{i+1} = \dfrac{N_i}{2}$. A minimum integer solution is then, for every $i \in \{1, \ldots, n\}$, $N_i^\star = 2^{n-i}$ and the size of the minimum expansion is in $\mathcal{O}(2^n)$. Thus, its size might be exponential and using the expansion might not be suitable for a wide class of graphs.

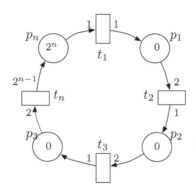

FIGURE 6.7: The number of vertices of the minimum expansion of $\mathcal{G}_n$ is exponential.

6.5 Periodic Schedule of a Normalized Timed Marked WEG

This section is dedicated to the presentation of polynomial algorithms to check the existence, and to compute the optimal throughput, of a periodic schedule for a normalized timed marked WEG. These results were first presented in [3]. From a practical point of view, the limitation to periodic schedules is often considered by many authors (see [5, 8, 27]) to get schedules easier to implement.

The complexity of these two problems is unknown for general (non periodic) cyclic schedules. One of the interests of this limitation will be to derive polynomial sufficient conditions of liveness and an upper bound on the optimum throughput for a normalized timed marked WEG.

Periodic schedules are first formally defined. Then, for every place $p =$

(t_i, t_j), a condition on the starting time of the first execution of t_i and t_j is expressed to fulfill the precedence relations induced by P. A polynomial algorithm checking the existence and computing a periodic schedule is then deduced. A simple example is lastly presented to illustrate that the throughput of a periodic schedule may be quite far from the optimal throughput.

6.5.1 Periodic Schedules

DEFINITION 6.7 *A schedule σ is periodic if each transition $t_i \in T$ has a period w_i^σ such that*

$$\forall k \geqslant 1, S_{<t_i,k>}^\sigma = s_i^\sigma + (k-1) \cdot w_i^\sigma.$$

s_i^σ *is the starting time of the first firing of t_i. The other firings of t_i are then repeated every w_i^σ time units.*

6.5.2 Properties of Periodic Schedules

LEMMA 6.7
Let us consider a place $p = (t_i, t_j) \in P$ and the two integer values $k_{min} = \dfrac{\max\{u(p) - v(p), 0\} - M_0(p)}{gcd_p}$ and $k_{max} = \dfrac{u(p) - M_0(p)}{gcd_p} - 1.$

1. *If p induces a precedence relation between the firings $< t_i, \nu_i >$ and $< t_j, \nu_j >$ then there exists $k \in \{k_{min}, \ldots, k_{max}\}$ such that $u(p) \cdot \nu_i - v(p) \cdot \nu_j = k \cdot gcd_p$.*

2. *Conversely, for any $k \in \{k_{min}, \ldots, k_{max}\}$, there exists an infinite number of tuples $(\nu_i, \nu_j) \in \mathbb{N}^{*2}$ such that $u(p) \cdot \nu_i - v(p) \cdot \nu_j = k \cdot gcd_p$ and p induces a precedence relation between the firings $< t_i, \nu_i >$ and $< t_j, \nu_j >$.*

PROOF

1. Since $gcd_p = gcd(v(p), u(p))$, for any tuple $(\nu_i, \nu_j) \in \mathbb{N}^{*2}$ there exists $k \in \mathbb{Z}$ such that $u(p) \cdot \nu_i - v(p) \cdot \nu_j = k \cdot gcd_p$. Now, if there is a precedence relation between $< t_i, \nu_i >$ and $< t_j, \nu_j >$, we get by Lemma 6.1, assuming by Lemma 6.3 that $M_0(p)$ is a multiple of gcd_p,

$$u(p) - M_0(p) > u(p) \cdot \nu_i - v(p) \cdot \nu_j \geqslant \max\{u(p) - v(p), 0\} - M_0(p),$$

   which is equivalent to

$$u(p) - M_0(p) - gcd_p \geqslant k \cdot gcd_p \geqslant \max\{u(p) - v(p), 0\} - M_0(p).$$

So we get $k_{min} \leqslant k \leqslant k_{max}$.

2. Conversely, there exists $(a, b) \in \mathbb{Z}^2$ such that $a \cdot u(p) - b \cdot v(p) = gcd_p$. Then for any $k \in \{k_{min}, \dots, k_{max}\}$, and any integer $q \geqslant 0$, the couple of integers $(\nu_i, \nu_j) = (k \cdot a + q \cdot v(p), k \cdot b + q \cdot u(p))$ is such that $u(p) \cdot \nu_i - v(p) \cdot \nu_j = k \cdot gcd_p$. Thus p induces a precedence relation between $< t_i, \nu_i >$ and $< t_j, \nu_j >$, which achieves the proof.

□

THEOREM 6.4

Let $\mathcal{G}$ be a normalized timed marked WEG. For any periodic schedule σ, there exists a rational $K^\sigma \in \mathbb{Q}^{+*}$ such that, for any couple of transitions $(t_i, t_j) \in T^2$, $\dfrac{w_i^\sigma}{Z_i} = \dfrac{w_j^\sigma}{Z_j} = K^\sigma$. Moreover, the precedence relations associated with any place $p = (t_i, t_j)$ are fulfilled by σ iff

$$s_j^\sigma - s_i^\sigma \geqslant \ell(t_i) + K^\sigma \cdot (Z_j - M_0(p) - gcd_p).$$

PROOF Consider a place $p = (t_i, t_j) \in P$ inducing a precedence relation between the firings $< t_i, \nu_i >$ and $< t_j, \nu_j >$. Then,

$$S_{<t_i, \nu_i>}^\sigma + \ell(t_i) \leqslant S_{<t_j, \nu_j>}^\sigma.$$

Since σ is periodic, we get

$$s_i^\sigma + (\nu_i - 1) \cdot w_i^\sigma + \ell(t_i) \leqslant s_j^\sigma + (\nu_j - 1) \cdot w_j^\sigma.$$

Then, by Lemma 6.7, there exists $k \in \{k_{min}, \dots, k_{max}\}$ such that $\nu_j = \dfrac{u(p) \cdot \nu_i - k \cdot gcd_p}{v(p)}$ and

$$s_j^\sigma - s_i^\sigma \geqslant \ell(t_i) + w_j^\sigma - w_i^\sigma + \nu_i \cdot w_i^\sigma - \dfrac{u(p) \cdot \nu_i - k \cdot gcd_p}{v(p)} \cdot w_j^\sigma.$$

So, $s_j^\sigma - s_i^\sigma \geqslant \ell(t_i) + \left(w_i^\sigma - \dfrac{u(p)}{v(p)} \cdot w_j^\sigma \right) \cdot \nu_i + \left(1 + \dfrac{k \cdot gcd_p}{v(p)} \right) \cdot w_j^\sigma - w_i^\sigma$. This inequality must be true for any value $\nu_i \in \mathbb{N}^*$, so $w_i^\sigma - \dfrac{u(p)}{v(p)} \cdot w_j \leqslant 0$ and then $\dfrac{w_i^\sigma}{u(p)} \leqslant \dfrac{w_j^\sigma}{v(p)}$. As $\mathcal{G}$ is normalized, $u(p) = Z_i$ and $v(p) = Z_j$. Since $\mathcal{G}$ is unitary, it is strongly connected and thus, for any place $p = (t_i, t_j)$, $\dfrac{w_i^\sigma}{Z_i} = \dfrac{w_j^\sigma}{Z_j}$. So,

there exists a value $K^\sigma \in \mathbb{Q}^*$ such that, for any transition $t_i \in T$, $\dfrac{w_i^\sigma}{Z_i} = K^\sigma$.
Then, the previous inequality becomes

$$s_j^\sigma - s_i^\sigma \geqslant \ell(t_i) + K^\sigma \cdot Z_j \cdot \left(1 + \frac{k \cdot gcd_p}{Z_j}\right) - K^\sigma \cdot Z_i$$

and thus

$$s_j^\sigma - s_i^\sigma \geqslant \ell(t_i) + K^\sigma \cdot (Z_j - Z_i + k \cdot gcd_p).$$

Now, the right term grows with k and according to Lemma 6.7 there exists $(\nu_i, \nu_j) \in \mathbb{N}^{*2}$ such that $k = k_{max}$, thus the precedence relation holds iff

$$s_j^\sigma - s_i^\sigma \geqslant \ell(t_i) + K^\sigma \cdot (Z_j - Z_i + Z_i - M_0(p) - gcd_p)$$

which is equivalent to

$$s_j^\sigma - s_i^\sigma \geqslant \ell(t_i) + K^\sigma \cdot (Z_j - M_0(p) - gcd_p).$$

Conversely, assume this last inequality and that $\forall t_i \in T$, $\dfrac{w_i^\sigma}{Z_i} = K^\sigma$.
Then, for any integers ν_i and ν_j with $u(p) \cdot \nu_i - v(p) \cdot \nu_j = k \cdot gcd_p$ for $k \in \{k_{min}, \ldots, k_{max}\}$, we can prove that σ checks the precedence relation between $< t_i, \nu_i >$ and $< t_j, \nu_j >$. $\qquad\square$

6.5.3   Existence of Periodic Schedules

The constraints expressed by Theorem 6.4 may be modeled by a valued graph $G = (X, A)$ built as follows:

1. the set of vertices is the set of transitions, *i.e.*, $X = T$;

2. to each place $p = (t_i, t_j)$ is associated an arc $a = (t_i, t_j)$ valued by $v(a, K^\sigma) = \ell(t_i) + K^\sigma \cdot (Z_j - M_0(p) - gcd_p)$. Following the notation of Chapter 5, we set $L(a) = \ell(t_i)$, $H(a) = M_0(p) + gcd_p - Z_j$ to obtain $v(a, K^\sigma) = L(a) - K^\sigma \cdot H(a)$.

For a given value $K^\sigma \in \mathbb{Q}^+$, the set of inequalities on the starting times of the first firings of the transitions is a difference constraint system as defined by Lemma 5.3 in Chapter 5. By extension, for every path μ of G, we set

$$L(\mu) = \sum_{a \in \mu} L(a) \text{ and } H(\mu) = \sum_{a \in \mu} H(a).$$

Since $L(a) > 0$ for every arc $a \in A$, the following theorem is easily deduced from Lemma 5.3 in Chapter 5:

THEOREM 6.5
Let $\mathcal{G}$ be a normalized timed WEG. There exists a periodic schedule iff, for every circuit c of G, $H(c) > 0$.

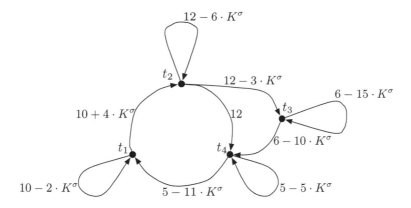

FIGURE 6.8: Valued graph $G = (X, A)$ associated with the normalized marked WEG pictured by Figure 6.4.

Surprisingly, this condition is similar to a sufficient condition of liveness proved in [24]. An algorithm in $\mathcal{O}(\max\{nm, m \max_{t_i \in T} \log Z_i\})$ to evaluate this condition can be found in this paper. It is also proved in [24] that this condition is a necessary and sufficient condition of liveness for circuits composed by two transitions. So, the following corollary is easily deduced:

COROLLARY 6.1
Let $\mathcal{G}$ be a normalized timed marked WEG composed by a circuit of two transitions. $\mathcal{G}$ is live iff $\mathcal{G}$ has a periodic schedule.

This corollary is not true anymore for circuits with three transitions. For example, let us consider the normalized timed WEG $\mathcal{G}$ presented by Figure 6.9 with no particular assumption on firing durations. The sequence of firings $s = t_3 t_1 t_1 t_1 t_2 t_3 t_1 t_1 t_1 t_1 t_2 t_2$ can be repeated infinitely, so it is live.
However, for the circuit $c = t_1 t_2 t_3 t_1$ we get:

$$H(c) = \sum_{i=1}^{3} M_0(p_i) + \sum_{i=1}^{3} gcd_{p_i} - \sum_{i=1}^{3} Z_i = 28 + 12 - 41 < 0$$

so the condition of Theorem 6.5 is false and this circuit has no periodic schedule.

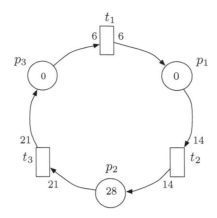

FIGURE 6.9: $\mathcal{G}$ is live but has no periodic schedule.

6.5.4 Optimal Periodic Schedule

Assume here that $\mathcal{G}$ is a normalized timed WEG which fulfills the condition expressed by Theorem 6.5. Then, the minimum value of K^σ is

$$K^{opt} = \max_{c \in C(G)} \frac{L(c)}{H(c)}$$

where $C(G)$ is the set of the circuits of G. K^{opt} is the minimum cycle mean of G, as defined by [16]. The computation of K^{opt} and the determination of a corresponding constraint graph was discussed in Chapter 5.

Now, we can observe that the throughput of a periodic schedule may be quite far from the optimum. For example, let us consider a marked normalized timed WEG circuit composed by two places $p_1 = (t_1, t_2), p_2 = (t_2, t_1)$ such that $gcd_{p_1} = gcd_{p_2} = 1$, $M_0(p_1) = Z_2 + Z_1 - 1$ and $M_0(p_2) = 0$ (see Figure 6.10).

It fulfills the condition stated by Theorem 6.5:

$$M_0(p_1) + M_0(p_2) + gcd_{p_1} + gcd_{p_2} - Z_2 - Z_1 = 1 > 0$$

The associated bi-valued graph G is then pictured by Figure 6.11.

We get $K^{opt} = \max\left\{\frac{\ell(t_1)}{Z_1}, \frac{\ell(t_2)}{Z_2}, \ell(t_1) + \ell(t_2)\right\} = \ell(t_1) + \ell(t_2)$ and the throughput of transitions for the optimum periodic schedule σ is $\tau_1^\sigma = \dfrac{1}{w_1^\sigma} =$

$\dfrac{1}{Z_1 \cdot (\ell(t_1) + \ell(t_2))}$ and $\tau_2^\sigma = \dfrac{1}{w_2^\sigma} = \dfrac{1}{Z_2 \cdot (\ell(t_1) + \ell(t_2))}$.

Let us consider now the earliest schedule σ' of the latter example. Since the total number of tokens in the circuit is $Z_1 + Z_2 - 1$, transitions t_1 and t_2 will never be fired simultaneously by σ'. Moreover, if we denote by n_1 (*resp.*

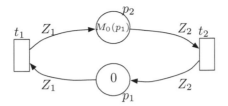

FIGURE 6.10: A normalized timed WEG with two places with $gcd_{p_1} = gcd_{p_2} = 1$ and $M_0(p_1) = Z_2 + Z_1 - 1$.

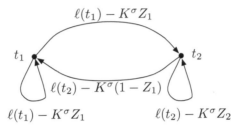

FIGURE 6.11: Bi-valued graph G associated with the normalized timed WEG with two places.

n_2) the number of firings of t_1 (*resp.* t_2) such that the system will return to its initial state (*i.e.*, with $Z_1 + Z_2 - 1$ tokens in p_1 and 0 token in p_2), then we must have $n_1 \cdot Z_1 - n_2 \cdot Z_2 = 0$. Thus, there exists $k \in \mathbb{N}^*$ with $n_1 = k \cdot Z_2$ and $n_2 = k \cdot Z_1$. The throughput of transitions t_1 and t_2 for the earliest schedule is then $\tau_1^{\sigma'} = \dfrac{Z_2}{Z_2 \cdot \ell(t_1) + Z_1 \cdot \ell(t_2)}$ and $\tau_2^{\sigma'} = \dfrac{Z_1}{Z_2 \cdot \ell(t_1) + Z_1 \cdot \ell(t_2)}$. Let us define now the ratio

$$R = \frac{\tau_1^{\sigma'}}{\tau_1^{\sigma}} = \frac{\tau_2^{\sigma'}}{\tau_2^{\sigma}} = \frac{Z_1 \cdot Z_2 \cdot (\ell(t_1) + \ell(t_2))}{Z_2 \cdot \ell(t_1) + Z_1 \cdot \ell(t_2)}.$$

Assume without loss of generality that $Z_1 \geqslant Z_2$, then

$$R = Z_1 \cdot \left(\frac{Z_2 \cdot \ell(t_1) + Z_1 \cdot \ell(t_2) - (Z_1 - Z_2) \cdot \ell(t_2)}{Z_2 \cdot \ell(t_1) + Z_1 \cdot \ell(t_2)} \right).$$

So,

$$R = Z_1 \cdot \left(1 - \frac{(Z_1 - Z_2) \cdot \ell(t_2)}{Z_2 \cdot \ell(t_1) + Z_1 \cdot \ell(t_2)} \right) < Z_1.$$

The ratio R is then maximum when $\ell(t_1)$ tends to infinity and the bound $\max\{Z_1, Z_2\}$ is asymptotically reached.

6.6 Conclusion

This chapter presented some basic recent advances on timed WEGs. It has focused on two open questions for such systems, namely the development of efficient algorithms for checking the liveness and computing the maximal throughput. These questions are fundamental for a practical point of view, since most of optimization problems expressed on these systems need to solve them efficiently, in order to evaluate the solutions obtained.

As we pointed out in this chapter, the complexity of the previous two problems is still open and remains a challenging theoretical issue. If no polynomial algorithm exists, the computation of another lower bound for the maximum throughput should also be investigated to improve those presented here.

The mathematical tools introduced in this chapter also allow us to polynomially solve two optimization problems: the computation of an initial live marking minimizing the places capacities is developped in [23] based on the sufficient condition of liveness expressed by Theorem 6.5. An approximation algorithm was also developed in [22], which maximizes the throughput for place capacities at most twice the minimum. These two algorithms illustrate that efficient polynomial algorithms may be obtained for some particular problems on WEG, even if the general problem does not necessarily belong to $\mathcal{NP}$.

Acknowledgments. We wish to thank Pascal Urard from STMicroelectronics for his encouragements and his interest in our work. We are also very grateful to Claire Hanen and Mohamed Benazouz for their numerous constructive remarks on this chapter.

References

[1] M. Adé. Data Memory Minimization for Synchronous Data Flow Graphs Emulated on DSP-FPGA Targets. Ph.D. thesis, Université Catholique de Louvain, 1997.

[2] M. Adé, R. Lauwereins, and J. A. Peperstraete. Buffer memory requirements in DSP applications. In *IEEE 5th International Workshop on Rapid System Prototyping*, pages 108–123, Grenoble, France, 1994.

[3] A. Benabid, C. Hanen, O. Marchetti, and A. Munier-Kordon. Periodic schedules for unitary timed weighted event graphs. In *ROADEF 2008 conference*, Clermont-Ferrand, France, 2008.

[4] S. S. Bhattacharyya, P. K. Murthy, and E. A. Lee. *Software Synthesis from Dataflow Graphs*. Kluwer, first edition, 1996.

[5] S. S. Bhattacharyya, P. K. Murthy, and E. A. Lee. Synthesis of embedded software from synchronous dataflow specifications. *Journal of VLSI Signal Processing*, 21(2):151–166, 1999.

[6] P. Chrzastowski-Wachtel and M. Raczunas. Liveness of weighted circuits and the diophantine problem of Frobenius. In *Proceeding of FCT'93*, volume 710 of *Lecture Notes in Computer Science*, pages 171–180. Springer, 1993.

[7] T. Cormen, C. Leiserson, and R. Rivest. *Introduction to Algorithms*. MIT Press, 1990.

[8] M. Čubrić and P. Panangaden. Minimal memory schedules for dataflow networks. In *CONCUR '93, 4th International Conference on Concurrency Theory*, volume 715 of *Lecture Notes in Computer Science*, pages 368–383, 1993.

[9] M. R. Garey and D. S. Johnson. *Computers and Intractability. A Guide to the Theory of NP-Completeness*. W. H. Freeman & Co, San Francisco, 1979.

[10] S. Gaubert. An algebraic method for optimizing resources in timed event graphs. In *Proceedings of the 9th conference on Analysis and Optimization of Systems*, volume 144 of *Lecture Notes in Computer Science*, pages 957–966. Springer-Verlag, 1990.

[11] M. Geilen, T. Basten, and S. Stuijk. Minimising buffer requirements of synchronous dataflow graphs with model cheking. In *Proocedings of the DAC*, 2005.

[12] A. Ghamarian, M. Geilen, T. Basten, A. Moonen, M. Bekooij, B. Theelen, and M. Mousavi. Throughput analysis of synchronous data flow graphs. In *Proocedings of the Sixth IEEE International Conference on Applications of Concurrency for System Design*, 2006.

[13] A. Giua, A. Piccaluga, and C. Seatzu. Firing rate optimization of cyclic timed event graphs. *Automatica*, 38(1):91–103, Jan. 2002.

[14] H. Hillion and J. Proth. Performance evaluation of a job-shop system using timed event graph. *IEEE Transactions on Automatic Control*, 34(1):3–9, 1989.

[15] M. Karczmarek. Constrained and Phased Scheduling of Synchronous Data Flow Graphs for StreamIt Language. Master's thesis, Massachussetts Institute of Technology, 2002.

[16] R. M. Karp. A characterization of the minimum cycle mean in a digraph. *Discrete Mathematics*, 23, 1978.

[17] R. M. Karp and R. E. Miller. Properties of a model for parallel computations: Determinacy, termination, queueing. *SIAM Journal on Applied Mathematics*, 14(6):1390–1411, 1966.

[18] S. Laftit, J. Proth, and X. Xie. Optimization of invariant criteria for event graphs. *IEEE Transactions on Automatic Control*, 37(5):547–555, 1992.

[19] E. A. Lee. Overview of the Ptolemy project. Technical report, University of California, Berkeley, 2003.

[20] E. A. Lee and D. G. Messerschmitt. Static scheduling of synchronous data flow programs for digital signal processings. *IEEE Transaction on Computers*, 36(1):24–35, 1987.

[21] E. A. Lee and D. G. Messerschmitt. Synchronous data flow. *Proceedings of the IEEE*, 75(9):1235–1245, 1987.

[22] O. Marchetti and A. Munier-Kordon. Complexity results for bi-criteria cyclic scheduling problems. In *17èmes Rencontres Francophones du Parallélisme (RenPar'2006), Perpignan, France*, pages 17–24, 2006.

[23] O. Marchetti and A. Munier-Kordon. Minimizing place capacities of weighted event graphs for enforcing liveness. *Discrete Event Dynamic Systems*, 18(1):91–109, 2008.

[24] O. Marchetti and A. Munier-Kordon. A sufficient condition for the liveness of weighted event graphs. *European Journal of Operational Research*, 197(2):532–540, 2009.

[25] A. Munier. Régime asymptotique optimal d'un graphe d'événements temporisé généralisé: application à un problème d'assemblage. *RAIRO-Automatique Productique Informatique Industrielle*, 27(5):487–513, 1993.

[26] A. Munier. The basic cyclic scheduling problem with linear precedence constraints. *Discrete Applied Mathematics*, 64(3):219–238, 1996.

[27] P. K. Murthy, S. S. Bhattacharyya, and E. A. Lee. Joint minimization of code and data for synchronous dataflow programs. *Formal Methods in System Design*, 11(1):41–70, 1997.

[28] P. M. Murthy. Scheduling Techniques for Synchronous and Multidimensional Synchronous Dataflow. Ph.D. thesis, University of California at Berkeley, 1996.

[29] J.-L. Peterson. *Petri Net Theory and the Modeling of Systems*. Prentice Hall PTR, 1981.

[30] N. Sauer. Marking optimization of weighted marked graph. *Discrete Event Dynamic Systems*, 13(3):245–262, 2003.

[31] S. Stuijk, M. Geilen, and T. Basten. Exploring trade-offs in buffer requirements and throughput constraints for synchronous dataflow graphs. In *DAC '06: Proceedings of the 43rd Annual Conference on Design automation*, pages 899–904, New York, NY, USA, 2006. ACM.

[32] E. Teruel, P. Chrzastowski-Wachtel, J. M. Colom, and M. Silva. On weighted T-systems. In *Proocedings of the 13th Internationnal Conference on Application and Theory of Petri Nets 1992*, volume 616 of *Lecture Notes in Computer Science*. Springer, 1992.

[33] W. Thies, M. Karczmarek, M. Gordon, D. Z. Maze, J. Wong, H. Hoffman, M. Brown, and S. Amarasinghe. StreamIt: A compiler for streaming applications. Technical report, Massachussetts Institute of Technology, 2001.

Chapter 7

Steady-State Scheduling

Olivier Beaumont

INRIA

Loris Marchal

CNRS and Université de Lyon

Abstract   In this chapter, we propose a general framework for deriving efficient (polynomial-time) algorithms for steady-state scheduling. In the context of large scale platforms (grids or volunteer computing platforms), we show that the characteristics of the resources (volatility, heterogeneity) limit their use to large regular applications. Therefore, steady-state scheduling, that consists in optimizing the number of tasks that can be processed per time unit when the number of tasks becomes arbitrarily large, is a reasonable setting. In this chapter, we concentrate on bag-of-tasks applications, and on scheduling collective communications (broadcast and multicast). We prove that efficient schedules can be derived in the context of steady-state scheduling, under realistic communication models that take into account both the heterogeneity of the resources and the contentions in the communication network.

7.1 Introduction

Modern computing platforms, such as Grids, are characterized by their large scale, their heterogeneity and the variations in the performance of their resources. These characteristics strongly influence the set of applications that can be executed using these platforms. First, the running time of the application has to be large enough to benefit from the platform scale, and to minimize the influence of start-up times due to sophisticated middlewares. Second, an application executed of such a platform typically consists in many small tasks, mostly independent. This allows to minimize the influence of variations in resource performance and to limit the impact of resource failures. From a scheduling point of view, the set of applications that can be efficiently executed on Grids is therefore restricted, and we concentrate in this chapter on "embarrassingly parallel" applications consisting in many independent tasks. In this context, makespan minimization, i.e, minimizing the minimal time to process a given number of tasks, is usually intractable. It is thus more reasonable to focus on throughput maximization, i.e., to optimize the number of tasks that can be processed within T time units, when T becomes arbitrarily large.

Contrary to the application model, the platform model may be rather sophisticated. Indeed, these platforms, made of the aggregation of many resources owned by different entities, are made of strongly heterogeneous computing resources. Similarly, due to long distance communications and huge volume of transmitted data, the cost of communications has to be explicitly taken into account. Some computation nodes within the same cluster may be able to communicate very quickly, whereas the communications between two nodes on both sides of the Atlantic may take much longer. Of course, predicting the exact duration of a 1GB communication through a transatlantic backbone is unreachable, but, as it is advocated in Chapter 11, the difficulties in estimating communication times should not lead us to neglect communications and assume a homogeneous network without congestion!

Therefore, when we consider scheduling issues on heterogeneous platforms, we need to cope with a rather complicated communication model (described in Section 7.2.1), but a rather simple application model (described in Section 7.2.2). Our goal in this chapter is to take advantage of the regularity of the applications; as we consider that the applications are made of a large number of identical operations, we relax the scheduling problem and consider the *steady-state* operation: we assume that after some transient initialization phase, the throughput of each resource will become stable. The idea behind Steady-State Scheduling is to relax the scheduling problem in order to only deal with resource activities and to avoid problems related to integral number of tasks. The question is therefore the following: Do scheduling problems become harder because we consider more sophisticated communication models or do they become simpler because we target simpler applications? In Section 7.3, we propose to represent a schedule by a weighted set of allocations (representing the way to transmit and execute a task) and by a set of weighted valid patterns (representing the way to organize communications and computations). In Section 7.4, we prove that it is possible to re-build a valid schedule from compatible sets of allocations and valid patterns. In Section 7.5, we prove that for many different applications under many different communication models, it is possible to find both the optimal solution (i.e., the optimal throughput) and to build a distributed schedule that achieves this throughput in strongly polynomial time, thus answering positively to the above question. At last, since the general framework proposed in Section 7.5 is based on the Ellipsoid method for solving linear programs and may lead to very expensive algorithms, we propose in Section 7.6 several efficient polynomial time algorithms for solving several Steady-State Scheduling problems.

7.2 Problem Formulation

In this section, we detail the modeling of the platform and of the target applications.

7.2.1 Platform Model

A platform is represented by a *graph* $G = (V, E)$, where vertices correspond to processors and edges to communication links. For the sake of generality, we also introduce the concept of *resource*, that can either represent a processor (computation resource) or a link (communication resource). We denote by $\mathscr{R}$ the set of all resources.

A processor P_u is characterized by its computing speed s_u, measured in flops (floating point operations per second). A communication link l_e is characterized by its bandwidth bw_e, measured in byte per second (B/s). For the sake of generality, we also extend the concept of *speed* to any resource r: s_r corresponds either to the computing speed if the resource is a processor, or to the bandwidth, if the resource is a communication link. Figure 7.1(a) gives an example of such a platform graph.

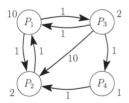 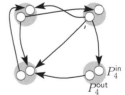

(a) Example of a platform graph. Processors and links are labelled with their speed. $P_{\text{source}} = P_1$.

(b) Example of bag-of-tasks application. Edges (files) and nodes (tasks) are labelled with their size.

(c) Bipartite communication graph for the bidirectional one-port model.

FIGURE 7.1: Example of platform graph and of bag-of-tasks applications.

Note that we do not take latencies into account: the communication of a message of size L on a link l_e takes L/bw_e time-units. The interaction between simultaneous communications are taken into account specifically by each model. In all cases, communications and computations can be overlapped.

Bidirectional one-port model: At a given time step, a processor can simultaneously be involved in two communications: sending a message and receiving another message.

Unidirectional one-port model: At a given time step, a processor can be involved in only one communication: either sending or receiving a message.

Bounded multi-port model: At given time step, a processor can be involved in several sending and receiving operations, provided that the total bandwidths used for incoming and outgoing communications does not exceed the capacity of the communication links. This model is close to the one proposed by Hong and Prasanna in [14].

We consider these different communication models as an illustration that steady-state scheduling can be used for a large range of models. However, in Section 7.6.2, we will more specifically study the bidirectional one-port model, which takes into account the fact that communications are usually serialized, and that in nowadays networks, most links are bidirectional (full-duplex).

7.2.2   Applications

Steady-State Scheduling can be applied to a large range of applications, going from collective communications to structured applications. We present here three applications that will be detailed in the rest of this chapter.

The first application (and the simplest) is called a **bag-of-tasks** application. We consider an application consisting of a large number of independent and similar tasks. Although simple, this application models most of the embarrassingly parallel applications such as parameter sweep applications [9] or BOINC-like computations [8]. Each task of the bag consists in a message "data" containing the information needed to start the computation, a computational task denoted by "comp", and a message "result" containing the information produced by the computation. Initially, all data messages are held by a given processor P_{source}. Similarly, all result messages must eventually be sent to P_{source}. All data messages have the same size (in bytes), denoted by δ_{data}, and all result messages have size δ_{result}. The cost of the each computational task comp (in flops) is w_{comp}. The simple application graph associated to the bag-of-tasks application is depicted in Figure 7.1(b). This application is close to the one presented in the chapter devoted to Divisible Load Scheduling (Chapter 8); however, even if, like in Divisible Load Scheduling, we relax the problem to work with rational number of tasks, our ultimate goal is to build a schedule where task atomicity is preserved.

The second application is a collective communication operation between processors. It consists in broadcasting a message from one processor P_{source} to all other processors of the platform. No computation is induced by this operation, so that the only parameter to deal with is the size of the message data, δ_{data}. Broadcasting in computer networks is the subject of a wide literature, as parallel algorithms often require to send identical data to the whole computing platform, in order to disseminate global information (see [5] for extensive references). In Steady-State Scheduling, we concentrate on the **series of broadcast** problem: the source processor P_{source} has a large number of same-sized messages to broadcast to all other processors. We can symmetrically assume that the source processor has a message of large size to broadcast, which is divided into small chunks, and we target the pipelined scheduling of the series of chunks.

The third application considered here is the multicast operation, which is very close to the previous one: instead of broadcasting a message to all processors in the platform, we target only a given subset of the nodes denoted by $\mathcal{P}_{\text{target}}$. As previously, we focus on the **series of multicast** problem, where a large number of same-sized messages has to be broadcast to target processors.

Just as we gather processors and communication links under the word *"resource"*, we sometimes do not distinguish between computational tasks and files, and we call them *"operations"*. Each operation is characterized by its size, which can either be the size of the corresponding message (in bytes) or the cost (in flops) of the corresponding computational task. The size of operation o is denoted by δ_o. Thus, the duration of operation o on resource r is δ_o/s_r. Note that this formulation forbids us to deal with *unrelated* computation models, where the execution time of a given task can vary arbitrarily among processors. It would be possible to study such complex computation models under the steady-state assumption, but at the cost of complicating the notations. Therefore, in this chapter, we limit our study to *related* computation models to keep notations simple.

At last, we denote by $\mathcal{O}$ the set of all operations in the considered application.

7.3   Compact Description of a Schedule

Scheduling an application on a parallel platform requires to describe *where* and *when* each task will be processed and each transfer will be executed. Since we concentrate on the problem of scheduling a large number of similar jobs, this description may well have a very large size. Indeed, providing the explicit answer to above questions for each task would lead to a description of size $\Omega(n)$, thus pseudo-polynomial in the size of the input. Indeed, since all n tasks are identical, the size of the input is of order $\log n$ and not n. Therefore, we are looking for a compact (i.e., polynomial in the size of the input) description of the schedule.

In this section, we separate temporal and spatial descriptions of such a schedule. We first explain how to get a compact description of the spatial aspect of the schedule (*where* operations are done); then we focus on the temporal aspect (*when* they are done).

7.3.1   Definition of the Allocations

In order to introduce the concept of allocation, let us consider the bag-of-tasks application and the platform graph represented on Figure 7.1(a). We also assume that the source processor is P_1. The problem consists in scheduling a large number of similar tasks. We first concentrate on a *single task* in the series, and study *where*, i.e., on which processor this task can be processed and on which links the transfers of the data and result messages can be done.

The computational task can be computed on any processor: P_1, P_2, P_3 or P_4. Once we have decided where this particular task will be processed, say P_2, we have to determine how the data (and the result) are sent from P_source to P_2 (respectively from P_2 to P_source). Any path may be taken to bring the data from the source to the computing processor: $P_1 \rightarrow P_2$, $P_1 \rightarrow P_3 \rightarrow P_2$ or $P_1 \rightarrow P_3 \rightarrow P_4 \rightarrow P_2$. To fully describe where a task is done, an allocation should provide (i) the path taken by the data, (ii) the processor computing the task, and (iii) the path taken by the result. Figure 7.2 presents some valid allocations for the bag-of-tasks application on the previous platform.

To cope with the different applications and to be able to schedule both communication primitives and complex scheduling problems, we propose the following general definition of an allocation.

DEFINITION 7.1 (allocation)   *An allocation A is a function which associates a set of platform resources to each operation such that all constraints of the particular application are satisfied.*

To complete the definition for the three applications introduced above, we have to detail their respective constraints.

- For the **bag-of-tasks** application, the constraints on the allocation are the following:
    - The set of resources $A(\mathsf{comp})$ is a single processor;
    - The set of resource $A(\mathsf{data})$ is a path from processor P_source to

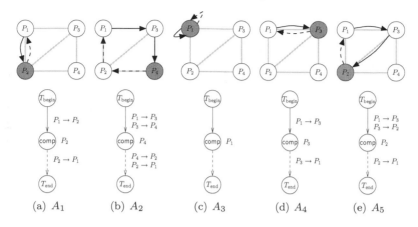

FIGURE 7.2: Examples of allocations for a bag-of-tasks application.

processor $A(\mathsf{comp})$;
   – The set of resource $A(\mathsf{result})$ is a path from $A(\mathsf{comp})$ to P_{source}.

- For the **series of broadcast** and the **series of multicast** applications, there is a single operation, which consists in the message data to be broadcast. The set of resources $A(\mathsf{data})$ must be a tree made of edges from the platform graph G, rooted at P_{source}. Furthermore, for the broadcast application, this tree has to span the whole platform graph, whereas for the multicast operation, the tree has to span a subset of the processors that contains all destination nodes.

Note that several resources may be associated to a single operation, like in the broadcast application (where several links are associated to the transfer of the data message). In this case, we assume that there is no need for a simultaneous use of these resources. For example, assume some legacy code used in a task imposes that four processors must be enrolled to process this task: such a task cannot be modeled with our framework, since we are unable to guarantee that the four processors allocated to the task will be available at the very same time for this particular task. On the contrary, in the case of the broadcast operation, each link of a given route can be used one after the other, by storing the message on intermediate nodes. Of course there is a precedence constraint (the links have to be used in a given order), and we are able to deal with this case in our model.

Allocations characterize where the computations and the transfers take place for a single task. Let us come back to the series of tasks problem: each task of the series has its own allocation. However, the number of allocations is finite: for instance, in the case of the bag-of-tasks application, there are at most $H^2 \times P$ different possible allocations, where P is the number of processors, and H the number of paths (without cycles) in the platform graph.

We call $\mathscr{A}$ the set of possible allocations.

For each allocation $A \in \mathscr{A}$, we focus on the number of times this allocation is used, that is the number of tasks among the series that will be transfered and processed according to this allocation. Yet, as we do not want to depend on the total number of tasks in the series, we rather compute the *average throughput* of a given allocation: we denote by x_a the (fractional) number of times the allocation A_a is used per time-units (say, seconds).

7.3.2   Definition of Valid Patterns

We have studied how to describe *where* the operations of a given application are executed. We now focus on the temporal aspect of the schedule, that is *when* all the operations involved by the allocations can take place. More specifically, we are looking for sets of operations (communications or computations) that can take place simultaneously. Such a set corresponds to a *valid pattern*.

DEFINITION 7.2 (valid pattern)   *A valid pattern π is a set of operations (communications and/or computations) that can take place simultaneously according to a given model.*

We denote by Π the set of all possible valid patterns. According to the communication model we consider, a valid pattern corresponds to different structures in the platform graph.

Unidirectional one-port model. In this model, a processor can be involved in at most one communication, but computations and communications can be overlapped. The communication links involved in a simultaneous pattern of communications constitutes a matching in the platform graph. A valid pattern is therefore a matching in G plus any subset of computing processors. Some examples of valid patterns for this model are depicted in Figure 7.3.

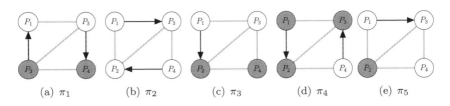

FIGURE 7.3: Examples of valid patterns for the unidirectional one-port model. Shaded processors and dark links are the resources taking part in the pattern.

Bidirectional one-port model. In this model, a processor can be involved in at most two communications, since it can simultaneously send and receive data. To better model this behavior, let us build the communication graph G_B obtained by splitting each processor P_u into two communicating processors: P_u^{out} is in charge of the sending messages while P_u^{in} is dedicated to receiving messages. All communication links are naturally translated into G_B: a link $P_u \to P_v$ is turned into a link $P_u^{\text{out}} \to P_v^{\text{in}}$ in G_B. G_B is a bipartite graph between P_*^{out} and P_*^{in} nodes, and the communications that can take place simultaneously in this model build up a matching in G_B. A valid pattern is thus made of a matching in G_B plus any subset of communicating processors. Figure 7.1(c) illustrates G_B for the platform graph described on Figure 7.1(a).

Bounded multi-port model. In this model, all operations (communications and computations) can take place simultaneously, provided that their aggregated throughput does not exceed platform capacity. Therefore, a valid pattern for this model is made of any subset of communication links and computing processors.

Since the computations are independent from the communications in all the models we consider, valid patterns are sometimes denoted "communication patterns" or even "matchings" when using the one-port communication models.

Each valid pattern may be used several times in a schedule. Rather than specifying for each task which pattern is used, we characterize each pattern π_p by its *average utilization time* y_p, that corresponds to the ratio between the time a given pattern is used and the total schedule time. We have considered an average throughput for the allocations; we similarly consider that a pattern π_p is used for a time y_p per time unit.

7.4 From Allocations and Valid Patterns to Schedules

7.4.1 Conditions and Weak Periodic Schedules

In this section, we describe how to move from a description made of weighted allocations and weighted valid patterns to an actual schedule. Let us therefore consider a set of allocations with their average throughput, and a set of valid patterns with their average utilization time.

We first give necessary conditions on throughputs and utilization times to build a schedule based on these two sets. The first condition concerns resource activity. We consider a given resource r; this resource might be used by several tasks corresponding to different allocations. We further focus on a given allocation A_a; this allocation makes use of resource r for an operation o

if and only if $r \in A(o)$. Since operation o has size δ_o and resource r has speed s_r, r needs δ_o/s_r time units to perform operation o. Assuming that allocation A_a has an average throughput x_a, the total time needed on resource r for this allocation during one time-unit is given by

$$\sum_{\substack{o \in \mathcal{O} \\ r \in A(o)}} x_a \frac{\delta_o}{s_r}.$$

We now focus on the time available on a given resource r. This resource is active and can be used during each valid pattern π_p where r appears. Assuming that the average utilization time of π_p is y_p, then the total availability time of resource r is given by

$$\sum_{\substack{\pi_p \in \Pi \\ r \in \pi_p}} y_p.$$

We are now able to write our first set of constraints: on each resource, the average time used by the allocations during one time-unit must be smaller than the average available time, i.e.,

$$\forall r \in \mathcal{R} \quad \sum_{A_a \in \mathcal{A}} \sum_{\substack{o \in \mathcal{O} \\ r \in A(o)}} x_a \frac{\delta_o}{s_r} \leqslant \sum_{\substack{\pi_p \in \Pi \\ r \in \pi_p}} y_p. \tag{7.1}$$

The second constraint comes from the definition of the availability times: y_p is the average time during which valid pattern π_p is used. During one time-unit, no more than one time-unit can be spent using all possible patterns, hence the second constraint

$$\sum_{\pi_p \in \Pi} y_p \leqslant 1. \tag{7.2}$$

These conditions must be satisfied by any pair of weighted sets of allocations and patterns that corresponds to a valid schedule. Figure 7.4 depicts an example of such sets, made of two allocations from Figure 7.2: A_1 with average throughput $1/8$, and A_2 with throughput $1/16$, and three valid patterns from Figure 7.3: π_1 with utilization time $3/8$, π_2 with utilization time $1/4$, and π_3 with utilization time $1/8$. Let us consider for example the communication link $\pi_2 \rightarrow \pi_1$. It is used by both allocations to ship the result message of size 2, and has speed 1. The total utilization time of this communication link is therefore $\frac{1}{8} \times \frac{2}{1} + \frac{1}{16} \times \frac{2}{1} = \frac{3}{8}$. This link belongs to valid pattern π_1, with utilization time $3/8$, thus Constraint (7.1) is satisfied on this link. We can similarly check that Constraint (7.1) is satisfied for each resource, and that the overall utilization time of valid patterns is given by $\frac{3}{8} + \frac{1}{4} + \frac{1}{8} \leqslant 1$, so that Constraint (7.2) is also satisfied.

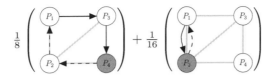

(a) Allocations A_1 and A_2 with their throughputs.

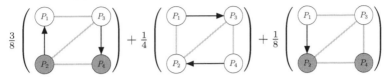

(b) Valid patterns π_1, π_2, and π_3 with their utilization times.

FIGURE 7.4: Example of allocations and valid patterns satisfying conditions 7.1 and 7.2.

7.4.2 Weak Periodic Schedules and Cyclic Scheduling

In this section, we present the basic block of a steady-state schedule, which is called a weak periodic schedule. We also describe how to build such a weak periodic schedule based on allocations and valid patterns that satisfy previous conditions.

DEFINITION 7.3 (*K*-weak periodic schedule of length *T*) *A K-weak periodic schedule of length T is an allocation of K instances of the application within time T that satisfies resource constraints (but not the dependencies between operations constituting an instance).*

Here by *instance*, we mean a single task in the bag-of-tasks applications, or a single message to broadcast in the series of broadcast (or multicast). In a weak periodic schedule, we do not take precedence constraints into account: for example a data message does not have to be received before a computation comp takes place. The precedence constraints will be taken care of in the next section, and fulfilled by the use of several consecutive weak periodic schedules.

We present here a greedy algorithm to build a weak periodic schedule. We assume that we have a set of allocations weighted by their average throughput, and a set of valid patterns weighted by the average utilization time, that satisfy both conditions (7.1) and (7.2). Furthermore, we assume that the values of average throughputs and average utilization times are rational numbers.

1. For each resource r, we do the following:

   (a) We consider the set $\Pi(r)$ of valid patterns which contains resource r;

   (b) Let us consider the set L_r of pairs (o, A) such that $A(o)$ contains resource r; for each element (o, A) of this set, we compute the time needed by this operation: $t(o, A, r) = x_a \frac{\delta_o}{s_r}$.

Thanks to Constraint (7.1), we know that

$$\sum_{(o,A)\in L_r} t(o, A, r) \leqslant \sum_{i=1}^{l} y_{p_i}$$

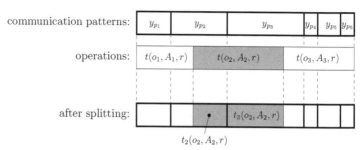

Then, we split each $t(o, A, r)$ into pieces so that they can fit into the intervals defined by the y_{p_i} (in any order), as illustrated above. We call $t_p(o, A, r)$ the time allocated for pair (o, A) using resource r on pattern π_p, and $n_p(o, A, r) = \frac{s_r}{\delta_o} \times t_p(o, A, r)$ the (fractional) number of instances of operation o for allocation A that can be computed in time $t_p(o, A, r)$.

2. We compute T, the least common multiple (lcm) of the denominators of all values $n_p(o, A, r)$, and we set $K = T \times \sum_{A_a \in \mathscr{A}} x_a$.

3. We build a K-weak periodic schedule of length T as follows. A weak periodic schedule is composed of $|\Pi|$ intervals of length $T \times y_p$. During the p-th interval, $\left[T \times \sum_{i=1}^{p-1} y_p \; ; \; T \times \sum_{i=1}^{p} y_p \right]$, the schedule "follows" the valid pattern π_p:

   - Resource r is active if and only if $r \in \pi_p$;
   - For each couple (o, A) such as $r \in A(o)$, r performs $T \times n_p(o, A, r)$ operations o for allocation A (in any order). Due to the definition of T, we know that $T \times n_p(o, A, r)$ is an integer.

Since Condition (7.2) is satisfied, the weak periodic schedule takes time $T \times \sum_{\pi_p \in \Pi} y_p \leqslant T$.

Figure 7.5 shows the construction of the weak periodic schedule corresponding to the allocations and valid patterns of Figure 7.4.

THEOREM 7.1

Given a K-weak periodic schedule of length T, we can build a cyclic schedule of period T, with throughput K/T.

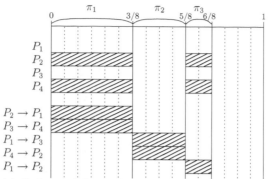

(a) Skeleton of the weak periodic schedule: valid patterns with their utilization times.

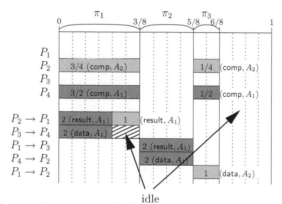

(b) After step 1: operations of allocations are mapped into the slots of the patterns. Numbers represents quantities $n_p(o, A, r)$.

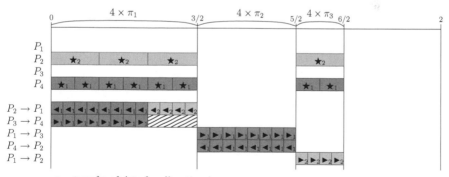

$\blacktriangleright_i$: transfer of **data** for allocation i
$\star_i$: computation of **comp** for allocation i
$\blacktriangleleft_i$: transfer of **result** for allocation i

(c) Final weak periodic schedule, after scaling ($T = 4$), with all operations labelled.

FIGURE 7.5: Example of the construction of a weak periodic schedule from allocations and valid patterns of Figure 7.4

The throughput represents the average number of instances of the application that are performed per one time-unit in steady state. To prove this result and formally define the throughput of a cyclic schedule, important concepts of cyclic scheduling theory must be introduced, which is not the aim of this chapter. We refer the interested reader to [2] where the theory is detailed, and all previous theorems are proved for the problem of scheduling series of workflows (DAGs) on a heterogeneous platform.

We now have all the pieces to build a schedule from the allocations and valid patterns.

THEOREM 7.2

Given a set of allocations $\mathscr{A}$ weighted by their average throughput x and a set of valid patterns Π weighted by their average utilization time y, where x and y are vectors of rational values, if these sets satisfy Constraints (7.1) and (7.2), then we can build a periodic schedule with throughput $\sum_{A_a \in \mathscr{A}} x_a$.

This result is a simple corollary of Theorem 7.1 applied to the weak periodic schedule builded from the sets of allocations and valid patterns.

7.5   Problem Solving in the General Case

In this section, we focus on the construction of allocations and valid patterns satisfying Constraints (7.1) and (7.2) and that maximize the overall throughput. We first gather this objective together with the constraints into a linear program:

$$\text{MAXIMIZE } \rho = \sum_{A_a \in \mathscr{A}} x_a, \text{ UNDER THE CONTRAINTS}$$

$$
\begin{cases}
\quad \sum_{\pi_p \in \Pi} y_p \leqslant 1, \\[2ex]
\forall r \in \mathscr{R} \quad \sum_{\substack{A_a \in \mathscr{A}}} \sum_{\substack{o \in \mathscr{O} \\ r \in A(o)}} x_a \frac{\delta_o}{s_r} \leqslant \sum_{\substack{\pi_p \in \Pi \\ r \in \pi_p}} y_p, \\[2ex]
\forall A_a \in \mathscr{A}, \quad x_a \geqslant 0, \\[1ex]
\forall \pi_p \in \Pi, \quad y_p \geqslant 0.
\end{cases}
\tag{7.3}
$$

This linear program is not directly tractable. Indeed, both the number of possible allocations and the number of possible valid patterns can be of exponential size in the size of the original problem. Therefore, the number of variables of this linear program may be huge. To simplify the notations, let us rewrite the previous linear program as follows

$$\text{MAXIMIZE } c^T \cdot X, \text{ UNDER THE CONSTRAINTS}$$

$$\begin{cases} A \cdot X \leqslant b, \\ X \geqslant 0. \end{cases} \tag{P}$$

Let us denote by $m = |\mathscr{A}| + |\Pi|$ the number of variables and by $n = |\mathscr{R}| + 1$ the number of non-trivial constraints, then matrix A has size $n \times m$, b is a vector of size n, and c a vector of size m. Variable X gathers both the x_a and the y_p: $X_i = x_i$ for $1 \leqslant i \leqslant |\mathscr{A}|$ and $X_{|\mathscr{A}|+i} = y_i$ for $1 \leqslant i \leqslant |\Pi|$. Matrix A is described below:

	for each allocation A_a	for each valid pattern π_p
	$0,0,\ldots,0$	$1,1,\ldots,1$
for each resource $r \in \mathscr{R}$	$\displaystyle\sum_{\substack{o \in \mathscr{O} \\ r \in A(o)}} \frac{\delta_o}{s_r}$	$(-1$ if $r \in \pi_p, 0$ otherwise$)$

7.5.1 Existence of a Compact Solution

A solution of the above linear program is *a priori* described by a huge number of weighted allocations and a huge number of valid patterns. In this section, we prove that there exists a compact optimal solution, i.e., an optimal solution with polynomial size in the size of the problem.

We write each δ_o and each s_r as irreducible fractions $\frac{\alpha_o}{\beta_o}$ and $\frac{\alpha'_r}{\beta'_r}$, and we compute $\Delta = \max\{\alpha_o, \beta_o, \alpha'_r, \beta'_r\}$. Thus, the encoding of each coefficient of matrix A has size at most $2 \log \Delta$.

The number of constraints in the linear program is not polynomial in the size of the problem. However, most of these constraints are trivial ($X \geqslant 0$), and only $|\mathscr{R}| + 1$ constraints need to be checked through computations. Thus, we assume that we have an oracle which, given a solution $(i_1, X_{i_1} = a_{i_1}/b_{i_1}), \ldots, (i_k, X_{i_k} = a_{i_k}/b_{i_k})$ (we implicitly assume that $X_i = 0$ on all other components) and for all $K \in \mathbb{Q}$, can determine if the following system of constraints is satisfied in time $O(k^\gamma \log(\max\{a_{i_j}, b_{i_j}\}))$:

$$\begin{cases} A \cdot X \leqslant b, \\ X \geqslant 0, \\ c^T \cdot X \geqslant K. \end{cases}$$

The last constraint enables to check if the throughput of the solution is above a given threshold K, as in the decision problem associated to the previous linear program.

DEFINITION 7.4 LIN-PROG-DEC   *Does there exists $X \in \mathbb{Q}^m$ which satisfies the constraints of Linear Program (P) and such that the value of the objective function verifies $c^T \cdot X \geqslant K$?*

Since the number of constraints to be checked is huge, this problem does not *a priori* belong to the $\mathcal{NP}$ class. We consider the following reduced decision problem.

DEFINITION 7.5 RED-LIN-PROG-DEC   *Does there exist $X \in \mathbb{Q}^m$ such that:*

  (i) X has at most n positive components;
  (ii) the positive components of X, $X_i = a_i/b_i$, satisfy:

$$\log(a_i) + \log(b_i) \leqslant B$$

(iii) X satisfies the constraints of Linear Program (P), and $c^T \cdot X \geqslant K$?

The complexity bound B depends on Δ, and is set to $B = 2n(\log n + 2n \log \Delta) + n \log \Delta$ for technical reasons. We are able to prove that these two versions of the decision problem are equivalent, but the second one belongs to $\mathcal{NP}$.

THEOREM 7.3
RED-LIN-PROG-DEC belongs to the $\mathcal{NP}$ class, and if there exists a solution X of LIN-PROG-DEC, then there exists a solution Y of RED-LIN-PROG-DEC, and Y is also solution of LIN-PROG-DEC.

PROOF   We briefly present the idea of the proof. The detailed proof can be found in [18].

It is easy to check that RED-LIN-PROG-DEC belongs to $\mathcal{NP}$: given a solution that satisfies the conditions of the decision problem, we can apply the oracle desribed above to this solution, and check that it is a valid solution in polynomial time.

To prove the second part of the theorem, we consider a solution X of LIN-PROG-DEC. We know that there exists an optimal solution Y of the linear program that corresponds to a vertex of the convex polyhedron defined by the constraints of (P). Y can be obtained by solving a linear system of size $m \times m$ extracted from the constraints of (P). Since A has size $n \times m$, the linear systems contains at most n rows from matrix A, and at least $m - n$ rows $X_i = 0$. This means that at most n components of Y are different from zero (point (i)).

To prove that the positive components of Y have a bounded complexity (point (ii) of RED-LIN-PROG-DEC), we can go further on the computation of Y with the linear system, and consider that each component of Y is obtained with Cramer formulae [11] as a fraction of two determinants extracted from A. This allows to bound the value of the numerator and denominator of all elements (point (iii)).

Finally, since Y is an optimal solution, its throughput is larger that the throughput of X. ⬚

This proves that our expression of the problem makes sense: there exist compact, easily described optimal solutions, and we now aim at finding them.

7.5.2 Resolution with the Ellipsoid Method

We now focus on throughput optimization. The method we propose is based on the Ellipsoid method for linear programming, introduced by Khachiyan [16] and detailed in [19]. To apply this method, we consider the dual linear program of (P)

MINIMIZE $b^T U$, UNDER THE CONTRAINTS

$$\begin{cases} A^T \cdot U \geqslant c, \\ U \geqslant 0. \end{cases} \tag{D}$$

Let us analyze this linear program. There is one constraint per allocation A_a, and one per valid pattern π_p, whereas there is one variable (U_1) corresponding to Constraint (7.1) and one variable (U_{r+1}) for each constraint of type (7.2), i.e., one for each resource r. The previous linear program can also be written:

MINIMIZE $b^T U$, UNDER THE CONTRAINTS

$$\begin{cases} (I) \quad \forall A_a \in \mathscr{A} \quad \displaystyle\sum_{\substack{r \in \mathscr{R} \\ r \in A\ a(o)}} \sum_{o \in \mathscr{O}} \frac{\delta_o}{s_r} U_{r+1} \geqslant 1, \\[2em] (II) \quad \forall \pi_p \in \Pi \quad \displaystyle\sum_{r \in \pi_p} U_{r+1} \leqslant U_1, \\[1em] (III) \qquad\qquad U \geqslant 0. \end{cases} \tag{D_2}$$

Given an optimization problem on a convex and compact set K of $\mathbb{Q}^k$, we consider the following two problems. The first one is the optimization problem.

DEFINITION 7.6 OPT(K, v)   *Given a convex and compact set K and a vector $v \in \mathbb{Q}^k$, find a vector $x \in K$ which maximizes $v^T \cdot x$, or prove that K is empty.*

The second problem is a separation problem, which consists in deciding whether the convex K contains a vector v and, if not, in finding an hyperplane separating v from K.

DEFINITION 7.7 SEP(K, v)   *Given a convex and compact set K and a vector $v \in \mathbb{Q}^k$, decide if $v \in K$ and if not, find a vector x such that $x^T \cdot v >$ $\max\{x^T \cdot y, \ y \in K\}$.*

The Ellipsoid method is based on the equivalence between above two problems, as expressed by the following result [12, Chapter 6].

THEOREM 7.4
Each of the two problems, $OPT(K, v)$ and $SEP(K, v)$ can be solved in polynomial time for each well described polyhedron if we know a polynomial-time oracle for the other problem.

A convex polyhedron is well described if it can be encoded in polynomial size, which is possible in our case (see [18] for details).

THEOREM 7.5 (Theorem 6.5.15 in [12])
There exists a polynomial-time algorithm that, for $x \in \mathbb{Q}^k$ and a well described polyhedron (corresponding to the dual (D)) given by a polynomial-time separation oracle,
(i) finds a solution to the primal problem (P).
(ii) or proves that (P) has unbounded solutions or no solution at all.

To solve the optimization problem using Ellipsoid method, it is therefore enough to find a polynomial-time algorithm that solves the separation problem in (D) (or (D_2)).

7.5.3   Separation in the Dual Linear Program

Given a vector $U \in \mathbb{Q}^k$ (with $k = |\mathscr{R}| + 1$), we can check whether U satisfies all the constraints of (D_2), and if not, find a constraint which is violated. This unsatisfied constraint provides a hyperplane separating vector U from the convex. There are three types of constraints in the linear program. The last constraints (type *(III)*) are easy to check, as the number of variables is small. The other constraints can be linked with allocations (type *(I)*) or valid patterns (type *(II)*).

7.5.3.1 Allocation Constraints

Let us first consider constraints of type *(I)*. To each resource r (processor or link), we associate a weight $w_r = \sum_{o \in \mathcal{O}} \frac{\delta_o}{s_r} U_{r+1}$. The constraint for allocation A_a is satisfied if the weighted sum of the resources taking part in the allocation is larger or equal to one. We specify this for the three examples of applications:

Broadcast. An allocation is a spanning tree in the platform graph, rooted at P_{source}. The constraint for allocation A_a states that the weight of the tree is larger than one. In order to check all constraints, we compute (in polynomial time) a spanning tree of minimum weight $A_{\min}$ [10]. If its weight is larger or equal to one, then all spanning trees satisfy this constraint. On the contrary, if its weight is smaller than one, then the corresponding constraint is violated for this minimal weight tree.

Multicast. The allocations for the multicast operation are very closed to the ones of the broadcast: an allocation is a tree rooted at P_{source} but spanning only the nodes participating to the multicast operation. Once the platform graph is weighted as above, checking all the constraints can be done by searching for a minimum weight multicast spanning tree. However, finding a minimum tree spanning a given subset of the nodes in a graph is a well known NP-complete problem, also called the Steiner problem [15]. This suggests that the steady-state multicast problem is difficult, but does not provide any proof. Nevertheless, finding the optimal throughput for the series of multicast problem happens to be NP-complete [4].

Bag-of-tasks application. For this problem, allocations are a little more complex than simple trees, but finding the allocation with smallest weight can still be done in polynomial time: each allocation consists in a computing processor P_{comp}, a path from the source to the computing processor $P_{\text{source}} \rightsquigarrow P_{\text{comp}}$, and a path back $P_{\text{comp}} \rightsquigarrow P_{\text{source}}$. Note that the choices of the two paths are made independently and they may have some edges in common: an edge (and its weight) is thus counted twice if it is used by both paths. A simple method to compute the allocation of minimum weight is to consider iteratively each processor: for each processor P_u, we can find in polynomial time a path of minimum weight from P_{source} to P_u [10], and another minimum weight path from P_u to P_{source}. By taking the minimum of the total weight among all computing processors, we obtain a minimum weight allocation.

7.5.3.2 Valid Pattern Constraints

The constraints of type *(II)* are quite similar of the constraints of type *(I)*: for each valid pattern, we have to check that the sum of the weights of the resources involved in this pattern is smaller than a given threshold U_1, where the weight of a resource r is U_{r+1}. We proceed very similarly to the previous case: we search a valid pattern with maximum weight, and check if it satisfies the constraint: if it does, all other valid patterns (with smaller weights) lead to satisfied constraints, and if it does not, it constitutes a violated constraint. Finding a pattern with maximum weight depends on the communication model in use.

Unidirectional one-port model. A valid pattern in this model corresponds to a matching in the communication graph plus any subset of computing nodes. To maximize its weight, all computing processors can be chosen, and the problem turns into finding a maximum weighted matching in the graph, what can be done in polynomial time [10].

Bidirectional one-port model. In this model, a valid pattern is a matching in the bipartite communication graph G_B, plus any subset of computing nodes. Again, the problem turns into finding a maximum weighted matching (in a bipartite graph in this context).

Bounded multi-port model. In that case, all communications and computations can be done simultaneously. A valid pattern with maximum weight is thus the one including all processors and all communication links.

Thus, we can solve the separation problem in the dual (D_2) in polynomial time. Using Theorem 7.5, we can solve the original linear program (7.3), i.e.,

- Compute the optimal steady-state throughput;
- Find weighted sets of allocations and valid patterns to reach this optimal throughput;
- Construct a periodic schedule realizing this throughput, with the help of Theorem 7.2.

The framework we have developed here is very general. It can thus be applied to a large number of scheduling problems, under a large number of communication models, as soon as we can explicit the allocations and the valid patterns, and solve the separation problem for these allocations and patterns. This allows us to prove that many scheduling problems have a polynomial complexity when studied in the context of steady-state scheduling.

7.6 Toward Compact Linear Programs

7.6.1 Introduction

In previous sections, we have proposed a general framework to analyze the complexity of Steady-State Scheduling problems. This framework is based on the polynomial-time resolution of linear programs (see Linear Program (7.3)). The variables in (7.3) are the set of all possible allocations and valid patterns and thus may both be of exponential size in the size of the platform. Therefore, the Ellipsoid method is the only possible mean to solve these linear programs in polynomial time, since all other methods require to explicitly write the linear program. We have proved that this method enables the design of polynomial-time algorithms for a large number of problems under a large number of communication schemes. Nevertheless, the Ellipsoid method is known to have a prohibitive computational cost, so that it cannot be used in practice.

Our goal in this section is to provide efficient polynomial-time algorithms for several steady-state scheduling problems. More precisely, we prove in Section 7.6.2, that the set of variables corresponding to valid patterns can be replaced by a much smaller set of variables and constraints dealing with local congestion only. In Section 7.6.3, we show that for a large set of problems, the set of variables dealing with allocations can also be replaced by a more compact set of variables and constraints. In this latter case, optimal steady-state schedules can be efficiently solved in practice, since corresponding linear programs only involve polynomial number of variables and constraints (with low degree polynomials). Nevertheless, for some problems, especially those under the unidirectional one-port model, no efficient polynomial algorithm is known, whereas those problems lie in $\mathcal{P}$, as assessed in the previous section.

7.6.2 Efficient Computation of Valid Patterns under the Bidirectional One-Port Model

Steady-State Scheduling problems can be formulated as linear programs where there is one variable per allocation and per valid pattern (see Linear Program (7.3)). In the case of bidirectional one port model, the variables corresponding to valid patterns can be replaced by variables and constraints dealing with local congestion only. Indeed, let us denote by $t_{u,v}$ the occupation time of the communication link from P_u to P_v induced by the allocations. Then,

$$\forall r = (P_u, P_v) \in \mathcal{R} \quad t_{u,v} = \sum_{A_a \in \mathcal{A}} \sum_{\substack{o \in \mathcal{O} \\ r \in A(o)}} x_a \frac{\delta_o}{s_r}.$$

In the bidirectional one-port model, incoming and outgoing communications at a node P_u can take place simultaneously, but all incoming (respectively outgoing) communications at P_u must be sequentialized. Therefore, we can define t_u^{in} (resp. t_u^{out}), the time when P_u is busy receiving (resp. sending) messages as

$$t_u^{\mathsf{in}} = \sum_{(P_v, P_u) \in E} t_{v,u} \quad \text{and} \quad t_u^{\mathsf{out}} = \sum_{(P_u, P_v) \in E} t_{u,v},$$

and both quantities must satisfy $t_u^{\mathsf{in}} \leqslant 1$ and $t_u^{\mathsf{out}} \leqslant 1$. Therefore, when considering only constraints on communications, we can replace the original linear program by the following one,

$$\text{MAXIMIZE } \rho = \sum_{A_a \in \mathscr{A}} x_a, \text{ UNDER THE CONTRAINTS}$$

$$
\begin{cases}
\forall r = (P_u, P_v) \in \mathscr{R} & t_{u,v} = \displaystyle\sum_{A_a \in \mathscr{A}} \sum_{\substack{o \in \mathscr{O} \\ r \in A(o)}} x_a \frac{\delta_o}{s_r}, \\[2em]
& t_u^{\text{in}} = \displaystyle\sum_{(P_v, P_u) \in E} t_{v,u}, \quad t_u^{\text{in}} \leqslant 1, \\[1.5em]
\forall r = P_u \in \mathscr{R} & t_u^{\text{out}} = \displaystyle\sum_{(P_u, P_v) \in E} t_{u,v}, \quad t_u^{\text{out}} \leqslant 1,
\end{cases}
\tag{7.4}
$$

where the set of variables dealing with valid patterns, which had an exponential size *a priori*, have been replaced by $|E| + 2|V|$ variables and constraints. The following theorem states that this modification in the linear program does not affect the optimal throughput.

THEOREM 7.6
On an application using only communications (and not computations), both Linear Programs (7.3) and (7.4) provide the same optimal objective value.

PROOF   Let us first consider a solution of Linear Program (7.3). Then,

$$\forall r = (P_u, P_v) \in \mathscr{R}, \quad t_{u,v} = \sum_{A_a \in \mathscr{A}} \sum_{\substack{o \in \mathscr{O} \\ r \in A(o)}} x_a \frac{\delta_o}{s_r} \leqslant \sum_{\substack{P = \pi_p \in \Pi \\ (P_u, P_v) \in \pi_p}} y_p$$

and therefore

$$t_u^{\text{out}} = \sum_{(P_u, P_v) \in E} t_{u,v} \leqslant \sum_{(P_u, P_v) \in E} \sum_{\substack{\pi_p \in \Pi \\ (P_u, P_v) \in \pi_p}} y_p.$$

Moreover, the edges (P_u, P_v) and (P_u, P_k) cannot appear simultaneously in a valid pattern π_p, so that each π_p appears at most once in the above formula. Therefore,

$$t_u^{\text{out}} \leqslant \sum_{\pi_p \in \Pi} y_p \leqslant 1.$$

Similarly, $t_i^{\text{in}} \leqslant 1$ holds true and the t_u's satisfy the constraints of Linear Program (7.4).

On the other hand, let us consider an optimal solution of Linear Program (7.4) and let us build the bipartite graph G_B (see Section 7.3.2) representing communications, where the weight of the edge between the outgoing port of P_u (denoted by P_u^{out}) and the incoming port of P_v (denoted by P_v^{in}) is given by $t_{u,v}$. This bipartite graph can be decomposed into a set of matchings, using the refined version of the Edge Coloring Lemma [19, vol. A, Chapter 20]:

Corollary 20.1a in [19, vol. A, Chapter 20]. *Let $G_B = (V', E', t_{u,v})$ be a weighted bipartite graph. There exist K matchings $M_1, \ldots, M_K$ with weights $\mu_1, \ldots, \mu_K$ such that*

$$\forall u, v, \quad \sum_{i=1}^{K} \mu_i \chi_{u,v}(M_i) = t_{u,v},$$

where $\chi_{u,v}(M_i) = 1$ if $(P_u^{out}, P_v^{in}) \in M_i$ and 0 otherwise, and

$$\sum_{i=1}^{K} \mu_i = \max\left(\max_u \sum_v t_{u,v}, \max_v \sum_u t_{u,v} \right).$$

Moreover, the matchings can be found in strongly polynomial time and by construction $K \leqslant |E'|$.

We build valid patterns directly from the matchings: for M_i, if $\chi_{u,v}(M_i) = 1$, then we include the communication link (P_u^{out}, P_v^{in}) in the pattern π_i. Therefore, we can build from the solution of Linear Program (7.4) a set of valid patterns (the matchings M_i) to organize the communications. Thus, both linear programs provide the same optimal objective value. □

For the sake of simplicity, we have not considered processing resources in this section. To take processing into account, we can bound the occupation time of a processor with a new constraint in the linear program:

$$\forall r = P_u \in \mathscr{R} \quad \sum_{A_a \in \mathscr{A}} \sum_{\substack{o \in \mathscr{O} \\ r \in A(o)}} x_a \frac{\delta_o}{s_r} \leqslant 1. \tag{7.5}$$

Then, since communications and computations do not interfere, we can schedule the computations of processor P_u, which last $t_u = \sum_o x_a \delta_o / s_u$, during the first t_u time units to build the valid patterns.

Therefore, in the case of the bidirectional one-port model, it is possible to replace the (exponential size) set of variables representing valid patterns by a much smaller (polynomial size) set of variables and constraints dealing with local congestion. This holds also true in the case of the bounded multi-port model, since building the valid patterns from local transfers is also equivalent to pealing a graph into a sum of weighted edge subsets. On the other hand, in the case of the unidirectional one-port model, the corresponding graph is not bipartite and it is therefore not possible to express the maximal throughput via local congestion constraints only. In fact, the design of efficient polynomial time algorithms for optimizing steady-state throughput is still completely open (for bag-of-tasks application, broadcast, ...).

7.6.3   Efficient Computation of Allocations

In the previous section, we have shown how the set of variables dealing with valid patterns can be replaced by a much smaller set of variables and constraints in the case of the bidirectional one-port (and bounded multi-port) model. However, an exponential set of variables still remains, describing the allocations.

In this section, we will concentrate on the bidirectional one-port model and prove how the (exponential size) set of variables dealing with allocations can also be replaced by a polynomial size set of variables and constraints. For simplicity, we will focus only on independent task distribution, but the same framework also applies to broadcasting [5] and independent task graphs scheduling [6] (under some conditions). The approach we propose is closely related to the pioneering work of Bertsimas and Gamarnik [7].

Let us consider the **bag-of-tasks** application problem described in Section 7.2.2, where a source (master) node P_{source} holds a large number of identical independent tasks to be processed by distant nodes (workers). All data messages have the same size (in bytes), denoted by δ_{data}, and all result messages have common size δ_{result}. The cost of any computational task comp is w_{comp} (in flops). We denote by s_u the computational speed of P_u and by $s_{u,v}$ the speed of the link between P_u and P_v.

Let us first concentrate on worker P_u and let us derive a set of equations corresponding to the communications and the processing induced by the execution of α_u tasks per time unit on P_u. First, P_u should be able to process the tasks, i.e., $\frac{\alpha_u \times w_{\text{comp}}}{s_u} \leqslant 1$. Let us now introduce the new variables $\alpha^{i,j}_{u,\text{data}}$ (respectively $\alpha^{i,j}_{u,\text{result}}$) that represent the number of data (resp. result) messages for the tasks processed by P_u which are transmitted through link (P_i, P_j). Clearly, all data (resp. result) messages should leave (resp. reach) P_{source} and reach (resp. leave) P_u and therefore,

$$\sum_j \alpha^{\text{source},j}_{u,\text{data}} = \sum_i \alpha^{i,u}_{u,\text{data}} = \alpha_u = \sum_j \alpha^{u,j}_{u,\text{result}} = \sum_i \alpha^{i,\text{source}}_{u,\text{result}}$$

and no messages should be created or lost at any node different from P_{source} and P_u

$$\forall P_i, \ P_i \neq P_{\text{source}}, \ P_i \neq P_u, \quad \begin{cases} \displaystyle\sum_j \alpha^{i,j}_{u,\text{data}} = \sum_j \alpha^{j,i}_{u,\text{data}}, \\[2ex] \displaystyle\sum_j \alpha^{i,j}_{u,\text{result}} = \sum_j \alpha^{j,i}_{u,\text{result}}. \end{cases}$$

Let us now consider the transfers corresponding to all messages to all workers simultaneously. P_i is busy receiving data and result messages during

$$t^{\text{in}}_i = \sum_u \sum_j \frac{\alpha^{j,i}_{u,\text{data}} \times \delta_{\text{data}}}{s_{j,i}} + \sum_u \sum_j \frac{\alpha^{j,i}_{u,\text{result}} \times \delta_{\text{result}}}{s_{j,i}}$$

and P_i is busy sending data and result messages during

$$t_i^{\text{out}} = \sum_u \sum_j \frac{\alpha_{u,\text{data}}^{i,j} \times \delta_{\text{data}}}{s_{i,j}} + \sum_u \sum_j \frac{\alpha_{u,\text{result}}^{i,j} \times \delta_{\text{result}}}{s_{i,j}}.$$

Therefore, the following linear program provides an upper bound on the possible number of tasks that can be processed, at steady state, during one time-unit

MAXIMIZE $\rho = \sum_u \alpha_u$, UNDER THE CONTRAINTS

$$\begin{cases} \forall P_u, \quad \alpha_u \geqslant 0, \quad \dfrac{\alpha_u \, w_{\text{comp}}}{s_u} \leqslant 1, \\[2mm] \forall P_u, \quad \displaystyle\sum_j \alpha_{u,\text{data}}^{\text{source},j} = \sum_j \alpha_{u,\text{data}}^{j,u} = \alpha_u = \sum_j \alpha_{u,\text{result}}^{u,j} = \sum_j \alpha_{u,\text{result}}^{j,\text{source}}, \\[4mm] \forall P_i, \; P_i \neq P_{\text{source}}, \; P_i \neq P_u, \quad \begin{cases} \displaystyle\sum_j \alpha_{u,\text{data}}^{i,j} = \sum_j \alpha_{u,\text{data}}^{j,i}, \\[3mm] \displaystyle\sum_j \alpha_{u,\text{result}}^{i,j} = \sum_j \alpha_{u,\text{result}}^{j,i}, \end{cases} \\[6mm] t_i^{\text{in}} = \displaystyle\sum_u \sum_j \dfrac{\alpha_{u,\text{data}}^{j,i} \times \delta_{\text{data}}}{s_{j,i}} + \sum_u \sum_j \dfrac{\alpha_{u,\text{result}}^{j,i} \times \delta_{\text{result}}}{s_{j,i}}, \\[4mm] t_i^{\text{out}} = \displaystyle\sum_u \sum_j \dfrac{\alpha_{u,\text{data}}^{i,j} \times \delta_{\text{data}}}{s_{i,j}} + \sum_u \sum_j \dfrac{\alpha_{u,\text{result}}^{i,j} \times \delta_{\text{result}}}{s_{i,j}}, \\[4mm] \forall P_i, \quad t_i^{\text{in}} \leqslant 1, \quad t_i^{\text{out}} \leqslant 1. \end{cases}$$

$$(7.6)$$

From the solution of the linear program, the set of valid patterns can be determined using the general framework described in Section 7.6.2. In order to build the set of allocations, we can observe that the set of values $\alpha_{u,\text{data}}^{i,j}$ (resp. $\alpha_{u,\text{result}}^{i,j}$) defines a flow of value α_u between P_{source} and P_u (resp. P_u and P_{source}). Each of these flows can be decomposed into a weighted set of at most $|E|$ disjoint paths [19, vol. A, Chapter 10]. Thus, it is possible to find at most $2|E|$ weighted allocations that represent the transfer of data from P_{source} to P_u, the processing on P_u and the transfer of results from P_u to P_{source}. Therefore, under the bidirectional one-port model, the Linear Program (7.6) provides the optimal throughput for the **bag-of-tasks** application problem. Moreover, this linear program is compact since it only involves $\Theta(|V||E|)$ variables and constraints.

7.7   Conclusion

Figure 7.7 summarizes the complexity of the problem from the point of view of steady-state scheduling. All NP-complete problems of this table are NP-complete for any of the communication models that we have considered. For the problems with polynomial complexity, we do not know a better algorithm than the one using the Ellipsoid method (see Section 7.5.2) for the unidirectional one-port model, whereas we have efficient algorithms under the bidirectional one-port model (as for the bag-of-tasks application in Section 7.6.3).

	NP-complete problems	problems with polynomial complexity
collective operations	multicast and prefix computation [4]	broadcast [5], scatter and reduce [17]
scheduling problems	general DAGs [3]	bags of tasks, DAGs with bounded dependency depth [3]

Table 7.1: Complexity results for steady-state problems.

In this chapter, we have shown that changing the metric, from makespan minimization to throughput maximization, enables to derive efficient polynomial time algorithms for a large number of scheduling problems involving heterogeneous resources, even under realistic communication models where contentions are explicitly taken into account. Nevertheless, from an algorithmic point of view, large scale platforms are characterized by their heterogeneity and by the dynamism of their components (due to processor or link failures, workload variation,...). Deriving efficient scheduling algorithms that can self-adapt to variations in platforms characteristics is still an open problem. We strongly believe that in this context, one needs to inject some static knowledge into scheduling algorithms, since greedy algorithms are known to exhibit arbitrarily bad performance in presence of heterogeneous resources. For instance, He, Al-Azzoni, and Down propose in [13] to base dynamic mapping decisions in resource management systems on the solution of a linear program, and Awerbuch and Leighton [1] show how to derive fully distributed approximation algorithms for multi-commodity flow problems based on potential queues whose characteristics are given by the pre-computation of the optimal solution.

References

[1] B. Awerbuch and T. Leighton. Improved approximation algorithms for the multi-commodity flow problem and local competitive routing in dynamic networks. In *26-th Annual ACM Symposium on Theory of Computing*, pages 487–496, 1994.

[2] O. Beaumont. *Nouvelles méthodes pour l'ordonnancement sur plates-formes hétérogènes*. Habilitation à diriger des recherches, Université de Bordeaux 1, Dec. 2004.

[3] O. Beaumont, A. Legrand, L. Marchal, and Y. Robert. Assessing the impact and limits of steady-state scheduling for mixed task and data parallelism on heterogeneous platforms. Research report 2004-20, LIP, ENS Lyon, France, Apr. 2004. Also available as INRIA Research Report RR-5198.

[4] O. Beaumont, A. Legrand, L. Marchal, and Y. Robert. Complexity results and heuristics for pipelined multicast operations on heterogeneous platforms. In *Proceedings of the 33rd International Conference on Parallel Processing (ICPP)*. IEEE Computer Society Press, 2004.

[5] O. Beaumont, A. Legrand, L. Marchal, and Y. Robert. Pipelining broadcasts on heterogeneous platforms. *IEEE Transactions on Parallel Distributed Systems*, 16(4):300–313, 2005.

[6] O. Beaumont, A. Legrand, L. Marchal, and Y. Robert. Steady-state scheduling on heterogeneous clusters. *International Journal of Foundations of Computer Science*, 16(2):163–194, 2005.

[7] D. Bertsimas and D. Gamarnik. Asymptotically optimal algorithm for job shop scheduling and packet routing. *Journal of Algorithms*, 33(2):296–318, 1999.

[8] Berkeley Open Infrastructure for Network Computing. http://boinc.berkeley.edu.

[9] H. Casanova and F. Berman. *Grid Computing: Making The Global Infrastructure a Reality*, chapter Parameter Sweeps on the Grid with APST. John Wiley, 2003. Hey, A. and Berman, F. and Fox, G., editors.

[10] T. Cormen, C. Leiserson, and R. Rivest. *Introduction to Algorithms*. MIT Press, 1990.

[11] G. H. Golub and C. F. V. Loan. *Matrix Computations*. Johns Hopkins, 1989.

[12] M. Grötschel, L. Lovász, and A. Schrijver. *Geometric Algorithm and Combinatorial Optimization.* Algorithms and Combinatorics 2. Springer-Verlag, 1994. Second corrected edition.

[13] Y.-T. He, I. Al-Azzoni, and D. Down. MARO-MinDrift affinity routing for resource management in heterogeneous computing systems. In *CASCON '07: Proceedings of the 2007 Conference of the Center for Advanced Studies on Collaborative Research*, pages 71–85, New York, 2007. ACM.

[14] B. Hong and V. Prasanna. Distributed adaptive task allocation in heterogeneous computing environments to maximize throughput. In *International Parallel and Distributed Processing Symposium IPDPS'2004*. IEEE Computer Society Press, 2004.

[15] R. Karp. Reducibility among combinatorial problems. In R. E. Miller and J. W. Thatcher, editors, *Complexity of Computer Computations (Proc. Sympos., IBM Thomas J. Watson Res. Center, Yorktown Heights, NY, 1972)*, pages 85–103. Plenum, NY, 1972.

[16] L. G. Khachiyan. A polynomial algorithm in linear programming. *Doklady Akademiia Nauk SSSR*, 224:1093–1096, 1979. English Translation: *Soviet Mathematics Doklady*, Volume 20, pp. 191–194.

[17] A. Legrand, L. Marchal, and Y. Robert. Optimizing the steady-state throughput of scatter and reduce operations on heterogeneous platforms. *Journal of Parallel and Distributed Computing*, 65(12):1497–1514, 2005.

[18] L. Marchal. *Communications collectives et ordonnancement en régime permanent sur plates-formes hétérogènes.* Thèse de doctorat, École Normale Supérieure de Lyon, France, Oct. 2006.

[19] A. Schrijver. *Combinatorial Optimization: Polyhedra and Efficiency*, volume 24 of *Algorithms and Combinatorics*. Springer-Verlag, 2003.

Chapter 8

Divisible Load Scheduling

Matthieu Gallet

École normale supérieure de Lyon and Université de Lyon

Yves Robert

École normale supérieure de Lyon, Institut Universitaire de France, and Université de Lyon

Frédéric Vivien

INRIA and Université de Lyon

8.1 Introduction

In this chapter, we focus on the problem of scheduling a large and compute-intensive application on parallel resources, typically organized as a master-worker platform. We assume that we can arbitrarily split the total work, or load, into chunks of arbitrary sizes, and that we can distribute these chunks to an arbitrary number of workers. The job has to be perfectly parallel, without any dependence between sub-tasks. In practice, this model is a reasonable relaxation of an application made up of a large number of identical, fine-grain parallel computations. Such applications are found in many scientific areas, like processing satellite pictures [14], broadcasting multimedia contents [2, 3], searching large databases [11, 10], or studying seismic events [12]. This model is known as the *Divisible Load* model and has been widespread by Bharadwaj et al. in [9]. Steady-State Scheduling, detailed in Chapter 7, is another relaxation, more sophisticated but well-suited to complex applications. In [15], Robertazzi shows that the Divisible Load model is a tractable approach, which applies to a great number of scheduling problems and to a large variety of platforms, such as bus-shaped, star-shaped, and even tree-shaped platforms.

In this chapter, we motivate the Divisible Load model using the example of a seismic tomography application. We solve this example first with the classical approach in Section 8.1.2, and then using the Divisible Load model in Section 8.2. After this complete resolution, we use weaker assumptions to study more general but harder problems in Section 8.3. Finally, we conclude in Section 8.4.

8.1.1 Motivating Example

We use a specific example of application and platform combination as a guideline, namely an Earth seismic tomography application deployed on a star-shaped platform of processors. The application is used to validate a model for the internal structure of the Earth, by comparing for every seismic event the propagation time of seismic waves as computed by the model with the time measured by physical devices. Each event is independent from the others and there is a large number of such events: 817,101 events were recorded for the sole year of 1999. The master processor owns all items, reads them, and scatters them among m active workers. Then each worker can process the data received from the master independently. The objective is to minimize the total completion time, also known as the makespan. The simplified code of the program can be written as:

```
if (rank = MASTER)
    raydata ← read W_total lines from data file;
MPI_Scatter(raydata,W_total/m,...,rbuff,...,MASTER,MPI_COMM_WORLD);
compute_work(rbuff);
```

This application fulfills all assumptions of the Divisible Load model, since it is made of a very large number of fine grain computations, and these computations are independent. Indeed, we do not have any dependence, synchronization, nor communication between tasks.

Throughout this chapter we will consider two different types of platforms. The first one is a bus-shaped master-worker platform, where all workers are connected to the master through identical links, and the second one is a star-shaped master-worker platform, where workers are connected to the master through links of different characteristics.

8.1.2 Classical Approach

In this section, we aim at solving the problem in a classical way. The target platform is a bus-shaped network, as shown in Figure 8.1. Workers are *a priori* heterogeneous, hence they have different computation speeds. We enforce the full one-port communication model: the master can communicate to at most one worker at a time. This model corresponds to the behavior of two widespread MPI libraries, IBM MPI and MPICH, which serialize asynchronous communications as soon as message sizes exceed a few tens of kilobytes [16]. Finally, each worker receives its whole share of data in a single message.

We will use the following notations, as illustrated in Figure 8.2:

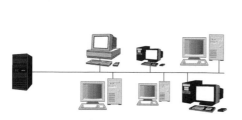

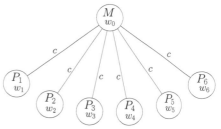

FIGURE 8.1: Example of bus-shaped network.

FIGURE 8.2: Theoretical model of a bus-shaped network.

- M is the master processor, which initially holds all the data to process.
- There are m workers, denoted as $P_1, \ldots, P_m$. In order to simplify some equations, P_0 denotes the master processor M.
- Worker P_i takes a time w_i to execute a task. $M = P_0$ requires a time w_0 to process such a load.
- Any worker P_i needs c time units to receive a unit-size load from the master. Recall that all workers communicate at the same speed with the master.
- M initially holds a total load of size W_{total}, where W_{total} is a very large integer.
- M will allocate n_i tasks to worker P_i. n_i is an integer, and since all tasks have to be processed, we have $\sum_{i=0}^{p} n_i = W_{total}$ (we consider that the master can also process some tasks).
- By definition, processor P_i processes n_i tasks in time $C_i = n_i \cdot w_i$.
- The completion time T_i corresponds to the end of the computations of P_i.

We allow the overlap of communications by computations on the master, i.e., the master can send data to workers while computing its own data. A worker cannot begin its computations before having received all its data. The objective is to minimize the total completion time T needed to compute the load W_{total}. We have to determine values for the n_i's that minimize T. Let us compute the completion time T_i of processor P_i: If we assume that processors are served in the order $P_1, \ldots, P_m$, equations are simple:

- P_0: $T_0 = n_0 \cdot w_0$,
- P_1: $T_1 = n_1 \cdot c + n_1 \cdot w_1$,
- P_2: $T_2 = n_1 \cdot c + n_2 \cdot c + n_2 \cdot w_2$,
- P_i: $T_i = \sum_{j=1}^{i} n_j \cdot c + n_i \cdot w_i$ for $i \geqslant 1$.

These equations are illustrated by Figures 8.3, 8.4, and 8.5. If we let $c_0 = 0$ and $c_i = c$ for $i \geqslant 1$, we render the last equation more homogeneous: $T_i = \sum_{j=0}^{i} n_j \cdot c_j + n_i \cdot w_i$ for $i \geqslant 0$.

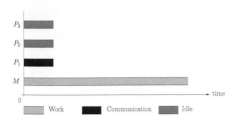

FIGURE 8.3: M computes and sends data to P_1.

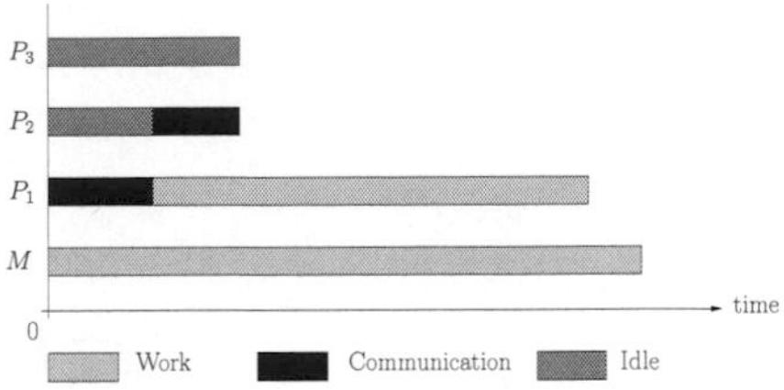

FIGURE 8.4: M and P_1 compute, M sends data to P_2.

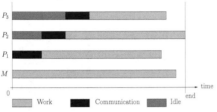

FIGURE 8.5: Complete schedule.

By definition, the total completion time T is equal to:

$$T = \max_{0 \leqslant i \leqslant m} \left(\sum_{j=0}^{i} n_j \cdot c_j + n_i \cdot w_i \right). \tag{8.1}$$

If we rewrite Equation 8.1 as

$$T = n_0 \cdot c_0 + \max \left(n_0 \cdot w_0, \max_{1 \leqslant i \leqslant m} \left(\sum_{j=1}^{i} n_j \cdot c_j + n_i \cdot w_i \right) \right),$$

we recognize an optimal sub-structure for the distribution of $W_{total} - n_0$ tasks among processors P_1 to P_m. This remark allows to easily find a solution for $n_0, \ldots, n_p$ using dynamic programming. Such a solution is given by Algorithm 8.1.[1]

Nevertheless, this solution is not really satisfying and suffers from several drawbacks. First, we do not have a closed form solution (neither for the n_i's nor for T). Moreover, the order of the processors during the distribution is arbitrarily fixed (the master communicates with P_1, then with P_2, P_3, and so on). Since processor speeds are *a priori* different, this order could be suboptimal, and this solution does not help us to find the right order among the $m!$ possible orders. There are by far too many possible orders to try

---

[1]Algorithm 8.1 builds the solution as a list. Hence the use of the constructor *cons* that adds an element at the head of a list.

Algorithm 8.1: Solution for the classical approach, using dynamic programming

---

$solution[0, m] \leftarrow cons(0, NIL)$
$cost[0, m] \leftarrow 0$
for $d \leftarrow 1$ **to** W_{total} **do**
   | $solution[d, m] \leftarrow cons(d, NIL)$
   | $cost[d, m] \leftarrow d \cdot c_m + d \cdot w_m$
for $i \leftarrow m - 1$ **downto** 0 **do**
   | $solution[0, i] \leftarrow cons(0, solution[0, i + 1])$
   | $cost[0, i] \leftarrow 0$ **for** $d \leftarrow 1$ **to** W_{total} **do**
   | | $(sol, min) \leftarrow (0, cost[d, i + 1])$
   | | **for** $e \leftarrow 1$ **to** d **do**
   | | | $m \leftarrow e \cdot c_i + \max(e \cdot w_i, cost[d - e, i + 1])$
   | | | **if** $m < min$ **then**
   | | | | $(sol, min) \leftarrow (e, m)$
   | | $solution[d, i] \leftarrow cons(sol, solution[d - sol, i + 1])$
   | | $cost[d, i] \leftarrow min$
return $(solution[W_{total}, 0], cost[W_{total}, 0])$

---

an exhaustive search. Furthermore, the time complexity of this solution is $W_{total}^2 \times m$, so the time to decide for the right values of the n_i's can be greater than the time of the actual computation! Finally, if W_{total} is slightly changed, we cannot reuse the previous solution to obtain a new distribution of the n_i's and we have to redo the entire computation.

Even if we have an algorithm, which gives us a (partial) solution in pseudo-polynomial time, we can still look for a better way to solve the problem. Let us consider the Divisible Load model. If we have around 800,000 tasks for 10 processors, there are on average 80,000 tasks on each processor. So, even if we have a solution in rational numbers (i.e., a non-integer number of tasks on each processor), we easily afford the extra-cost of rounding this solution to obtain integer numbers of tasks and then a valid solution in practice. The new solution will overcome all previous limitations.

8.2 Divisible Load Approach

The main principle of the Divisible Load model is to relax the integer constraint on the number of tasks on each worker. This simple idea can lead to high-quality results, even if we loose a little precision in the solution. Now, let us instantiate the problem using this relaxation: instead of an integer number

n_i of tasks, processor P_i (with $0 \leqslant i \leqslant m$) will compute a fraction α_i of the total load W_{total}, where $\alpha_i \in \mathbb{Q}$. The number of tasks allocated to P_i is then equal to $n_i = \alpha_i W_{total}$ and we add the constraint $\sum_{i=0}^{m} \alpha_i = 1$, in order to ensure that the whole workload will be computed.

8.2.1   Bus-Shaped Network

In this paragraph, we keep exactly the same platform model as before: i.e., a bus-shaped network with heterogeneous workers, and data is distributed to workers in a single message, following a linear cost model. Equation 8.1 can easily be translated into Equation 8.2.

$$T = \max_{0 \leqslant i \leqslant m} \left(\sum_{j=0}^{i} \alpha_j \cdot c_j + \alpha_i \cdot w_i \right) W_{total}. \tag{8.2}$$

Using this equation, we can prove several important properties of optimal solutions:

- all processors participate ($\forall i, \alpha_i > 0$) and end their work at the same time (Lemma 8.1),
- the master processor should be the fastest computing one but the order of other processors is not important (Lemma 8.2).

LEMMA 8.1
In an optimal solution, all processors participate and end their processing at the same time.

PROOF   Consider any solution, such that at least two workers do not finish their work at the same time. Without any loss of generality, we can assume that these two processors are P_i and P_{i+1} (with $i \geqslant 0$) and such that either P_i or P_{i+1} finishes its work at time T (the total completion time). In the original schedule, P_i finishes at time T_i, P_{i+1} finishes at time T_{i+1} and $\max(T_i, T_{i+1}) = T$. We transfer a fraction δ of work from P_{i+1} to P_i (Figure 8.6), assuming that transfering a negative amount of work corresponds to a transfer from P_i to P_{i+1}. In this new schedule, P_i finishes at time T_i' and P_{i+1} finishes at time T_{i+1}'. We want P_i and P_{i+1} to finish at the same time, thus we have the following equations:

$$T_i' = T_{i+1}'$$
$$\Leftrightarrow \left(\sum_{j=0}^{i} \alpha_j c_j \right) + \delta c_i + \alpha_i w_i + \delta w_i = \left(\sum_{j=0}^{i} \alpha_j c_j \right)$$
$$+ \delta c_i + (\alpha_{i+1} - \delta)(c_{i+1} + w_{i+1})$$
$$\Leftrightarrow \qquad \delta c_i + \alpha_i w_i + \delta w_i = \delta c_i + \alpha_{i+1}(c_{i+1} + w_{i+1})$$
$$- \delta(c_{i+1} + w_{i+1})$$
$$\Leftrightarrow \qquad \delta(w_i + c_{i+1} + w_{i+1}) = \alpha_{i+1}(c_{i+1} + w_{i+1}) - \alpha_i w_i$$

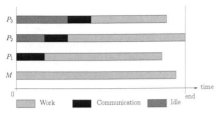

(a) P_1 finishes earlier than P_2.

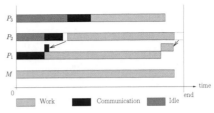

(b) A fraction of the load allocated to P_2 is given to P_1.

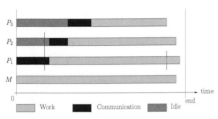

(c) This new schedule is strictly better.

FIGURE 8.6: Discharging some work from P_2 to P_1 to obtain a better schedule.

Therefore,

$$T'_i = T'_{i+1} \quad \Leftrightarrow \quad \delta = \frac{\alpha_{i+1}\,(c_{i+1} + w_{i+1}) - \alpha_i w_i}{w_i + c_{i+1} + w_{i+1}}.$$

We have:

$$T'_{i+1} - T_{i+1} = \delta\,(c_i - c_{i+1} - w_{i+1})$$
$$T'_i - T_i = \delta\,(c_i + w_i)$$

We know that $c_{i+1} \geqslant c_i \geqslant 0$ (because $c_i = c$ if $i \geqslant 1$ and $c_0 = 0$). Thus, if δ is positive, then we have $T = T_{i+1} > T'_{i+1} = T'_i > T_i$, and $T = T_i > T'_i = T'_{i+1} > T_{i+1}$ if δ is negative. In both cases, processors P_i and P_{i+1} finish strictly earlier than T. Finally, δ cannot be zero, since it would mean that $T_i = T'_i = T'_{i+1} = T_{i+1}$, contradicting our initial hypothesis.

Moreover, there are at most $m - 1$ processors finishing at time T in the original schedule. By applying this transformation to each of them, we exhibit a new schedule, strictly better than the original one. In conclusion, any solution such that all processors do not finish at the same time can be strictly improved, hence the result.

□

LEMMA 8.2
If we can choose the master processor, it should be the fastest processor, but the order of other processors has no importance.

PROOF   Let us consider any optimal solution $(\alpha_0, \alpha_1, \ldots, \alpha_m)$. By using Lemma 8.1, we know that $T = T_0 = T_1 = \ldots = T_m$. Hence the following equations:

- $T = \alpha_0 \cdot w_0 \cdot W_{total}$,
- $T = \alpha_1 \cdot (c + w_1) \cdot W_{total}$ and then $\alpha_1 = \frac{w_0}{c+w_1}\alpha_0$,
- $T = (\alpha_1 \cdot c + \alpha_2 \cdot (c + w_2)) W_{total}$ and then $\alpha_2 = \frac{w_1}{c+w_2}\alpha_1$,
- for all $i \geqslant 1$, we derive $\alpha_i = \frac{w_{i-1}}{c+w_i}\alpha_{i-1}$.

Thus, we have $\alpha_i = \prod_{j=1}^{i}\frac{w_{j-1}}{c+w_j}\alpha_0$ for all $i \geqslant 0$. Since $\sum_{i=0}^{m}\alpha_i = 1$, we can determine the value of α_0, and thus we have closed formulas for all the α_i's:

$$\alpha_i = \frac{\prod_{j=1}^{i}\frac{w_{j-1}}{c_j+w_j}}{\sum_{k=0}^{m}\left(\prod_{j=1}^{k}\frac{w_{j-1}}{c_j+w_j}\right)}.$$

Using these formulas, we are able to prove the lemma. Let us compute the work done in time T by processors P_i and P_{i+1}, for any i, $0 \leqslant i \leqslant m - 1$. As in Section 8.1.2, we let $c_0 = 0$ and $c_i = c$ for $1 \leqslant i \leqslant m$, in order to have homogeneous equations. Then, the two following equations hold true:

$$T = T_i = \left(\left(\sum_{j=0}^{i-1}\alpha_j \cdot c_j\right) + \alpha_i \cdot w_i + \alpha_i \cdot c_i\right) \cdot W_{total} \qquad (8.3)$$

$$T = T_{i+1} = \left(\left(\sum_{j=0}^{i-1}\alpha_j \cdot c_j\right) + \alpha_i \cdot c_i + \alpha_{i+1} \cdot w_{i+1} + \alpha_{i+1} \cdot c_{i+1}\right) \cdot W_{total}$$
$$(8.4)$$

With $K = \frac{T - W_{total} \cdot \left(\sum_{j=0}^{i-1}\alpha_j \cdot c_j\right)}{W_{total}}$, we have

$$\alpha_i = \frac{K}{w_i + c_i} \qquad \text{and} \qquad \alpha_{i+1} = \frac{K - \alpha_i \cdot c_i}{w_{i+1} + c_{i+1}}.$$

The total fraction of work processed by P_i and P_{i+1} is equal to

$$\alpha_i + \alpha_{i+1} = \frac{K}{w_i + c_i} + \frac{K}{w_{i+1} + c_{i+1}} - \frac{c_i \cdot K}{(w_i + c_i)(w_{i+1} + c_{i+1})}.$$

Therefore, if $i \geqslant 1$, $c_i = c_{i+1} = c$, and this expression is symmetric in w_i and w_{i+1}. We can then conclude that the communication order to the workers has no importance on the quality of a solution, even if it is contrary to the intuition. However, when $i = 0$ the amount of work done becomes:

$$\alpha_0 + \alpha_1 = \frac{K}{w_0} + \frac{K}{w_1 + c}.$$

Since $c > 0$, this value is maximized when w_0 is smaller than w_1. Hence, the root processor should be the fastest computing ones, if possible. $\square$

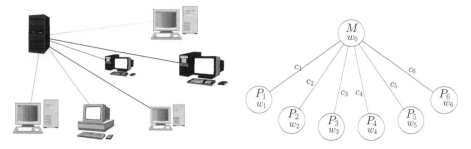

FIGURE 8.7: Example of star-shaped network.

FIGURE 8.8: Theoretical model of a star-shaped network.

The previous analysis can be summarized as:

THEOREM 8.1

For Divisible Load applications on bus-shaped networks, in any optimal solution, the fastest computing processor is the master processor, the order of the communications to the workers has no impact on the quality of a solution, and all processors participate in the work and finish simultaneously. A closed-form formula gives the fraction α_i of the load allocated to each processor:

$$\forall i \in \{0, \ldots, p\}, \alpha_i = \frac{\prod_{j=1}^{i} \frac{w_{j-1}}{c_j + w_j}}{\sum_{k=0}^{m} \left(\prod_{j=1}^{k} \frac{w_{j-1}}{c_j + w_j} \right)}.$$

8.2.2 Star-Shaped Network

The bus-shaped network model can be seen as a particular case, with homogeneous communications, of the more general star-shaped network. Now, we focus our attention on such star-shaped networks: every worker P_i is linked to M through a communication link of different capacity as shown in Figure 8.7, and processors have different speeds.

Notations are similar to the previous ones and are illustrated by Figure 8.8: a master processor M, and m workers $P_1, \ldots, P_m$. The master sends a unit-size message to P_i (with $1 \leqslant i \leqslant m$) in time c_i, and P_i processes it in time w_i. The total workload is W_{total}, and P_i receives a fraction α_i of this load (with $\alpha_i \in \mathbb{Q}$ and $\sum_{i=1}^{m} \alpha_i = 1$).

We assume that the master does not participate in the computation, because we can always add a virtual processor P_0, with the same speed w_0 as the master and with instantaneous communications: $c_0 = 0$. As before, M sends a single message to each worker, and it can communicate to at most one worker at a time, following the one-port model.

This new model seems to be a simple extension of the previous one, and

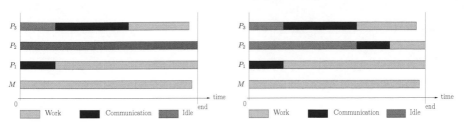

FIGURE 8.9: Some work is added to P_k.

we will check whether our previous lemmas are still valid. We show that the following lemmas are true for any optimal solution:

- all processors participate in the work (Lemma 8.3),
- all workers end their work at the same time (Lemma 8.4),
- all active workers must be served in the non-decreasing order of the c_i's (Lemma 8.5),
- an optimal solution can be computed in polynomial time using a closed-form expression (Lemma 8.6).

LEMMA 8.3
In any optimal solution, all processors participate in the work, i.e., $\forall i, \alpha_i > 0$.

PROOF   Consider an optimal solution $(\alpha_1, \ldots, \alpha_p)$ and assume that at least one processor P_i remains idle during the whole computation ($\alpha_i = 0$). Without any loss of generality, we can also assume that communications are served in the order $(P_1, \ldots, P_m)$. We denote by k the greatest index such that $\alpha_k = 0$ (i.e., P_k is the last processor which is kept idle during the computation). We have two cases to consider:

- $k < m$

  By definition, P_m is not kept idle and thus $\alpha_m \neq 0$. We know that P_m is the last processor to communicate with the master, and then there is no communication during the last $\alpha_m \cdot w_m \cdot W_{total}$ time units. Therefore, once P_m as received its share of work, we can send $\frac{\alpha_m \cdot w_m \cdot W_{total}}{c_k + w_k} > 0$ load units to the processor P_k, as illustrated in Figure 8.9, and it would finish its computation at the same time as P_m. So, we can exhibit a strictly better solution than the previous optimal one, in which P_k was kept idle.

- $k = m$

  We modify the initial solution to give some work to the last processor, without increasing the total completion time. Let k' be the greatest index such that $\alpha_{k'} \neq 0$. By definition, since $P_{k'}$ is the last served processor, there is no communication with the master during at least $a_{k'}.w_{k'} \cdot W_{total}$ time units. As previously, we can give $\frac{\alpha_{k'} \cdot w_{k'} \cdot W_{total}}{c_m + w_m} > 0$

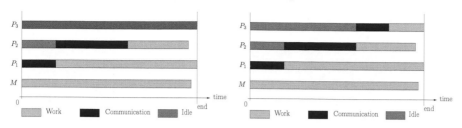

FIGURE 8.10: Some work is added to P_m.

load units to P_m and then exhibits a strictly better solution than the previous optimal one, as represented by Figure 8.10.
Then, in all cases, if at least one processor remains idle, we can build a solution processing strictly more work within the same time. Then, scaling everything, this leads to a solution processing the same amount of work in strictly less time. This proves that in any optimal solution, all processors participate in the work. ∎

LEMMA 8.4
In any optimal solution, all workers end their work at the same time.

PROOF   Since this proof requires more technical arguments, the reader may want to skip it. Consider any optimal allocation. Without loss of generality, we suppose the processors to be numbered according to the order in which they receive their shares of the work. We denote our optimal solution $(\alpha_1, \ldots, \alpha_m)$. Thanks to Lemma 8.3, we know that all workers participate in the computation and then all α_i's are strictly positive. Consider the following linear program:

$$\text{MAXIMIZE } \sum_{i=1}^m \beta_i \text{ SUBJECT TO}$$
$$\begin{cases} \forall i, \beta_i \geqslant 0, \\ \forall i, \sum_{k=1}^i \beta_k c_k + \beta_i w_i \leqslant T \end{cases}$$

The α_i's obviously satisfy the set of constraints above, and from any set of β_i's satisfying the set of inequalities, we can build a valid schedule that processes exactly $\sum_{i=1}^m \beta_i$ units of load. Let $(\beta_1, \ldots, \beta_m)$ be any optimal solution to this linear program. By definition, we have $\sum_{i=1}^m \alpha_i = \sum_{i=1}^m \beta_i$.
We know that one of the extremal solutions $\mathcal{S}_1$ of the linear program is one of the vertices of the convex polyhedron $\mathcal{P}$ induced by the inequalities [17]: this means that in the solution $\mathcal{S}_1$, there are at least m equalities among the $2 \cdot m$ inequalities, m being the number of variables. If we use Lemma 8.3, we know that all the β_i's are positive. Then this vertex is the solution of the

following (full rank) linear system

$$\forall i \in (1, \ldots, m), \sum_{k=1}^{i} \beta_k c_k + \beta_i w_i = T.$$

Thus, we derive that there exists at least one optimal solution, such that all workers finish simultaneously.

Now, consider any other optimal solution $\mathcal{S}_2 = (\delta_1, \ldots, \delta_m)$, different from $\mathcal{S}_1$. Similarly to $\mathcal{S}_1$, $\mathcal{S}_2$ belongs to the polyhedron $\mathcal{P}$. Now, consider the following function f:

$$f : \begin{cases} \mathbb{R} \to \mathbb{R}^m \\ x \mapsto \mathcal{S}_1 + x(\mathcal{S}_2 - \mathcal{S}_1) \end{cases}$$

By construction, we know that $\sum_{i=1}^{m} \beta_i = \sum_{i=1}^{m} \delta_i$. Thus, with the notation $f(x) = (\gamma_1(x), \ldots, \gamma_m(x))$:

$$\forall i \in (1, \ldots, m), \gamma_i(x) = \beta_i + x(\delta_i - \beta_i),$$

and therefore

$$\forall x, \sum_{i=1}^{m} \gamma_i(x) = \sum_{i=1}^{m} \beta_i = \sum_{i=1}^{m} \delta_i.$$

Therefore, all the points $f(x)$ that belong to $\mathcal{P}$ are extremal solutions of the linear program.

Since $\mathcal{P}$ is a convex polyhedron and both $\mathcal{S}_1$ and $\mathcal{S}_2$ belong to $\mathcal{P}$, then $\forall x, 0 \leqslant x \leqslant 1, f(x) \in \mathcal{P}$. Let us denote by x_0 the largest value of $x \geqslant 1$ such that $f(x)$ still belongs to $\mathcal{P}$. x_0 is finite, otherwise at least one of the upper bounds or one of the lower bounds would be violated. At least one constraint of the linear program is an equality in $f(x_0)$, and this constraint is not satisfied for $x > x_0$. We know that this constraint cannot be one of the upper bounds: otherwise, this constraint would be an equality along the whole line $(\mathcal{S}_1, f(x_0))$, and would remain an equality for $x > x_0$. Hence, the constraint of interest is one of the lower bounds. In other terms, there exists an index i, such that $\gamma_i(x_0) = 0$. This is in contradiction with Lemma 8.3 and with the fact that the γ_i's correspond to an optimal solution of the problem.

We can conclude that $\mathcal{S}_1 = \mathcal{S}_2$, and thus that for a given order of the processors, there exists a unique optimal solution and that in this solution all workers finish simultaneously their work.   ☐

LEMMA 8.5

In any optimal solution all active workers must be served in the non-decreasing order of the c_i's.

PROOF   We use the same method as for proving Lemma 8.2: we consider two workers P_i and P_{i+1} (with $1 \leqslant i \leqslant m - 1$) and we check whether the total work which can be done by them in a given time T is dependent of their order.

For these two processors, we use Lemma 8.4 to obtain the following Equations 8.5 and 8.6, similar to Equations 8.3 and 8.4 derived in the case of a bus-shaped network:

$$T = T_i = \left(\left(\sum_{j=1}^{i-1} \alpha_j \cdot c_j \right) + \alpha_i(w_i + c_i) \right) \cdot W_{total} \qquad (8.5)$$

$$T = T_{i+1} = \left(\left(\sum_{j=1}^{i-1} \alpha_j \cdot c_j \right) + \alpha_i \cdot c_i + \alpha_{i+1}(w_{i+1} + c_{i+1}) \right) \cdot W_{total} \qquad (8.6)$$

With $K = \frac{T - W_{total} \cdot \left(\sum_{j=1}^{i-1} \alpha_j \cdot c_j \right)}{W_{total}}$, Equations 8.5 and 8.6 can be respectively rewritten as:

$$\alpha_i = \frac{K}{w_i + c_i} \text{ and } \alpha_{i+1} = \frac{K - \alpha_i \cdot c_i}{w_{i+1} + c_{i+1}}$$

The time needed by communications can be written as:

$$(\alpha_i \cdot c_i + \alpha_{i+1} \cdot c_{i+1}) \cdot W_{total} =$$

$$\left(\left(\frac{c_i}{w_i + c_i} + \frac{c_{i+1}}{w_{i+1} + c_{i+1}} \right) - \frac{c_i \cdot c_{i+1}}{(w_i + c_i)(w_{i+1} + c_{i+1})} \right) \cdot K \cdot W_{total}.$$

We see that the equation is symmetric, thus that the communication time is completely independent of the order of the two communications.

The total fraction of work is equal to:

$$\alpha_i + \alpha_{i+1} = \left(\frac{1}{w_i + c_i} + \frac{1}{w_{i+1} + c_{i+1}} \right) \cdot K - \frac{K \cdot c_i}{(w_i + c_i)(w_{i+1} + c_{i+1})}.$$

If we exchange P_i and P_{i+1}, then the total fraction of work is equal to:

$$\alpha_i + \alpha_{i+1} = \left(\frac{1}{w_i + c_i} + \frac{1}{w_{i+1} + c_{i+1}} \right) \cdot K - \frac{K \cdot c_{i+1}}{(w_i + c_i)(w_{i+1} + c_{i+1})}.$$

The difference between both solutions is given by

$$\Delta_{i,i+1} = (c_i - c_{i+1}) \frac{K}{(w_i + c_i)(w_{i+1} + c_{i+1})}.$$

Contrary to the communication time, the fraction of the load done by processors P_i and P_{i+1} depends on the order of the communications. If we serve the fastest-communicating processor first, the fraction of the load processed by P_i and P_{i+1} can be increased without increasing the communication time for other workers. In other terms, in any optimal solution, participating workers should be served by non-decreasing values of c_i. □

LEMMA 8.6
When processors are numbered in non-decreasing values of the $c_i's$, the following closed-form formula for the fraction of the total load allocated to processor P_i ($1 \leqslant i \leqslant m$) defines an optimal solution:

$$\alpha_i = \frac{\frac{1}{c_i+w_i} \prod_{k=1}^{i-1}\left(\frac{w_k}{c_k+w_k}\right)}{\sum_{i=1}^{p} \frac{1}{c_i+w_i} \prod_{k=1}^{i-1} \frac{w_k}{c_k+w_k}}.$$

PROOF   Thanks to the previous lemmas, we know that all workers participate in the computation (Lemma 8.3) and have to be served by non-decreasing values of c_i (Lemma 8.5) and that all workers finish simultaneously (Lemma 8.4).

Without loss of generality, we assume that $c_1 \leqslant c_2 \leqslant \ldots \leqslant c_m$. By definition, the completion time of the i-th processor is given by:

$$T_i = T = W_{total}\left(\sum_{k=1}^{i} \alpha_k c_k + \alpha_i w_i\right)$$

$$= \alpha_i W_{total}(w_i + c_i) + \sum_{k=1}^{i-1}(\alpha_k c_k W_{total}). \quad (8.7)$$

Then, an immediate induction solves this triangular system and leads to:

$$\alpha_i = \frac{T}{W_{total}} \frac{1}{c_i + w_i} \prod_{k=1}^{i-1}\left(\frac{w_k}{c_k + w_k}\right). \quad (8.8)$$

To compute the total completion time T, we just need to remember that $\sum_{i=1}^{m} \alpha_i = 1$, hence

$$\frac{T}{W_{total}} = \frac{1}{\sum_{i=1}^{p} \frac{1}{c_i+w_i} \prod_{k=1}^{i-1} \frac{w_k}{c_k+w_k}}.$$

So, we can have the complete formula for the α_i's:

$$\alpha_i = \frac{\frac{1}{c_i+w_i} \prod_{k=1}^{i-1}\left(\frac{w_k}{c_k+w_k}\right)}{\sum_{i=1}^{p} \frac{1}{c_i+w_i} \prod_{k=1}^{i-1} \frac{w_k}{c_k+w_k}}.$$

$\square$

We summarize the previous analysis as:

THEOREM 8.2
For Divisible Loads applications on star-shaped networks, in an optimal solution:

- *workers are ordered by non-decreasing values of c_i,*
- *all participate in the work,*
- *they all finish simultaneously.*

Closed-form formulas give the fraction of the load allocated to each processor.

We conclude this section with two remarks.

- If the order of the communications cannot be freely chosen, Lemma 8.3 is not always true; for instance, when sending a piece of work to a processor is more expensive than having it processed by the workers served after that processor [12].
- In this section, we considered that the master processor did not participate in the processing. To deal with cases where we can enroll the master in the computation, we can easily add a virtual worker, with the same processing speed as the master, and a communication time equal to 0. Finally, if we also have the freedom to choose the master processor, the simplest method to determine the best choice is to compute the total completion time in all cases: we have only m different cases to check, so we can still determine the best master efficiently (although not as elegantly as for a bus-shaped platform).

8.3 Extensions of the Divisible Load Model

In the previous section, we have shown that the Divisible Load approach enables to solve scheduling problems that are hard to solve under classical models. With a classical approach, we are often limited to solving simpler problems, featuring homogeneous communication links and linear cost functions for computations and communications. In Section 8.2, we have seen how the Divisible Load model could help solve a problem that was already difficult even with homogeneous resources.

Realistic platform models are certainly intractable, and crude ones are not likely to be realistic. A natural question arises: how far can we go with Divisible Load theory? We present below several extensions to the basic framework described so far. In Section 8.3.1, we extend the linear cost model for communications by introducing *latencies*. Next, we distribute chunks to processors in *several* rounds in Section 8.3.2. In the latter two extensions, we still neglect the cost of *return messages*, assuming that data is only sent from the master to the workers, and that results lead to negligible communications. To conclude this section, we introduce return messages in Section 8.3.3.

8.3.1 Introducing Latencies

In the previous sections, we used a simple linear cost model for communications. The time needed to transmit a data chunk was perfectly proportional to its size. On real-life platforms, there is always some latency to account for. In other words, an affine communication and affine computation model would be more realistic. We have to introduce new notations for these latencies: C_i denotes the communication latency paid by worker P_i for a communication from the master, and W_i denotes the latency corresponding to initializing a computation. If P_i has to process a fraction α_i of the total load, then its communication time is equal to $C_i + c_i \alpha_i W_{total}$ and its computation time is equal to $W_i + w_i \alpha_i W_{total}$.

This variant of the problem is NP-complete [19] and thus much more difficult than the previous one. However, some important results can be shown for any optimal solution:

- Even if communication times are fixed (bandwidths are supposed to be infinite), the problem remains NP-complete, as shown by Yang et al. in [19].
- All participating workers end their work at the same time (Lemma 8.7).
- If the load is large enough, all workers participate in the work and must be served in non-decreasing order of the c_i's (Lemma 8.9).
- An optimal solution can be found using a mixed linear program (Lemma 8.8).

LEMMA 8.7
In any optimal solution, all participating workers have the same completion time.

PROOF   This lemma can be shown using a proof similar to the proof of Lemma 8.4. A detailed proof is given in [4].   ▯

The following mixed linear program aims at computing an optimal distribution of the load among workers.

MINIMIZE T_f SUBJECT TO

$$
\begin{cases}
(1) & \forall i, 1 \leqslant i \leqslant m, \quad \alpha_i \geqslant 0 \\
(2) & \qquad\qquad\qquad \sum_{i=1}^{m} \alpha_i = 1 \\
(3) & \forall j, 1 \leqslant j \leqslant m,, \quad y_j \in \{0,1\} \\
(4) & \forall i,j, 1 \leqslant i,j \leqslant m, \quad x_{i,j} \in \{0,1\} \\
(5) & \forall j, 1 \leqslant i \leqslant m, \quad \sum_{i=1}^{m} x_{i,j} = y_j \\
(6) & \forall i, 1 \leqslant i \leqslant m, \quad \sum_{j=1}^{m} x_{i,j} \leqslant 1 \\
(7) & \forall j, 1 \leqslant i \leqslant m, \quad \alpha_j \leqslant y_j \\
(8) & \forall i, 1 \leqslant i \leqslant m, \quad \sum_{k=1}^{i-1} \sum_{j=1}^{m} x_{k,j}(C_j + \alpha_j c_j W_{total}) \\
& \qquad\qquad + \sum_{j=1}^{m} x_{i,j}(C_j + \alpha_j c_j W_{total} + W_j + \alpha_j w_j W_{total}) \leqslant T_f
\end{cases}
$$

LEMMA 8.8

An optimal solution can be found using the mixed linear program above (with a potentially exponential computation cost).

PROOF   In [5], the authors added the resource selection issue to the original linear program given by Drozdowski in [11]. To address this issue, they added two notations: y_j, which is a binary variable equal to 1 if, and only if, P_j participates in the work, and $x_{i,j}$, which is a binary variable equal to 1 if, and only if, P_j is chosen for the i-th communication from the master. Equation 5 implies that P_j is involved in exactly y_j communication. Equation 6 states that that at most one worker is used for the i-th communication. Equation 7 ensures that non-participating workers have no work to process. Equation 8 implies that the worker selected for the i-th communication must wait for the previous communications before starting its own communication and then its computation.

This linear program always has a solution, which provides the selected workers and their fraction of the total load in an optimal solution.   ⬜

LEMMA 8.9

In any optimal solution, and if the load is large enough, all workers participate in the work and must be served in non-decreasing order of communication time c_i.

PROOF   We want to determine the total amount of work which can be done in a time T. Let us consider any valid solution to this problem. The set of the k active workers is denoted $\mathcal{S} = \{P_{\sigma(1)}, \ldots, P_{\sigma(k)}\}$, where σ is a one-to-one mapping from $[1 \ldots k]$ to $[1 \ldots m]$ and represents the order of communications. Let n_{TASK} denote the maximum number of processed units of load using this set of workers in this order.

- Consider the following instance of our problem, with k workers $P'_{\sigma(1)}$, ..., $P'_{\sigma(k)}$, such that $\forall i \in \{1, \ldots, k\}, C'_i = 0, W'_i = 0, c'_i = c_i, w'_i = w_i$ (in fact, we are just ignoring all latencies). The total number of work units n'_{TASK} which can be executed on this platform in the time T is greater than the number n_{TASK} of tasks processed by the original platform:

$$n_{TASK} \leqslant n'_{TASK}.$$

Using Theorem 8.2, we know that n'_{TASK} is given by a formula of the following form:

$$n'_{TASK} = f(\mathcal{S}, \sigma) \cdot T.$$

The main point is that n'_{TASK} is proportional to T.

- Now we will determine the number of works units n''_{TASK}, which could be done in a time $T'' = T - \sum_{i \in \mathcal{S}}(C_i + W_i)$. n''_{TASK} is clearly smaller than n_{TASK} since it consists in adding all latencies before the beginning of the work:

$$n''_{TASK} \leqslant n_{TASK}.$$

The previous equality still stands:

$$n''_{TASK} = f(\mathcal{S}, \sigma) \left(T - \sum_{i \in \mathcal{S}} (C_i + W_i) \right).$$

We have $n''_{TASK} \leqslant n_{TASK} \leqslant n'_{TASK}$ and then

$$f(\mathcal{S}, \sigma) \left(1 - \frac{\sum_{i \in \mathcal{S}} (C_i + W_i)}{T} \right) \leqslant \frac{n_{TASK}}{T} \leqslant f(\mathcal{S}, \sigma).$$

Therefore, when T becomes arbitrarily large, the throughput of the platform becomes close to the theoretical model without any latency. Thus, when T is sufficiently large, in any optimal solution, all workers participate in the work, and chunks should be sent on the ordering of non-decreasing communication times c_i.

Without any loss of generality, we can assume that $c_1 \leqslant \ldots \leqslant c_p$ and then the following linear system returns an asymptotically optimal solution:

MINIMIZE T SUBJECT TO
$$\begin{cases} \sum_{i=1}^{m} \alpha_i = 1 \\ \forall i \in \{1, \ldots, m\}, \sum_{k=1}^{i} (C_k + c_k \alpha_k W_{total}) + W_i + w_i \alpha_i W_{total} = T \end{cases}$$

Moreover, when T is sufficiently large, this solution is optimal when all c_i are different, but determining the best way to break ties among workers with the same communication speed remains an open question.

$\square$

8.3.2 Multi-Round Strategies

So far, we only used models with a strict one-port communication scheme, and data were transmitted to workers in a single message. Therefore, each worker had to wait while previous one were communicating with the master before it could begin receiving and processing its own data. This waiting time can lead to a poor utilization of the platform. A natural solution to quickly distribute some work to each processor is to distribute data in multiple rounds: while the first processor is computing its first task, we can distribute data to other workers, and then resume the distribution to the first processor, and so on. By this way, we hope to overlap communications with computations, thereby increasing platform throughput (see Figure 8.11). This idea seems promising, but we have two new questions to answer:

1. How many rounds should we use to distribute the whole load?

2. Which size should we allocate to each round?

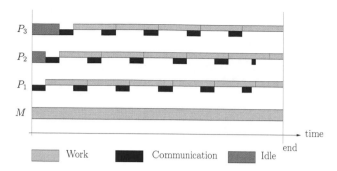

FIGURE 8.11: Multi-round execution over a bus-shaped platform.

If we follow the lesson learned using a single-round distribution, we should try to solve the problem without latencies first. However, it turns out that the linear model is not interesting in a multi-round framework: the optimal solution has an infinite number of rounds of size zero. On the contrary, when latencies are added to the model, they prevent solutions using such an infinite number of rounds. However, the problem becomes NP-complete. In the following, we present asymptotically optimal solutions. Finally, we assess the gain of multi-round strategies with respect to one-round solutions.

8.3.2.1 Homogeneous Bus-Shaped Platform, Linear Cost Model

In this first case, we assume a linear cost model for both communications and computations and we demonstrate that the corresponding optimal solution is obvious but unrealistic, since it requires an infinite number of rounds.

THEOREM 8.3
Let us consider any homogeneous bus-shaped master-worker platform, following a linear cost model for both communications and computations. Then any optimal multi-round schedule requires an infinite number of rounds.

PROOF
We prove this theorem by contradiction. Let $\mathcal{S}$ be any optimal schedule using a finite number K of rounds, as illustrated in Figure 8.12. We still have m workers in our platform, and the master allocates a fraction $\alpha_i(k)$ to worker i $(1 \leqslant i \leqslant m)$ during the k-th round. Without any loss of generality, we suppose that workers are served in a *Round-Robin* fashion, in the order $P_1, \ldots, P_m$: if you want P_2 to be served before P_1, just use a round distributing data only to P_2 (and a null fraction to other processors) followed by a round distributing data to other processors.

Since we consider any schedule, we need general notations: if we have

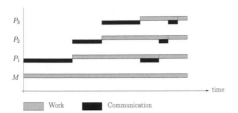

FIGURE 8.12: The original schedule, using two rounds.

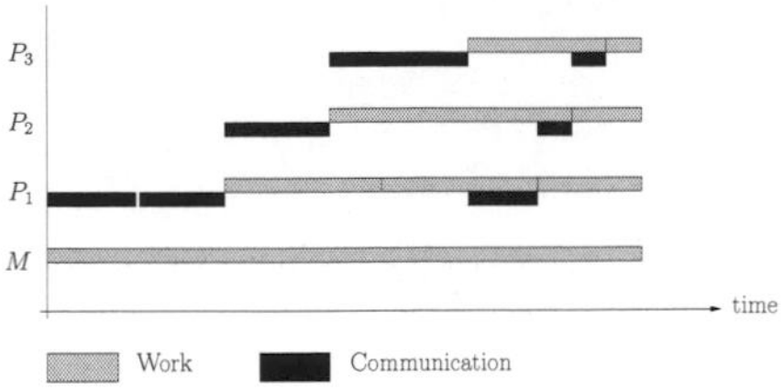

FIGURE 8.13: We split the first communication in two parts.

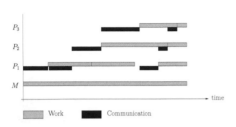

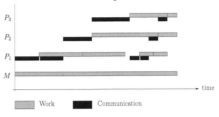

FIGURE 8.14: P_1 can start its work earlier.

FIGURE 8.15: We split the second communication in two parts, P_1 can finish its total work strictly earlier.

$1 \leqslant i \leqslant m$ and $1 \leqslant k \leqslant K$, $Comm_i^{start}(k)$ (respectively $Comm_i^{end}(k)$, $Comp_i^{start}(k)$, $Comp_i^{end}(k)$) denotes the start of the communication from the master to worker P_i of the data for the k-th round (respectively the end of the communication, the start and the end of the computation). We assume $Comm_i^{end}(0) = 0$ and $Comp_i^{end}(0) = 0$ for all $1 \leqslant i \leqslant m$, to simplify some equations. With these new notations, any valid schedule has to respect the following constraints:

- Two successive communications cannot overlap:

$$\begin{aligned}\forall i, k, 2 \leqslant i \leqslant m, 1 \leqslant k \leqslant K, \ Comm_i^{start}(k) &\geqslant Comm_{i-1}^{end}(k)\\ \forall k, 1 \leqslant k \leqslant K, \qquad\qquad Comm_1^{start}(k) &\geqslant Comm_p^{end}(k-1)\end{aligned} \quad (8.9)$$

- A computation cannot begin before the end of the corresponding communication:

$$\forall i, k, \ 1 \leqslant i \leqslant m, 1 \leqslant k \leqslant K, Comp_i^{start}(k) \geqslant Comm_i^{end}(k) \quad (8.10)$$

We force any computation to begin after the end of the previous one:

$$\forall i, k \leqslant i \leqslant m, 2 \leqslant k \leqslant K, Comp_i^{start}(k) \geqslant Comp_i^{end}(k-1) \quad (8.11)$$

This constraint is not necessary to obtain a valid schedule, but it allows to simplify equations and do not change the completion time of schedules, since the order of computations allocated to a worker has no importance.

We can easily modify the proof of Lemma 8.1 to show that in any optimal schedule, all workers finish their work simultaneously: basically, we just have to modify the last chunk. We want to show that we can build a new schedule $\mathcal{S}'$ based on $\mathcal{S}$, such that P_1 ends its work earlier than before, and other processors end their work at the same time as before.

As P_1 initially holds no data, it stays temporarily idle (i.e., it does not compute anything) at some points during schedule execution, waiting for some data to process. Let k_0 be the last round during which P_1 stays idle. Therefore, P_1 receives a positive load fraction during this round (otherwise it would also be idle at the beginning of round $k_0 + 1$). From the schedule $\mathcal{S}$, we build $\mathcal{S}'$, mostly identical to $\mathcal{S}$ except that we replace the k_0-th round by two successive rounds, as shown in Figure 8.13. In fact, we only send the fraction allocated to the first worker during round k_0 in two parts. Formally, using the same notations for $\mathcal{S}'$ than for $\mathcal{S}$, but adding prime symbols, $\mathcal{S}'$ is defined as follows:

- The first $k_0 - 1$ rounds are identical in both schedules:
  $\forall k, i, 1 \leqslant k \leqslant k_0 - 1, 1 \leqslant i \leqslant m, \alpha'_i(k) = \alpha_i(k)$,
- Round k_0 is defined as follows:
  $\alpha'_1(k_0) = \frac{1}{2}\alpha_1(k_0)$ and $\forall i, 2 \leqslant i \leqslant m, \alpha'_i(k_0) = 0$.
- Round $k_0 + 1$ is defined as follows:
  $\alpha'_1(k_0 + 1) = \frac{1}{2}\alpha_1(k_0)$ and $\forall i, 2 \leqslant i \leqslant m, \alpha'_i(k_0 + 1) = \alpha_i(k_0)$.
- Finally, the last rounds are identical in both schedules:
  $\forall i, k, 1 \leqslant i \leqslant m, k_0 + 1 \leqslant k \leqslant K, \alpha'_i(k + 1) = \alpha_i(k)$.

Now, focus on start and finish times for all workers, except the first one:

- The first $k_0 - 1$ rounds are unchanged, hence:

$$\forall k, i, 1 \leqslant k \leqslant k_0 - 1, 1 \leqslant i \leqslant m, Comm'^{start}_i(k) = Comm^{start}_i(k)$$
$$\forall k, i, 1 \leqslant k \leqslant k_0 - 1, 1 \leqslant i \leqslant m, Comm'^{end}_i(k) = Comm^{end}_i(k)$$
$$\forall k, i, 1 \leqslant k \leqslant k_0 - 1, 1 \leqslant i \leqslant m, Comp'^{start}_i(k) = Comp^{start}_i(k)$$
$$\forall k, i, 1 \leqslant k \leqslant k_0 - 1, 1 \leqslant i \leqslant m, Comp'^{end}_i(k) = Comp^{end}_i(k)$$

- We easily establish the following equations for round k_0:

$$i = 1 \ Comm'^{start}_1(k_0) = Comm'^{start}_1(k_0)$$
$$\forall i, 2 \leqslant i \leqslant m, Comm'^{start}_i(k_0) = Comm'^{end}_1(k_0)$$
$$i = 1 \ Comm'^{end}_1(k_0) = Comm'^{start}_1(k_0)$$
$$+ \tfrac{1}{2}\alpha_i(k_0) \cdot c \cdot W_{total}$$
$$\forall i, 2 \leqslant i \leqslant m, \ Comm'^{end}_i(k_0) = Comm'^{start}_1(k_0)$$
$$\forall i, 2 \leqslant i \leqslant m, \ Comp'^{start}_i(k_0) = Comp'^{end}_i(k_0 - 1)$$
$$\forall i, 2 \leqslant i \leqslant m, \ Comp'^{end}_i(k_0) = Comp'^{start}_i(k_0)$$

- And for the following round $k_0 + 1$:

$$i = 1 \; Comm'^{start}_1(k_0 + 1) = Comm'^{end}_1(k_0)$$
$$\forall i, 2 \leqslant i \leqslant m, \; Comm'^{start}_i(k_0 + 1) = Comm^{start}_i(k_0)$$
$$i = 1 \quad Comm'^{end}_1(k_0 + 1) = Comm'^{start}_1(k_0 + 1)$$
$$+ \tfrac{1}{2}\alpha_i(k_0) \cdot c \cdot W_{total}$$
$$\forall i, 2 \leqslant i \leqslant m, \quad Comm'^{end}_i(k_0 + 1) = Comm^{end}_i(k_0)$$
$$\forall i, 2 \leqslant i \leqslant m, \quad Comp'^{start}_i(k_0 + 1) = Comp^{start}_i(k_0)$$
$$\forall i, 2 \leqslant i \leqslant m, \quad Comp'^{end}_i(k_0 + 1) = Comp^{end}_i(k_0)$$

- Finally, for the last $K - k_0$ rounds:

$$\forall k, i, k_0 + 1 \leqslant k \leqslant K, 1 \leqslant i \leqslant m, \; Comm'^{start}_i(k + 1) = Comm^{start}_i(k)$$
$$\forall k, i, k_0 + 1 \leqslant k \leqslant K, 1 \leqslant i \leqslant m, \; Comm'^{end}_i(k + 1) = Comm^{end}_i(k)$$
$$\forall k, i, k_0 + 1 \leqslant k \leqslant K, 2 \leqslant i \leqslant m, \; Comp'^{start}_i(k + 1) = Comp^{start}_i(k)$$
$$\forall k, i, k_0 + 1 \leqslant k \leqslant K, 2 \leqslant i \leqslant m, \; Comp'^{end}_i(k + 1) = Comp^{end}_i(k)$$

Consider the first processor P_1. Using Equations 8.10 and 8.11 on the original schedule $\mathcal{S}$, we have:

$$Comp^{end}_1(k_0) = \max(Comp^{end}_1(k_0 - 1), Comm^{end}_1(k_0)) + \alpha_1(k_0) \cdot w_1 \cdot W_{total} \tag{8.12}$$

But we know that P_1 is idle just before the start of its computation, hence:

$$Comp^{end}_1(k_0 - 1) < Comm^{end}_1(k_0) = Comm^{start}_1(k_0) + \alpha_1 k_0 \cdot c \cdot W_{total}.$$

Using this remark, we expand Equation 8.12 to obtain:

$$Comp^{end}_1(k_0) = Comm^{start}_1(k_0) + \alpha_1 k_0 \cdot c \cdot W_{total} + \alpha_1(k_0) \cdot w_1 \cdot W_{total}$$
$$> Comp^{end}_1(k_0 - 1) + \alpha_1(k_0) \cdot w_1 \cdot W_{total} \tag{8.13}$$

Using Equations 8.10 and 8.11 on the new schedule $\mathcal{S}'$, we have:

$$Comp'^{end}_1(k_0) = \max(Comp'^{end}_1(k_0 - 1), Comm'^{end}_1(k_0))$$
$$+ \frac{1}{2}\alpha_1(k_0) \cdot w_1 \cdot W_{total} \tag{8.14}$$

$$Comp'^{end}_1(k_0 + 1) = \max(Comp'^{end}_1(k_0), Comm'^{end}_1(k_0 + 1))$$
$$+ \frac{1}{2}\alpha_1(k_0) \cdot w_1 \cdot W_{total} \tag{8.15}$$

Combining Equations 8.14 and 8.15, and noting that $Comp'^{end}_1(k_0 - 1) = Comp^{end}_1(k_0 - 1)$, leads to:

$$Comp'^{end}_1(k_0 + 1) = \frac{1}{2}\alpha_1(k_0) \cdot w_1 \cdot W_{total} + \max(Comm'^{end}_1(k_0 + 1),$$
$$\max(Comp^{end}_1(k_0 - 1), Comm'^{end}_1(k_0)) + \frac{1}{2}\alpha_1(k_0) \cdot w_1 \cdot W_{total}) \tag{8.16}$$

By definition, we have $Comm'^{end}_1(k_0) = Comm^{start}_1(k_0) + \frac{1}{2}\alpha_1(k_0) \cdot c \cdot W_{total}$ and thus $Comm'^{end}_1(k_0 + 1) = Comm^{start}_1(k_0) + \alpha_1(k_0) \cdot c \cdot W_{total}$.

Then we can expand Equation 8.16 into:

$$Comp'^{end}_1(k_0 + 1) =$$
$$\max \begin{cases} Comm^{start}_1(k_0) + \alpha_1(k_0) \cdot c \cdot W_{total} + \frac{1}{2}\alpha_1(k_0) \cdot w_1 \cdot W_{total}, \\ Comp^{end}_1(k_0 - 1) + \alpha_1(k_0) \cdot w_1 \cdot W_{total}, \\ Comm^{start}_1(k_0) + \frac{1}{2}\alpha_1(k_0) \cdot c \cdot W_{total} + \alpha_1(k_0) \cdot w_1 \cdot W_{total} \end{cases}$$

Since we have $\alpha_1(k_0) > 0$, we have respectively:

$$Comm^{start}_1(k_0) + \left(c + \frac{1}{2}w_1\right)\alpha_1(k_0)W_{total} <$$
$$Comm^{start}_1(k_0) + (c + w_1)\alpha_1(k_0)W_{total}$$

$$Comm^{start}_1(k_0) + \left(\frac{1}{2}c + w_1\right)\alpha_1(k_0)W_{total} <$$
$$Comm^{start}_1(k_0) + (c + w_1)\alpha_1(k_0)W_{total}$$

Using these inequalities and equation 8.13, we state that:

$$Comp'^{end}_1(k_0 + 1) < Comp^{end}_1(k_0).$$

In other words, if we split the previous distribution into two rounds, P_1 can finish its fraction of work strictly earlier, as shown in Figure 8.14; part of its idle time is switched from before round k_0-th to before round $k_0 + 1$ of the original schedule S (which is round $k_0 + 2$ of the new schedule S'). If we repeat this process, P_1 will be idle just before its last round, and, at the next step, will finish its work before its original finish time (see Figure 8.15).

We have previously shown that if a processor finish its work strictly before other workers, then we can build a better schedule (see Lemma 8.1). This is sufficient to prove that S is not an optimal schedule, and thus, any schedule with a finite number of rounds can be strictly improved. ☐

As a consequence of Theorem 8.3, we cannot use linear cost models in a multi-round setting. In the next section, we introduce latencies.

8.3.2.2 Bus-Shaped Network and Homogeneous Processors, Fixed Number of Rounds

The simplest case to explore is a bus-shaped network of homogeneous processors, i.e., with homogeneous communication links, and when the number of rounds to use is given to the algorithm.

Intuitively, rounds have to be small to allow a fast start of computations, but also have to be large to amortize the cost of latencies. These two contradictory objectives can be merged by using small rounds at the beginning, and then increasing them progressively to amortize paid latencies.

The first work on multi-round strategies was done by Bharadwaj, Ghose, and Mani using a linear cost model for both communications and computations [8]. This was followed by Yang and Casanova in [18], who used affine models instead of linear ones.

Since we only consider homogeneous platforms, we have for any worker i (with $1 \leqslant i \leqslant m$) $w_i = w$, $W_i = W$, $c_i = c$, and $C_i = C$. Moreover, R denotes the computation-communication ratio ($R = w/c$) and γ_j denotes the time to compute the chunk j excluding the computation latency: $\gamma_j = \alpha_j \cdot w \cdot W_{total}$. We assume that we distribute the whole load in M rounds of m chunks. For technical reasons, chunks are numbered in the reverse order, from $Mm - 1$ (the first one) to 0 (the last one).

Using these notations, we can write the recursion on the γ_j series:

$$\forall j \geqslant m, W + \gamma_j = \frac{1}{R}(\gamma_{j-1} + \gamma_{j-2} + \ldots + \gamma_{j-m}) + m \cdot C \qquad (8.17)$$

$$\forall 0 \leqslant j < m, W + \gamma_j = \frac{1}{R}(\gamma_{j-1} + \gamma_{j-2} + \ldots + \gamma_{j-m}) + j \cdot C + \gamma_0 \qquad (8.18)$$

$$\forall j < 0, \gamma_j = 0 \qquad (8.19)$$

Equation 8.17 expresses that a worker j must receive enough data to compute during exactly the time needed for the next m chunks to be communicated, ensuring no idle time on the communication bus. This equation is of course not true for the last m chunks. Equation 8.18 states that all workers have to finish their work at the same time. Finally, Equation 8.19 ensures the correctness of the two previous equations by setting out-of-range terms to 0.

This recursion corresponds to an infinite series in the γ_j, of which the first m values give the solution to our problem. Using generating functions, we can solve this recursion. Let $\mathcal{G}(x) = \sum_{j=0}^{\infty} \gamma_j x^j$ be the generating function for the series. Using Equations 8.17 and 8.18, the value of $\mathcal{G}(x)$ can be expressed as (see [18]):

$$\mathcal{G}(x) = \frac{(\gamma_0 - m \cdot C)(1 - x^m) + (m \cdot C - W) + C \cdot \left(\frac{x(1-x^{m-1})}{1-x} - (m-1)x^m \right)}{(1-x) - x(1-x^m)/R}$$

The rational expansion method [13] gives the roots of the polynomial denominator and then the correct values of the γ_j's, and finally, the values of the α_j's. The value of the first term γ_0 is given by the equation $\sum_{j=0}^{Mm-1} \gamma_j = W_{total} \cdot w$.

8.3.2.3   Bus-Shaped Network, Computing the Number of Rounds

In the previous section, we assumed that the number of rounds was given to the algorithm, thus we avoided one of the two issues of multi-round algorithms.

Now, we suppose that the number of chunks has to be computed by the scheduling algorithm as well as their respective sizes. As we said before, we have to find a good compromise between a small number of chunks, to reduce the overall cost of latencies, and a large one, to ensure a good overlap of communications by computations.

In fact, finding the optimal number of rounds for such algorithms and affine cost models is still an open question. Nonetheless, Yang, van der Raadt, and Casanova proposed the Uniform Multi-Round (UMR) algorithm [20]. This algorithm is valid in the homogeneous case as well as in the heterogeneous case, but we will only look at the homogeneous one for simplicity reasons. To simplify the problem, UMR assume that all chunks sent to workers during the same round have the same size. This constraint can limit the overlap of communications by computations, but it allows to find an optimal number of rounds.

In this section, α_j denotes the fraction of the load sent to any worker during the j-th round, and M denotes the total number of rounds. Then there are m chunks of size α_j, for a total of $m \cdot M$ chunks. The constraint of uniform sizes for chunks of the same round is not used for the last round, allowing the workers to finish simultaneously. To ensure a good utilization of the communication link, the authors force the master to finish sending work for round $j + 1$ to all workers when worker p finishes computing for round j. This condition can be written as:

$$W + \alpha_j \cdot w \cdot W_{total} = p \cdot (C + \alpha_{j+1} \cdot c \cdot W_{total}),$$

which leads to:

$$\alpha_j = \left(\frac{c}{m \cdot w}\right)^j (\alpha_0 - \gamma) + \gamma, \qquad (8.20)$$

where $\gamma = \frac{1}{w - m \cdot c} \cdot (m \cdot C - W)$. The case $w = m \cdot c$ is simpler and detailed in the original paper [20].

With this simple formula, we can give the makespan $\mathcal{M}$ of the complete schedule, which is the sum of the time needed by the worker m to process its data, the total latency of computations and the time needed to send all the chunks during the first round (the $\frac{1}{2}$ factor comes from the non-uniform sizes of the last round, since all workers finish simultaneously):

$$\mathcal{M}(M, \alpha_0) = \frac{W_{total}}{m} + M \cdot W + \frac{1}{2} \cdot m \cdot (C + c \cdot \alpha_0). \qquad (8.21)$$

The complete schedule needs to process the entire load, which can be written as:

$$\mathcal{G}(M, \alpha_0) = \sum_{j=0}^{M-1} p \cdot \alpha_j = W_{total}. \qquad (8.22)$$

Using these equations, the problem can be expressed as minimizing $\mathcal{M}(M, \alpha_0)$ subject to $\mathcal{G}(M, \alpha_0)$. The Lagrange Multiplier method [7] leads to a single equation, which cannot be solved analytically but only numerically. Several simulations showed that uniform chunks can reduce performance, when compared to the multi-round algorithm of the previous section, when latencies are small; but they lead to better results when latencies are large. Moreover, the UMR algorithm can be used on heterogeneous platforms, contrary to the previous multi-round algorithm.

Asymptotically optimal algorithm. Finding an optimal algorithm that distributes data to workers in several rounds, is still an open question. Nevertheless, it is possible to design asymptotically optimal algorithms. An algorithm is asymptotically optimal if the ratio of its makespan obtained with a load W_{total} over the optimal makespan with this same load tends to 1 as W_{total} tends to infinity. This approach is coherent with the fact that the Divisible Load model already is an approximation well suited to large workloads.

THEOREM 8.4
Consider a star-shaped platform with arbitrary values of computation and communication speeds and latencies, allowing the overlap of communications by computations. There exists an asymptotically optimal periodic multi-round algorithm.

PROOF   The main idea is to look for a periodic schedule: the makespan T is divided into k periods of duration T_p. The initialization and the end of the schedule are sacrificed, but the large duration of the whole schedule amortizes this sacrifice. We still have to find a good compromise between small and large periods. It turns out that choosing a period length proportional to the square-root of the optimal makespan T^* is a good trade-off. The other problem is to choose the participating workers. This was solved by Beaumont et al. using linear programming [4]. If $\mathcal{I} \subseteq \{1, \ldots, p\}$ denotes the selected workers, we can write that communication and computation resources are not exceeded during a period of duration T_p:

$$\sum_{i \in \mathcal{I}} (C_i + \alpha_i \cdot c_i W_{total}) \leqslant T_p,$$

$$\forall i \in \mathcal{I}, W_i + \alpha_i \cdot w_i \cdot W_{total} \leqslant T_p.$$

We aim to maximize the average throughput $\rho = \sum_{i \in \mathcal{I}} \frac{\alpha_i \cdot W_{total}}{T_p}$, where $\frac{\alpha_i \cdot W_{total}}{T_p}$ is the average number of load units processed by P_i in one time unit, under the following linear constraints:

$$\begin{cases} \forall i \in \mathcal{I}, & \frac{\alpha_i \cdot W_{total}}{T_p} w_i \leqslant 1 - \frac{W_i}{T_p} \quad \text{(overlap)}, \\ \sum_{i \in \mathcal{I}} \frac{\alpha_i \cdot W_{total}}{T_p} c_i \leqslant 1 - \frac{\sum_{i \in \mathcal{I}} C_i}{T_p} \quad \text{(1-port model)} \end{cases}$$

This set of constraints can be replaced by the following one, stronger but easier to solve:

$$\begin{cases} \forall i \in \{1, \ldots, p\}, & \frac{\alpha_i \cdot W_{total}}{T_p} w_i \leqslant 1 - \frac{\sum_{i=1}^{p} C_i + W_i}{T_p} \quad \text{(overlap)}, \\ \sum_{i=1}^{p} \frac{\alpha_i \cdot W_{total}}{T_p} c_i \leqslant 1 - \frac{\sum_{i=1}^{p} C_i + W_i}{T_p} \quad \text{(1-port model)} \end{cases} \quad (8.23)$$

Without any loss of generality, assume that $c_1 \leqslant c_2 \leqslant \ldots \leqslant c_p$ and let q be the largest index, such that $\sum_{i=1}^{q} \frac{c_i}{w_i} \leqslant 1$. Let ε be equal to $1 - \sum_{i=1}^{q} \frac{c_i}{w_i}$ if $q < p$, and to 0 otherwise. Then the optimal throughput for system 8.23 is realized with

$$\forall 1 \leqslant i \leqslant q, \quad \frac{\alpha_i \cdot W_{total}}{T_p} = \frac{1}{c_i} \left(1 - \frac{\sum_{i=1}^{p} C_i + W_i}{T_p} \right)$$

$$\frac{\alpha_{q+1} \cdot W_{total}}{T_p} = \left(1 - \frac{1}{T_p} \sum_{i=1}^{p} (C_i + W_i) \right) \left(\frac{\varepsilon}{c_{q+1}} \right)$$

$$\forall q + 2 \leqslant i \leqslant m \qquad \alpha_i = 0$$

and the throughput is equal to

$$\rho = \sum_{i=1}^{p} \frac{\alpha_i \cdot W_{total}}{T_p} = \left(1 - \frac{\sum_{i=1}^{p} C_i + W_i}{T_p} \right) \rho_{\text{opt}} \text{ with } \rho_{\text{opt}} = \sum_{i=1}^{q} \frac{1}{w_i} + \frac{\varepsilon}{c_{q+1}}.$$

To prove the asymptotic optimality of this algorithm, we need an upper bound on the optimal throughput ρ^*. This upper bound can be obtained by removing all latencies (i.e., $C_i = 0$ and $W_i = 0$ for any worker i). By definition, we have $\rho^* \leqslant \rho_{\text{opt}}$. If we call T^* the optimal time needed to process B load units, then we have

$$T^* \geqslant \frac{B}{\rho^*} \geqslant \frac{B}{\rho_{\text{opt}}}.$$

Let T denote the time needed by the proposed algorithm to compute the same workload B. The first period is dedicated to communications and is lost for processing, so $k = \left\lceil \frac{B}{\rho \cdot T_p} \right\rceil + 1$ periods are required for the whole computation.

We have $T = k \cdot T_p$, therefore:

$$T \leqslant \frac{B}{\rho} + 2 \cdot T_p \leqslant \frac{B}{\rho_{\text{opt}}} \left(\frac{1}{1 - \sum_{i=1}^{p} \frac{C_i + W_i}{T_p}} \right) + 2 \cdot T_p,$$

and, if $T_p \geqslant 2 \cdot \sum_{i=1}^{p} C_i + W_i$,

$$T \leqslant \frac{B}{\rho_{\text{opt}}} + 2 \cdot \frac{B}{\rho_{\text{opt}}} \sum_{i=1}^{p} \frac{C_i + W_i}{T_p} + 2 \cdot T_p$$

and if T_p is equal to $\sqrt{\frac{B}{\rho_{\text{opt}}}}$,

$$\frac{T}{T^*} \leqslant 1 + 2 \left(\sum_{i=1}^{p} (C_i + W_i) + 1 \right) \frac{1}{\sqrt{T^*}} = 1 + O \left(\frac{1}{\sqrt{T^*}} \right).$$

That suffices to show the asymptotic optimality of the proposed algorithm. ∎

Maximum benefit of multi-round algorithms. Using multi-round algorithms brings new difficulties to an already difficult problem. It is worthwhile to assess how much such algorithms can improve the solution. The answer is given by the following result:

THEOREM 8.5
Consider any star-shaped master-worker platform where communication cost and computation cost each follows either a linear or an affine model. Any multi-round schedule cannot improve an optimal single-round schedule by a factor greater than 2.

PROOF Let S be any optimal multi-round schedule, using a finite number K of rounds. We have m workers in our platform, and the master allocates a fraction $\alpha_i(k)$ to worker i ($1 \leqslant i \leqslant m$) during the k-th round. Let T denote the total completion time obtained by S. From S we build a new schedule S' which sends in a single message a fraction $\sum_{k=1}^{K} \alpha_i(k)$ to worker i (the messages are sent in an arbitrary order.) The master does not spend more time communicating under S' than under S. Therefore, no later than time T all workers will have finished to receive their work. No worker will spend more time processing its fraction of the load under S' than under S (the loads have same sizes). Therefore, none of them will spend more than T time-units to process its load under S'. Therefore, the makespan of the single round schedule S' is no greater than $2T$.

□

8.3.3 Return Messages

In previous sections, we assumed that computations required some input data but produced a negligible output, so we did not take its transmission back to the master into account. This assumption could be unduly restrictive for those computations producing large outputs, such as cryptographic keys. In this section, we incorporate return messages into the story. Which properties are still valid?

In the general case, there is no correlation between input and output sizes, but we simplify the problem by assuming the same size for input and output messages. In other words, if M needs $c_i \alpha_i W_{total}$ time units to send the input to worker P_i, P_i needs the same time to send the result back to M after having completed its computation. The communication medium is supposed to be bi-directional (as most of network cards are now full-duplex), so the master M can simultaneously send and receive data.

In our first framework with linear cost models and distribution in a single round, all workers participated in the work and we were able to find an optimal order to distribute data. If we allow return messages, we have two new

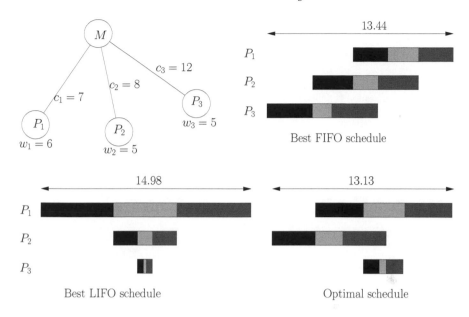

FIGURE 8.16: Optimal order can be neither FIFO nor LIFO.

issues: the order of return messages could be different from the distribution order and several workers could remain idle during the whole computation. Two simple ordering strategies are the FIFO strategy (return messages are sent in the same order as the input messages) and the LIFO strategy (return messages are sent in reverse order). In fact, several examples are exhibited in [6], where the optimal order for return messages is neither FIFO or LIFO, as illustrated by Figure 8.16, or where the optimal makespan is reached with some idle processors, as illustrated by Figure 8.17. The best FIFO and LIFO distribution are easy to compute, since all processors are involved in the work, they are served in non-decreasing values of communication times and do not remain idle between the initial message and the return message [6]. Furthermore, all FIFO schedules are optimal in the case of a bus-shaped platform [1].

Moreover, the distribution of data in a single round induces long waiting times, and a multi-round distribution could really improve this drawback. Regrettably, any optimal multi-round distribution for the linear cost model uses an infinite number of rounds. Thus, affine cost models are required to have realistic solutions, and the problem then becomes very hard to solve or even to approximate.

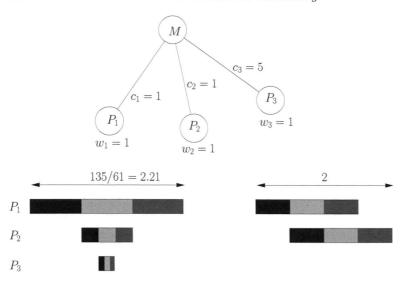

LIFO, best schedule with 3 processors   FIFO, optimal makespan with 2 processors

FIGURE 8.17: Example with an idle worker.

8.4   Conclusion

In this chapter, we have dealt with the Divisible Load model, a simple and useful relaxation to many scheduling problems. A general applicative example, the execution of a distributed computation made of independent tasks on a star-shaped platform, has been used as a guideline through this chapter. Without any relaxation, this example is a tractable problem, but the known solution to this problem is only partial and has a large computational complexity. Moreover, the linear cost function used for communications and computations, and the homogeneous communication model, limit the practical significance of this approach. We showed how to use the Divisible Load theory to simplify the problem and then solve it completely. Because we have simplified the problem with the Divisible Load relaxation, we can afford to use a more complicated model: we can deal with heterogeneous communication links and/or we can include latencies in communications and computations. This new model is more realistic, but still tractable thanks to the Divisible Load approach. However, it also has its own limits! Indeed, we have seen that problem complexity quickly increases as soon as we add latencies or return messages.

References

[1] M. Adler, Y. Gong, and A. Rosenberg. On "exploiting" node-heterogeneous clusters optimally. *Theory of Computing Systems*, 42(4):465–487, 2008.

[2] D. T. Altilar and Y. Paker. An optimal scheduling algorithm for parallel video processing. In *International Conference on Multimedia Computing and Systems*, pages 245–248, July 1998.

[3] D. T. Altilar and Y. Parker. Optimal scheduling algorithms for communication constrained parallel processing. In *Euro-Par 2002 Parallel Processing*, volume 2400 of *LNCS*, pages 197–206, 2002.

[4] O. Beaumont, H. Casanova, A. Legrand, Y. Robert, and Y. Yang. Scheduling divisible loads on star and tree networks: Results and open problems. Research report 4916, INRIA, 2003.

[5] O. Beaumont, H. Casanova, A. Legrand, Y. Robert, and Y. Yang. Scheduling divisible loads on star and tree networks: Results and open problems. *IEEE Transactions on Parallel and Distributed Systems*, 16(3):207–218, Mar. 2005.

[6] O. Beaumont, L. Marchal, and Y. Robert. Scheduling divisible loads with return messages on heterogeneous master-worker platforms. In *International Conference on High Performance Computing HiPC'2005*, pages 123–132. Springer-Verlag, 2005.

[7] D. P. Bertsekas. *Constrained optimization and Lagrange Multiplier methods*. Athena Scientific, 1996.

[8] V. Bharadwaj, D. Ghose, and V. Mani. Multi-installment load distribution in tree networks with delays. *IEEE Transactions on Aerospace and Electronic Systems*, 31(2):555–567, Apr. 1995.

[9] V. Bharadwaj, D. Ghose, V. Mani, and T. G. Robertazzi. *Scheduling Divisible Loads in Parallel and Distributed Systems*. IEEE Computer Society Press, 1996.

[10] J. Błażewicz, M. Drozdowski, and M. Markiewicz. Divisible task scheduling - concept and verification. *Parallel Computing*, 25(1):87–98, Jan. 1999.

[11] M. Drozdowski. Selected Problems of Scheduling Tasks in Multiprocessor Computer Systems. Ph.D. thesis, Instytut Informatyki Politechnika Poznanska, Poznan, 1997.

[12] S. Genaud, A. Giersch, and F. Vivien. Load-balancing scatter operations for grid computing. *Parallel Computing*, 30(8):923–946, 2004.

[13] R. Graham, D. Knuth, and O. Patashnik. *Concrete Mathematics: A Foundation for Computer Science.* Wesley, 1994.

[14] C.-K. Lee and M. Hamdi. Parallel image processing applications on a network of workstations. *Parallel Computing*, 21(1):137–160, 1995.

[15] T. G. Robertazzi. Ten reasons to use divisible load theory. *Computer*, 36(5):63–68, May 2003.

[16] T. Saif and M. Parashar. Understanding the behavior and performance of non-blocking communications in MPI. In *Euro-Par 2004 Parallel Processing*, volume 3149 of *LNCS*, pages 173–182. Springer, Dec. 2004.

[17] A. Schrijver. *Theory of Linear and Integer Programming.* John Wiley & Sons, New York, 1986.

[18] Y. Yang and H. Casanova. Extensions to the multi-installment algorithm: Affine cost and output data transfers. Technical Report CS2003-0754, Dept. of Computer Science and Eng., Univ. of California, San Diego, July 2003.

[19] Y. Yang, H. Casanova, M. Drozdowski, M. Lawenda, and A. Legrand. On the complexity of multi-round divisible load scheduling. Research Report 6096, INRIA, Jan. 2007.

[20] Y. Yang, K. van der Raadt, and H. Casanova. Multiround algorithms for scheduling divisible loads. *IEEE Transactions on Parallel and Distributed Systems*, 16(11):1092–1102, 2005.

Chapter 9

Multi-Objective Scheduling

Pierre-François Dutot
Université de Grenoble

Krzysztof Rzadca
Polish-Japanese Institute of Information Technology

Erik Saule
Université de Grenoble

Denis Trystram
Université de Grenoble

Abstract   This chapter considers multi-objective scheduling, i.e., scheduling with simultaneous optimization of several objectives. The main motivation to optimize more than one objective is the growing complexity of modern systems. Characterizing by one variable only the performance of a heterogeneous system with many users shows a narrow view of the system. In multi-objective scheduling, one can explicitly model, optimize, and find trade-offs between various performance measures.

We introduce multi-objective scheduling on three motivating problems to illustrate different approaches. In MAXANDSUM, a scheduler optimizes both the makespan and the sum of completion times. In EFFICIENTRELIABLE, the goal is to find the trade-off between the makespan and the reliability of schedule on failing processors. Finally, in TWOAGENTMINSUM, a processor must be shared fairly between jobs produced by two independent users.

We study the complexity of these three problems and propose different approaches to obtain an acceptable solution. In MAXANDSUM, an algorithm finds a single schedule that approximates both the minimal makespan and the minimal sum of completion times. In EFFICIENTRELIABLE, all the Pareto-optimal solutions are approximated by a linear number of schedules. In TWO-AGENTMINSUM, the multi-objective problem becomes NP-hard (whereas single objective problems are easy). Here, we propose an axiomatic way to characterize schedules "fair" to individual users.

9.1   Motivation

Optimizing one objective over a set of constraints has been widely studied for many combinatorial optimization problems including scheduling problems. There exists a host of problems related to scheduling. In parallel processing, the fundamental problem is, informally, to determine when and where to execute tasks of a program on a target computing platform. Scheduling is a hard problem that has been extensively studied in many aspects, theoretical as well as practical ones. Many efficient solutions have been proposed and implemented for a large spectrum of applications on most existing parallel and distributed platforms.

In this chapter, we focus on scheduling when several objectives have to be optimized simultaneously.

9.1.1   Once Upon a Time

Let us start by a short story that comes from an imaginary world. Some time ago, a manager of a software company asked a developer (let us call him *the hare*) to implement a program that runs fast on a given parallel supercomputer. The program was so smart that the hare was winning all the performance races during several years. Clients were delighted. Moreover, during many years, the hare succeeded to adapt the program to successive generations of parallel platforms. However, the systems gradually evolved to heterogeneous components linked by long distance connections. Hare's program was still the fastest one, but it was unable to cope with crashing processors, a more and more frequent situation on such large-scale heterogeneous computers.

Meanwhile, another developer (let us call her *the tortoise*) wrote a very simple program that was able to deliver a solution in almost all crash situations. The tortoise's program simply used only the most reliable processor. Of course, this solution was rather slow, but even so, clients were satisfied.

The hare worked hard and finally developed a new version of his program, a bit less reliable than the tortoise's, but still much faster. Clients were satisfied, as the new version was a good compromise between these two factors. After a few months, however, the main client started to work with critical applications that needed much higher reliability. This time, the hare was too tired to react and to develop yet another version of his code. Gradually, the hare's code fell into oblivion and the hare himself was forced to leave the company.

In the reminder of this chapter, we will formalize the main concepts hidden in this story. Using more classic vocabulary, schedules can be evaluated differently, depending on the goals and the context of the end user. These evaluations are usually modeled by objective functions, simple mathematical expressions that take into account the proposed schedule and the parameters of platform and jobs. Using some concrete examples, we will analyze a number of such situations, in which a scheduler optimizes several objective functions.

9.1.2 Diversity of Objectives

Although most existing studies concern optimizing the time the last job ends (the makespan), many other possible objectives exist. In this section, we recall the objective functions which are frequently used in the scheduling community (most of them are studied in other chapters of this book). We also present some exotic functions to give an insight on how to define different objectives which can be optimized simultaneously with a classic objective. Some of these objectives only make sense when used in a multi-objective setting, as they are trivial to optimize in a single objective problem.

The most common objectives use the completion time of each job C_i. This completion time can be considered on its own or coupled with another time parameter, such as the job's release date r_i or the job's deadline d_i. Completion time depends on the processing time of job i on processor j (denoted by $p_{i,j}$). It can also be considered together with an external parameter ω_i which is a weight associated to job i. All these values are related to a single job. To measure the efficiency of a schedule applied to all the jobs, we can either consider the maximum of values over all the jobs, the sum on all the jobs (with weights if necessary), or any other combination which may be relevant.

Below are listed the most popular examples of objectives involving completion time of jobs:

- Makespan: $C_{max} = \max_i C_i$,

- Minsum[1]: $\sum_i C_i$ or weighted minsum $\sum_i \omega_i C_i$ [12],

- Weighted sum flow: $\sum_i \omega_i (C_i - r_i)$,

---

[1]In numerous articles the minsum criteria and sum flow are confounded and called flow-time, since the (weighted) sum of release dates is a constant. However, for approximation purposes, we need to clearly distinguish both.

- Max stretch: $\max_i \frac{C_i - r_i}{p_{i,j}}$ [7],

- Number of late activities (or tardy jobs): $\sum_i U_i = \sum_i \delta(C_i - d_i)$ where $\delta(x)$ equals 0 when $x \leqslant 0$, and 1 otherwise.

In some problems, when optimizing an objective expressed as a function of time, completing a job is useless unless it meets its deadline. It is thus possible to reject some tasks. In this case, the number of rejected jobs and their associated costs can be optimized.

Even if some of these objectives are similar in nature, solving a scheduling problem can be easy for a certain objective, while NP-hard for a different one. For instance, minimizing the sum of completion times for scheduling independent jobs on a fixed number of machines is easy, while minimizing the makespan or the weighted sum is hard [9].

With more elaborate computational models, more exotic objectives can be formed that take into account more than the completion time. For example, taking into account a cost function which can depend on which processors are used, we can be interested in minimizing the overall budget [35]. In embedded systems, the energy consumption is a key issue to increase battery life. Different schedules can use more or less energy depending on how many processors are used and how processor frequencies are scaled [19, 10, 2]. Another possibility is to consider processor failures and to optimize the reliability, i.e., the probability that the job is completed successfully [34, 38].

Today, there is an increasing interest for considering a large variety of objectives. Indeed, the tremendous power of new computing platforms gave raise to new needs. The real challenge is to take into account the diversity of points of view into combinatorial optimization problems. The trend today is even to propose one's own objective which has not previously been considered!

9.1.3   Motivating Problems

In this chapter, we will illustrate all concepts and methods on three basic multi-objective scheduling problems that are presented below.

In MAXANDSUM [39], n independent jobs have to be scheduled on m identical processors. The problem is to minimize simultaneously the makespan ($C_{\max} = \max C_i$) and the sum of completion times $\sum C_i$. MAXANDSUM will serve as a basis for illustrating the basic definitions and it will be solved using zenith approximation.

In EFFICIENTRELIABLE (derived from [15]), n independent jobs have to be scheduled on m processors. Processors have the same speed, however, they do not have the same reliability. They are subject to crash faults, described by a failure rate λ_j. We denote by C^j the time when the last job is executed on P_j. The optimization problem is to minimize both the makespan and the reliability index of the schedule : $rel = \sum_j \lambda_j C^j$. The latter objective is directly related to the probability of failure of the schedule (see [15] for more details). This problem does not admit zenith approximation algorithm and will be solved using Pareto set approximation.

Finally, in TwoAgentMinSum [1], two *agents* A and B submit two sets of jobs to the same processor. Each agent aims at minimizing the sum of completion times of their jobs denoted respectively by $\sum C_i^A$ and $\sum C_i^B$. The notion of fairness will be illustrated on this problem.

9.1.4 Summary of Results on Single Objective Problems

We start by recalling some basic, well-established results on the example problems described in the previous section.

- Minimizing C_{max} for n independent jobs on m identical parallel machines.

  The C_{max} objective is the most popular objective and the most studied one for many variants of the scheduling problem. This problem has been proved to be NP-hard even for $m = 2$ [18]. However, the well-known Graham's list algorithm is a 2-approximation for any values of m (see the proof in Section 2.3.2).

- Minimizing $\sum C_i$ for n independent jobs on m identical machines.

  It is easy to prove that the algorithm that sorts the jobs by non-decreasing order of their processing times is optimal (see Section 4.4). This policy is known as the Shortest Processing Time (SPT).

- Minimizing the reliability index.

  This problem is easy since it exists a straightforward algorithm that computes the optimal solution: allocate all jobs on the most reliable machine.

Let us now briefly discuss how to obtain solutions for more general scheduling problems in a context of optimizing a single objective (without loss of generality, let us consider a minimization problem). Solutions may be obtained by exact methods, purely heuristic methods, or approximation methods. As most "interesting" problems are NP-hard, exact methods, like branch-and-bound, cannot be applied in practice. Thus, only approached solutions can be computed. Purely heuristic methods have no theoretical bounds that can be provided for all the instances of the problem. Moreover, they usually lack the formal framework for the analysis. The assessment is done by usage. That is why we prefer to focus on approximation algorithms (see Chapter 2) with a guaranteed worst-case performance regarding the optimal solutions.

9.1.5 Beyond the Scope of This Chapter

This chapter is focused on multi-objective optimization *for parallel systems*. Therefore, we left unaddressed many existing works on multi-objective scheduling in other contexts (such as job-shop scheduling). Different scheduling problems can be found in [42]. The authors study the complexity of these problems and provide some algorithms. However, approximation issues are not discussed.

In this chapter, we have a strong inclination towards the design of efficient algorithms, and therefore we will not provide non-polynomial algorithms, even if they find exact solutions.

There are also many studies in the context of networking and communication scheduling [24]. The problems are usually solved with *ad-hoc* solutions that are hard to generalize or to adapt to other problems.

From a different perspective we could look for solutions according to hidden objectives, either held by a single oracle which sorts solutions according to his/her preferences, or held by multiple players who compete for shared resources (with their own personal view of what is most desirable or profitable). The former setting is related to decision making, while the latter is related to cake division. In decision making only one solution is provided while we will always try to give a broader perspective of the possible solutions. In cake division, the classic approach focuses on preventing jealousy among users [41]. This can be very different from optimizing classic objectives as sometime giving less value to everyone prevents jealousy.

9.1.6   Chapter Organization

This section provided an informal introduction to multi-objective scheduling, along with some motivating problems. In the next section, we will consider the multi-objective optimization problem more formally. We define the multi-objective scheduling problem and the notion of Pareto-optimality.

We first discuss the main concepts and definitions of multi-objective optimization in Section 9.2.

Section 9.3 describes the existing approaches to multi-objective optimization. Subsection 9.3.1 discusses the most commonly used approaches for building algorithms solving multi-objective problems. Then, Subsection 9.3.2 discusses some complexity issues of such problems.

In Section 9.4 we describe an approximation algorithm for MAXANDSUM problem. The goal is to find a single solution that approximates the minimal makespan and, at the same time, the minimal sum of completion times.

Section 9.5 considers a different approach to solving a multi-objective problem, in which the result of the algorithm is an approximation of the whole set of Pareto-optimal solutions (instead of a single solution). The formal definition of the notion of approximate Pareto set is given (Section 9.5.2). Then, we present a general technique that produces an approximated Pareto set whose size is linear and we describe an algorithm that uses this approach for EFFICIENTRELIABLE problem in Section 9.5.3.

Finally, in Section 9.6 we consider how to ensure fairness by multi-objective optimization. Using TWOAGENTMINSUM, we assume that the system is used by many independent agents, who must be treated fairly. In the resulting multi-objective problem, the objectives correspond to performance measures of individual agents. Section 9.6.2 proposes the notion of axiomatic fairness, that describes a "fair" solution by axioms, that put further restrictions on Pareto-optimal solutions.

9.2 What Is Multi-Objective Optimization?

It is impossible to provide a simple definition of all the possible multi-objective optimization problems due on one hand to the variety of objectives and variables used to express them and on the other hand to the variety of problems which cannot be united into a single formulation. In this chapter, we propose a simplified definition which can be used for our motivating problems. Without loss of generality, we assume that all problems concern the minimization of objectives.

DEFINITION 9.1 *Multi-objective optimization problem*
Let m be the number of processors, denoted by $P_1, \ldots, P_m$. Let n be the number of jobs, denoted by $J_1, \ldots, J_n$, with processing time $p_{i,j}$ for job J_j on processor P_i.

The multi-objective optimization problem consists of finding starting times *S_j for all jobs and a function π that maps jobs to processors ($\pi(j) = i$ if J_j is scheduled on P_i), such that the processors compute jobs one at a time:*

$$\forall j, j' \text{ if } \pi(J_j) = \pi(J_{j'}) \text{ then } C_j \leqslant S_{j'} \text{ or } C_{j'} \leqslant S_j$$

and the objective functions are minimized:

$$\min \left(f_1(\pi, S_1, \ldots, S_n), \ldots, f_k(\pi, S_1, \ldots, S_n) \right)$$

We illustrate all the definitions of this section on an instance of MAXAND-SUM [39] problem.

Example 9.1
The instance is composed of three processors and nine tasks. All processors are identical, and thus tasks have the same processing time wherever they are scheduled (given in Table 9.1). The problem is to minimize the two following objective functions :

- $f_1(\pi, S_1, \ldots, S_n) = \max_i (S_i + p_i) = \max_i C_i$
- $f_2(\pi, S_1, \ldots, S_n) = \sum_i (S_i + p_i) = \sum_i C_i$

Table 9.1: Processing times of the nine tasks of our typical instance

Task number	1	2	3	4	5	6	7	8	9
Processing time	1	2	4	8	16	32	64	128	128

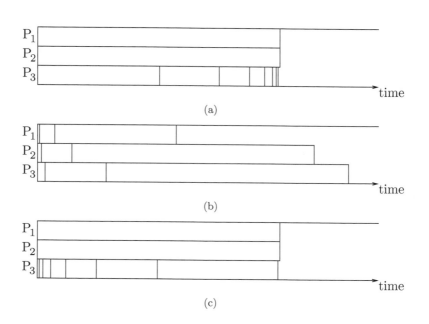

FIGURE 9.1: Some possible schedules obtained with classic algorithms (a) LPT (largest processing times first), (b) SPT (shortest processing times first) and (c) a modified version of LPT where tasks are first assigned to processors according to LPT and then reordered on each processor according to SPT.

Any feasible schedule is related to a k-tuple of values of objective functions. In the rest of the chapter, we may implicitly replace the schedule by its k-tuple. Conversely, if there exists a schedule for a given k-tuple, we may indifferently use either the k-tuple or one of the associated schedules. The schedules presented on Figure 9.1 can either be referred to as LPT, SPT, and LPT-SPT (see caption for details) or their objective values ((128, 1025), (164, 453), and (128, 503), respectively).

The notion of **Pareto-dominance** [44] formalizes the notion of best compromise solutions in multi-objective optimization. Solution σ^* Pareto-dominates solution σ if σ^* is as good as σ on all objective values and better on at least one objective.

DEFINITION 9.2 Pareto-dominance   *Let $\sigma = (\pi, S_1, \ldots, S_n)$ and $\sigma' = (\pi', S_1', \ldots, S_n')$ be two solutions to the multi-objective optimization problem.*

$$\sigma \ Pareto\text{-}dominates \ \sigma' \Leftrightarrow \forall l, f_l(\pi, S_1, \ldots, S_n) \leqslant f_l(\pi', S_1', \ldots, S_n').$$

Example 9.2
In our example, the LPT-SPT schedule (Fig. 9.1(c)) Pareto-dominates the LPT schedule (Fig. 9.1(a)).   ⬚

Note that the notion of Pareto-dominance defines a partial order on the set of all solutions. Indeed, two solutions are **Pareto-independent** when none of them Pareto-dominates the other one. Solution σ^* is **Pareto-optimal** if it is not dominated by any other solution.

Example 9.3
In our example, the SPT schedule (Fig. 9.1(b)) and the SPT-LPT schedule (Fig. 9.1(c)) are Pareto-independent.   ⬚

DEFINITION 9.3 Pareto-optimality
Let $\sigma = (\pi, S_1, \ldots, S_n)$ be a solution to the multi-objective optimization problem. Solution σ is Pareto-optimal if and only if:

$$\forall \sigma' = (\pi', S_1', \ldots, S_n') \neq \sigma$$
$$\exists k, \ f_k(\pi, S_1, \ldots, S_n) > f_k(\pi', S_1', \ldots, S_n')$$
$$\Rightarrow \exists l, \ f_l(\pi, S_1, \ldots, S_n) < f_l(\pi', S_1', \ldots, S_n')$$

Example 9.4
The schedule LPT-SPT with objective values $(128, 503)$ (Figure 9.1(c)) is Pareto-optimal since:

- No schedule can have a better makespan because the longest tasks requires 128 units of time;

- To achieve this makespan, each of the longest tasks must be placed alone on a processor. All the remaining tasks must be placed on the remaining processor;

- The remaining tasks are sorted according to the SPT rule which is known to be optimal for $\sum C_i$ on a single processor.

☐

Note that all Pareto-optimal solutions are Pareto-independent.

DEFINITION 9.4 Pareto set $\mathfrak{P}$
The **Pareto set** *(denoted by $\mathfrak{P}$) of a problem is the set of all Pareto-optimal solutions.*

$$\sigma \in \mathfrak{P} \Leftrightarrow \sigma \text{ is Pareto-optimal}$$

From all Pareto-optimal solutions, we can derive the optimal value achievable independently on each objective. Using these values, we construct the **zenith solution** that is optimal on all objectives. The zenith solution Pareto-dominates all the Pareto-optimal solutions. In almost all problems, such a solution is infeasible. However, this solution often serves as a reference to measure the quality of solutions returned by an optimization algorithm.

Example 9.5
The Pareto set is represented in Figure 9.2 with diamonds. It is composed of eight points. Note that not all the points are on the convex hull of the possible solutions. For one point, we marked by gray rectangles dominated solutions (in the upper right) and dominating solutions (in the lower left). Since there are no solutions in the dominating rectangle, the solution is Pareto optimal.
☐

9.3   Overview of the Various Existing Approaches

Existing studies concerning multi-objective scheduling optimization can be decomposed in two groups. The first one is algorithmic, based on the design and analysis of efficient heuristics to solve the multi-objective problems. The second group is more theoretical and centered on the characterization of the problems themselves, mainly on complexity issues. A nice survey about multi-objective scheduling has been written by Hoogeveen [21].

9.3.1   Algorithms Building One Trade-off Solution

The simplest algorithm analyses are done on simulations or experiments, assessing the efficiency of existing single objective algorithms on the different objectives (see for example [11]). The main asset of such studies are that they give an insight on the average behavior of the heuristics using a sample

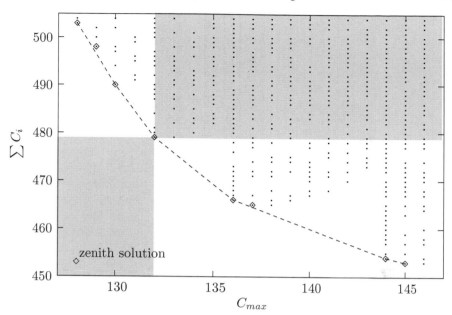

FIGURE 9.2: Some of the possible solutions for the instance used in this section. All possible solutions with makespan lower than 146 and $\sum C_i$ lower than 503 are included, each depicted as a single dot. The Pareto set is depicted by eight diamonds, while the convex hull of the possible solution set is drawn as a dashed line. Remark that it connects only six Pareto-optimal points. The zenith solution depicted in the lower left corner is not a feasible solution in this example.

of instances, as long as this sample is carefully selected to be representative of all possible instances. This selection is in itself a difficult problem that is partially addressed in [17].

A more sophisticated approach is to extend the classic single objective approach by aggregating several objective functions artificially. The first aggregations which come to mind is to optimize the sum of two objectives, or the maximum of both (see for instance [6, 5]). However, if the range of the objective values differ too much, some scaling has to be applied to compensate for the possible imbalance. The aggregation possibilities are infinite, and many have a specific meaning and need to be finely tuned to the considered problem. Many properties can be defined to put forward efficient aggregations [14]. Among the most classic one is the Pareto-optimality of the optimal solution for the aggregation. This property is for example true for all linear combinations of objectives. When representing the solution space as an k-dimensional space, some aggregation functions are closely related to Euclidean norms. There is an inherent limitation to the linear aggregation functions: the Pareto sets are not necessarily convex and thus, some Pareto-optimal solutions might not be reachable by any such linear functions. For example on Figure 9.2 only six solutions belong to the convex hull (solutions $(129, 498)$ and $(137, 465)$ are not). In the worst cases, only two extremal solutions can be found, the rest of the Pareto solutions being above the convex hull. Note that because usually Pareto sets are very close to the convex hull this technique is very useful in a first attempt to solve a multi-objective problem.

A natural way to map a multi-objective problem into a single objective one is to treat each objective separately, sorting the objective functions from the most important to the least important. In this approach the solutions are ranked according to the lexicographic order of their objective values, restoring a complete order on the k-tuples representing the solutions. An application of this technique can be found in [20].

Another possibility is to transform some objectives (usually, all but one) into constraints. Such an approach has been successfully used in [40]. Most of the time, the overall problem is solved using metaheuristic or integer linear programming in which adding constraints is easy. The main drawback of this approach is that the solving process is not really simplified. However, it can be natural for some problems. For instance, using the problem EFFICIENT-RELIABLE applied in the automotion field, the constraint on the makespan emerges from human braking response time.

Finally, integrated approaches aiming at approximating simultaneously all objectives have been proposed. For instance, an approximation algorithm for MAXANDSUM has been proposed in [39]. Most of the time, a trade-off parameter allows the user to tune the approximation ratio. Using such a technique, the system designer can control the degradation on each objective and make it fits into his/her applicative requirement. Another integrated approach is to look for solutions that fulfill constraints on a part of objective functions and that optimize the other objective functions. For instance, Shmoys and Tardös studied a problem of scheduling tasks on unrelated processors where executing a task on a processor induces a cost. They proposed an algorithm that, given a makespan threshold, finds a schedule that optimizes the cost of a schedule without doubling the makespan threshold [35]. These two techniques are described in more details in Section 9.4 and Section 9.5.

Let us remark that all the previous examples concern the optimization of two objectives. The problem of optimizing k (fixed) objectives does not seem to be so different (obviously, the analysis would become more technical). However, adding a new objective diminishes the chance to obtain a tractable solution for the global approximation of the Pareto set in a single point. Let us point out some papers dealing with three objectives [42, 28, 33, 24].

9.3.2   Complexity Issues

Remark first that in the multi-objective field, the pertinent problems are *optimization* problems, and not decision ones. However, like in classic single objective optimization, the decision version of an optimization problem gives insights about the hardness of the problem. It is straightforward that if the decision version of a single objective problem is NP-complete, then the multi-objective decision problem is NP-complete too. However, a multi-objective optimization problem could be hard while all single-objective optimization problems could be solved in polynomial time. For instance, the single objective counterparts of TWOAGENTMINSUM are both easy. Indeed, the SPT rule solves each single-objective problem optimally. However, the multi-objective decision problem associated to TWOAGENTMINSUM is NP-complete [1]. We now provide a sketch of the NP-completeness proof of the decision version of TWOAGENTMINSUM.

THEOREM 9.1
The decision version of TWOAGENTMINSUM *is NP-Complete [1].*

PROOF   The proof is a reduction from PARTITION.

First, let us remark that the decision problem belongs to $\mathcal{NP}$.

The decision version of TWOAGENTMINSUM is the problem of determining whether, for given sets of jobs and budgets Q^A and Q^B, there is a feasible schedule in which $\sum C_i^A \leqslant Q^A$ and $\sum C_i^B \leqslant Q^B$. Given an instance of PARTITION with set of elements $\mathcal{P} = \{s_i\}$, where the s_i are sorted in non-decreasing order, we construct an instance of TWOAGENTMINSUM, in which both agents have identical sets of jobs with $p_i^A = p_i^B = s_i$ and $Q = Q^A = Q^B = \frac{3}{2}S + 2(\sum_i(n-i)s_i)$, where $S = \sum_i s_i$.

First recall that a SPT schedule is a schedule that executes jobs in SPT order. Since both agents submit the same set of jobs each job has an identical twin in the other agent set. The goal of this reduction is to show that a schedule which does not exceed either budget is necessarily a SPT schedule where total processing time of jobs from agent A scheduled right before their counterpart from agent B is equal to the total processing time of jobs from agent A scheduled right after their counterpart.

Without loss of generality, we assume that all s_i are different. Note that there are 2^n SPT schedules in this instance. Firstly, in all SPT schedules, jobs J_i^A and J_i^B are both scheduled during time interval $[2\sum_{j<i} s_j; 2\sum_{j\leqslant i} s_j]$. Straightforwardly, the total sum of completion times of a SPT schedule is $\sum_i C_i^A + \sum_i C_i^B = 2Q$. Secondly, in any feasible schedule $\sum_i C_i^A \leqslant Q$ and $\sum_i C_i^B \leqslant Q$ and therefore $\sum_i C_i^A + \sum_i C_i^B \leqslant 2Q$. A swap argument proves that any feasible schedule is a SPT schedule.

In a schedule, let us define a marker $x(i)$, such that $x(i) = 0$ if J_i^A is executed before J_i^B, and $x(i) = s_i$ otherwise. Let $x = \sum_i x(i)$. Using x in such a schedule $\sum_i C_i^A = x + S + 2\sum_i(n-i)s_i$, and $\sum_i C_i^B = (S-x) + S + 2\sum_i(n-i)s_i$. As both budgets are lower or equal to Q, we have $x = \frac{S}{2}$, i.e., x must be equal to half of the sum of elements in PARTITION. Now, having a feasible schedule, it is straightforward to construct a solution to PARTITION:

if J_i^A is executed before J_i^B, s_i is added to $\mathcal{P}_1$, otherwise it is added to $\mathcal{P}_2$. Conversely, having $\mathcal{P}_1$ and $\mathcal{P}_2$ with sums of elements of $\frac{S}{2}$, we execute J_i^A before J_i^B iff $s_i \in \mathcal{P}_1$. $\qquad\qquad\qquad\qquad\qquad\qquad\qquad\qquad\qquad\qquad\qquad$ ▯

In practice, the central problem in multi-objective optimization is to choose a trade-off between all feasible solutions. Thus, one reasonable answer is to provide all the Pareto optimal solutions. We can distinguish three different questions. The first one is how many Pareto-optimal solutions there are. This number can be exponential in the size of the instance. For example, the number of Pareto-optimal solutions of an instance of TwoAgentMin-Sum can be in $\Omega(2^n)$. Problems that deliver the number of Pareto-optimal solutions are called **counting problems**. The second question is to provide all the Pareto-optimal solutions. Such problems are called **enumeration problems**. Finally, the last question is to approximate the Pareto set by a limited number of solutions. In the remainder of this section, we discuss successively these three questions.

The complexity class $\#P$ contains counting problems that can be solved in finite time where verifying if a solution is valid can be done in polynomial time [42]. Note that a problem belonging to $\#P$ needs to be solved in finite time, but not in bounded time. We can consider that $\#P$ is for counting problems what $\mathcal{NP}$ is for decision problems. $\mathcal{FP}$ is the sub-class of $\#P$ where counting problems are solvable in polynomial time.

Counting problems are only interesting to obtain insight about the difficulty of enumeration problems. From the point of view of complexity classes, the NPO class is generalized to $\mathcal{ENP}$ for enumeration problems [42]. A problem belongs to $\mathcal{ENP}$ if there exists an algorithm that checks whether a set of solutions is the Pareto set with a processing time bounded by a polynomial function of the instance's size and the number of Pareto-optimal solutions. A problem of $\mathcal{ENP}$ that can be solved in polynomial time (in the size of the instance and the number of Pareto-optimal solutions) belongs to $\mathcal{EP}$.

Let us note that such classes are not so useful in practice. The number of Pareto optimal solutions is not a criterion to classify problems. For instance, TwoAgentMinSum belongs to $\mathcal{EP}$, which is the most basic class. However, enumerating the Pareto set takes an exponential time since there can be an exponential number of Pareto optimal solutions.

More details about complexity classes for counting and enumeration problems can be found in [43].

Approximability issues are not well formalized in the context of the multi-objective optimization. It is often possible to derive some inapproximability results. Most of the time, such results are not complexity results (they do not use the $\mathcal{P} \neq \mathcal{NP}$ assumption) but are derived from the non-existence of solutions achieving a given objective value. To our knowledge, nobody investigated the extension of the classic $\mathcal{APX}$ or $\mathcal{PTAS}$ classes to multi-objective problems.

9.4 Zenith Approximation on MaxAndSum

This problem has already been introduced in Section 9.1.3 and used as an illustration for the general definitions of Section 9.2. Both objectives are directly derived from the completion times of the tasks. Considered separately, both objectives have been extensively studied, and approximation or optimal algorithms have been designed.

For the simple case of n independent jobs on m processors, let us first recall that shortest processing time (SPT) is an optimal algorithm on the sum of completion times and that List Scheduling is a 2-approximation algorithm for the makespan. SPT is a List Scheduling algorithm, therefore, $\sum C_i^{SPT} \leqslant \sum C_i^*$ and $C_{max}^{SPT} \leqslant 2C_{max}^*$. The SPT solution is a $(2,1)$-approximation of the zenith solution. Usually, such a result is stated as follows.

THEOREM 9.2
SPT is a $(2,1)$-approximation algorithm of MaxAndSum.

This simple result does not hold in more complex problems such as problems involving parallel tasks or in presence of release times. Moreover, it is not possible to derive from SPT an algorithm having better a guarantee on the makespan. In the remainder of this section, we expose a framework that was originally proposed by Stein [39]. The main idea behind this technique is to optimize the sum of completion times by focusing on the task at the beginning of the schedule, leaving the last task to be scheduled according to the makespan objective.

Consider you have two schedules namely, σ_{cmax} a ρ_{cmax}-approximated schedule on the makespan and σ_{minsum} a ρ_{minsum}-approximated schedule on the sum of completion times. The framework of Stein mixes these two schedules while not worsening both guarantees too much.

Let $l = \alpha C_{max}^{\sigma_{cmax}}$ where α is a positive constant. We now partition the tasks in two sets $T = T_{minsum} \cup T_{cmax}$ where $t \in T_{minsum}$ if t completes before l in σ_{minsum} and $t \in T_{cmax}$ otherwise. Tasks from T_{minsum} are scheduled at the beginning of the schedule according to σ_{minsum} (this step is called *truncation*). Then, the tasks from T_{cmax} are appended on the same processors on which they were executed in σ_{cmax} starting from time l (this step is called *composition*). Let us call σ the obtained schedule.

PROPOSITION 9.1
The maximum completion time of σ is bounded: $C_{max}^\sigma \leqslant (1+\alpha)\rho_{cmax} C_{max}^*$.

PROOF   No tasks are running at time l: each task either finishes before

l or begins after l (or exactly at l). Tasks scheduled after l are scheduled with σ_{cmax}. Such tasks finish before $l + C^{\sigma_{cmax}}_{max}$. Recall that $l = \alpha C^{\sigma_{cmax}}_{max}$ and that σ_{cmax} is a ρ_{cmax}-approximated schedule on the makespan. Thus, $C^{\sigma}_{max} \leqslant (1+\alpha)\rho_{cmax}C^{*}_{max}$. $\qquad\Box$

The bound on the makespan of σ enables to bound the degradation on the sum of completion times of tasks that have been scheduled accordingly with σ_{cmax}. This property is used for bounding the performance of σ on the sum of completion times.

PROPOSITION 9.2
The sum of completion times of σ is bounded: $\sum C^{\sigma}_i \leqslant \frac{1+\alpha}{\alpha}\rho_{minsum} \sum C^{*}_i$.

PROOF The proof is obtained by bounding the ratio $\frac{C^{\sigma}_i}{C^{\sigma_{minsum}}_i}$ for each task i. Either $i \in T_{minsum}$ and the ratio is exactly 1, or $i \in T_{cmax}$ and the ratio $\frac{C^{\sigma}_i}{C^{\sigma_{minsum}}_i} \leqslant \frac{(1+\alpha)C^{\sigma_{cmax}}_{max}}{\alpha C^{\sigma_{cmax}}_{max}} = \frac{1+\alpha}{\alpha}$ (i finishes before $l + C^{\sigma_{cmax}}_{max}$ in σ and after l in σ_{minsum}). Recall that σ_{minsum} is a ρ_{minsum} approximation on the sum of completion times. Thus, $\sum C^{\sigma}_i \leqslant \sum \frac{1+\alpha}{\alpha}C^{\sigma_{minsum}}_i \leqslant \frac{1+\alpha}{\alpha}\rho_{minsum} \sum C^{*}_i$. $\qquad\Box$

These two properties lead to the approximation ratio of the zenith solution of the MAXANDSUM problem by σ.

COROLLARY 9.1
σ is a $((1+\alpha)\rho_{cmax}, \frac{1+\alpha}{\alpha}\rho_{minsum})$ approximation of the zenith solution of MAXANDSUM.

Consider σ_{cmax} and σ_{minsum} are optimal solutions for the makespan and the sum of completion times. Then, σ is a $(1 + \alpha, \frac{1+\alpha}{\alpha})$-approximation of the zenith point. The interesting point is the existence of such good solutions. By selecting more precisely the cutting point in σ_{minsum}, it is possible to improve the result to a $(1 + \rho, \frac{e^{\rho}}{e^{\rho}-1})$-approximation of the zenith point $\forall \rho \in [0; 1]$ [4]. More recently, Rasala et al. showed that variants with release times do not admit an algorithm achieving a better approximation ratio since there can exist no such solution [30].

This technique is generic, and has been extended to other objective functions such as the sum of weighted completion times or maximum lateness [30]. This technique relies on the knowledge of two existing schedules each optimizing one of the objectives. It can be extended using two scheduling policies, where one dynamically uses the other to fill in specific time intervals. This has been done for the same pair of objectives, using the recursive doubling scheme [36] for moldable parallel tasks [16]. We cannot provide more than a brief sketch of the method here. The scheduling is produced by an algorithm which creates successive batches of doubling sizes in order to execute small jobs first (to obtain a reasonable sum of weighted completion times). These batches are scheduled with an off-line algorithm that optimizes the total weight scheduled within the alloted time (thus optimizing the makespan).

However, such techniques of combining schedules cannot be applied to all scheduling problems since both *truncation* and *composition* operations must generate valid schedules. For instance, *composition* can violate constraints in problems with hard deadlines.

Combining schedules is not the only way to obtain a Zenith approximation, for instance on a different problem Skutella [37] proposed to model the problem as a linear program depending on a parameter which provides a family of solutions with approximation ratio $(1 + \alpha, 1 + \frac{1}{\alpha})$ when the parameter α varies.

9.5 Pareto Set Approximation on EFFICIENTRELIABLE

9.5.1 Motivation

As discussed in Section 9.3.2, we are looking for all possibly interesting solutions of a multi-objective optimization problem. However, generating the Pareto set of an instance is usually a difficult problem: first, computing a single Pareto optimal solution can be an NP-complete problem and, second, the cardinality of the Pareto set can be exponential in the size of the instance. Thus, the only reasonable way is to determine an approximation of the Pareto set (with a polynomial number of solutions).

A first idea would be to determine a parametrized zenith approximation algorithm (like the one in the previous section) and to use it with different values of the parameter to build a set of interesting solutions. However such an algorithm does not always exist. This is illustrated on the EFFICIENT-RELIABLE problem. Recall that in this problem the two objectives are the maximum completion time and the reliability index $rel = \sum \lambda_j C^j$, where C^j is the completion time of the last task scheduled on processor j.

THEOREM 9.3
The zenith solution of the EFFICIENTRELIABLE *problem cannot be approximated within a constant factor.*

PROOF   The theorem is proved by exhibiting an instance of the problem on which no Pareto optimal solution (therefore, no solution at all) is at a constant factor of both optimal values.

Consider the instance composed of m processors and m unitary jobs (such that $p_i = 1$). The reliability index of the first processor $\lambda_1 = 1$ is smaller than the reliability indices of the other processors $\forall j > 1, \lambda_j = x, x > 1$.

The optimal schedule for the makespan consists in scheduling each job on a different processor. Thus $C^*_{\max} = 1$. The optimal schedule for the reliability consists in scheduling all the jobs on the most reliable processor (in this instance, on processor 1). Thus $rel^* = m$.

Now, suppose that there exists a (k, k)-approximation algorithm. The generated schedule is k-approximated on the makespan. Thus, at most k jobs are scheduled on processor 1 and $m - k$ jobs are scheduled on other processors. Then, the reliability of such a solution is such that $\sum_j C^j \lambda_j \geq k + (m - k)x$. Finally, the approximation ratio on the reliability is linear in x which is an unbounded value.

Therefore, there is no (k, k)-approximation algorithm for the EFFICIENT-RELIABLE problem. □

Remárk that this theorem is not a complexity result. It does not depend on the assumption $\mathcal{P} \neq \mathcal{NP}$. It is directly derived from the problem itself.

In the remainder of this section, we present techniques that help to deal with such problems. The concepts will be illustrated on the EFFICIENTRELIABLE problem.

9.5.2 Definition of Pareto Set Approximation

We need a notion close to approximation algorithm in single-objective optimization theory. Papadimitriou and Yannakakis [28] give a definition of an $(1 + \epsilon)$-approximation of the Pareto set. Informally, a set of solution P is an $(1 + \epsilon)$-approximation of the Pareto set, if each solution of the Pareto set is $(1 + \epsilon)$-dominated by at least a solution of P. The formal definition that follows is a slightly modified version of the definition in [28] but it is strictly equivalent.

P is a $\rho = (\rho_1, \dots, \rho_k)$-approximation of the Pareto set P^* if each solution $S^* \in P^*$ is ρ-approximated by a solution $S \in P$: $\forall S^* \in P^*, \exists S \in P, \forall i = 1, \dots k, f_i(S) \leq \rho_i f_i(S^*)$. Figure 9.3 illustrates the concept for two objective functions. Crosses are solutions of the scheduling problem represented in the $(f_1; f_2)$ space. Bold crosses are an approximated Pareto set. Each solution in this set (ρ_1, ρ_2)-dominates the space delimited in bold. All solutions are dominated by a solution of the approximated Pareto set as they are included in a (ρ_1, ρ_2)-dominated space. Thus, the bold set is a (ρ_1, ρ_2)-approximation of the Pareto set.

The following theorem states the existence of approximated set of reasonable size. It has first been proposed in [28].

THEOREM 9.4
For each multi-objective optimization problem whose decision version belongs

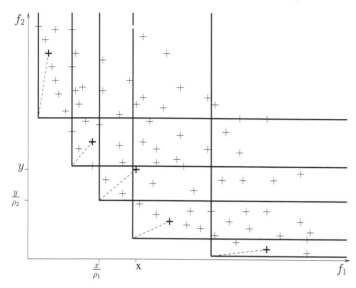

FIGURE 9.3: Bold crosses are a (ρ_1, ρ_2)-approximation of the Pareto set.

to $\mathcal{NP}$, for each instance I and $\epsilon > 0$, there exists an approximated Pareto set P_ϵ which is a $(1 + \epsilon)$-approximation of the Pareto set and P_ϵ has a cardinality polynomial in $|I|$ and $\frac{1}{\epsilon}$.

The detailed proof can be found in [28]. Remark that the cardinality of the approximated set is not polynomial in the number of objectives.

There are two main techniques for obtaining an approximation of the Pareto set. The first one [28] works on very restricted classes of objective functions and is computational intensive since it requires to run a huge amount of pseudo-polynomial algorithms. The second one [3] is presented in the next section.

9.5.3 The Thresholding Approach

In this section, we present a technique to approximate Pareto set using a thresholding approach. Some objectives are transformed into constraints in a way similar to dual approximation. Again the concepts are presented and illustrated on the EFFICIENTRELIABLE problem.

9.5.3.1 Notation

Most existing algorithms that solve a bi-objective problem set a threshold on the first objective and construct a ρ_2-approximation of the second objective. The schedule cannot exceed the threshold by more than a constant factor ρ_1. In the literature such algorithms are improperly said to be

(ρ_1, ρ_2)-approximation. This is a confusing notation, firstly since it is also used for zenith approximation (see Section 9.4), and secondly since the notation is symmetric while its meaning is not (the threshold is applied on the first objective and not on the second one).

For this purpose, a notation has been proposed that clarifies the notion of approximation in this context [22]. We denote an algorithm with the previous property as a $\langle \bar{\rho}_1, \rho_2 \rangle$-approximation algorithm. More formally,

DEFINITION 9.5 *Given a threshold value of the makespan ω, a $\langle \bar{\rho}_1, \rho_2 \rangle$-approximation algorithm delivers a solution whose $C_{\max} \leqslant \rho_1 \omega$ and $rel \leqslant \rho_2 rel^*(\omega)$ where $rel^*(\omega)$ is the best possible value of rel in schedules whose makespan is less than ω. The algorithm does not return a solution only if there is no solution fulfilling the makespan constraint.*

Tackling a problem through this kind of technique is similar to the ϵ-constraint problem defined in [42] where one should optimize one objective under a strict constraint on the other one. Thus, a $\langle \bar{1}, \rho \rangle$-approximation algorithm of the multi-objective problem is a ρ-approximation of the ϵ-constraint problem.

9.5.3.2 A $\langle \bar{2}, 1 \rangle$-Approximation Algorithm

We now present a $\langle \bar{2}, 1 \rangle$-approximation algorithm for the EFFICIENTRELIABLE problem called CMLF (for constrained min lambda first).

Let ω be a makespan value given as a parameter of CMLF. If this guess is too small, that is $\omega < \max(\frac{\sum p_i}{m}, \max p_i)$, the algorithm returns an error. Otherwise, the processors are considered in the order of non-decreasing values of λ_j. On each processor j, jobs are added until the total processing time of jobs on j becomes greater than ω.

THEOREM 9.5
CMLF is a $\langle \bar{2}, 1 \rangle$-approximation algorithm for the EFFICIENTRELIABLE problem. The time complexity of CMLF is in $O(m \log m + n)$.

PROOF In the case of $\omega < \frac{\sum p_i}{m}$, there is obviously no solution with a makespan better than ω. Thus, we are interested in the case where $\omega \geqslant \frac{\sum p_i}{m}$. First remark that the algorithm returns a valid schedule on m processors. Indeed, the makespan of the last processor cannot exceed ω while some more jobs are unscheduled. It would imply that $\sum p_i > m\omega$ which contradicts the hypothesis ($\omega \geqslant \frac{\sum p_i}{m}$). Secondly, the makespan is less than 2ω. Indeed, for a given processor j, no job begins after ω. Thus, $C_j \leqslant \omega + \max p_i$. The fact that $\max p_i \leq \omega$ proves the approximation on $C_{\max}$.

The proof of the approximation of the reliability is a bit harder. Recall that the optimal reliability is obtained by scheduling all the jobs on the most reliable processor. In schedules of makespan lower than ω, a lower bound on the best reliability $rel^*(\omega)$ is obtained by relaxing the problem to its preemptive version. With preemptions, we can schedule jobs on the most reliable processors up to the bound ω. When the bound ω is reached on a processor, the job is preempted. Its remaining part is scheduled on the next most reliable processor. The schedule generated by CMLF executes even more than ω units of work on the most reliable processors. Thus, its reliability is better than $rel^*(\omega)$. $\qquad\square$

Remark that a $\langle \bar{2}, 1 \rangle$-approximation algorithm was proposed by Shmoys and Tardös in [35] for scheduling jobs on unrelated machines in which a cost is paid when a job is scheduled on a processor. This result could also be used to deal with the EFFICIENTRELIABLE in the case of unrelated processors.

9.5.3.3 Framework

Algorithm 9.1, described below, constructs an approximation of the Pareto set of the problem by applying the $\langle \bar{2}, 1 \rangle$-approximation algorithm on a geometric sequence of makespan thresholds.

The set of efficient values of makespan is partitioned into chunks of exponentially increasing sizes. The ith chunk contains solutions whose makespan are in $[(1 + \frac{\epsilon}{2})^{i-1} C_{\max}^{\min}; (1 + \frac{\epsilon}{2})^i C_{\max}^{\min}]$, where $C_{\max}^{\min}$ is a lower bound of the optimal makespan. Similarly, $C_{\max}^{\max}$ is an upper bound of reasonable makespans, which is defined below. For each chunk i, the algorithm computes a solution π_i using CMLF. As we will show in Theorem 9.6, π_i approximates all the solutions of chunk i within a factor $(2 + \epsilon, 1)$.

Algorithm 9.1: Pareto set approximation algorithm

Input: ϵ a positive real number
Output: S a set of solutions
$S = \emptyset$;
for $i \in \left\{ 1, 2, \ldots, \left\lceil \log_{1 + \frac{\epsilon}{2}} \left(\frac{C_{\max}^{\max}}{C_{\max}^{\min}} \right) \right\rceil \right\}$ **do**
$\quad \omega_i = (1 + \frac{\epsilon}{2})^i C_{\max}^{\min}$;
$\quad \pi_i = CMLF(\omega_i)$;
$\quad S = S \cup \pi_i$
return S

The values of efficient makespan are bounded by the two following values. The lower bound $C_{\max}^{\min} = \frac{\sum_i p_i}{m}$ is obtained by considering a perfect load balancing between processors. The upper bound $C_{\max}^{\max} = \sum_i p_i$ is the makespan when all the jobs are scheduled on the most reliable processor. Such a solution

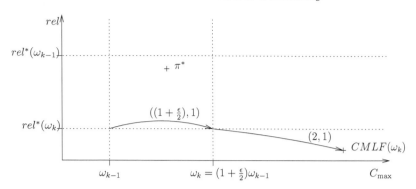

FIGURE 9.4: The Pareto-optimal solution π^* is dominated by $(\omega_{k-1}, rel^*(\omega_k))$. By construction, $(\omega_{k-1}, rel^*(\omega_k))$ is $(1 + \frac{\epsilon}{2}, 1)$-approximated by $(\omega_k, rel^*(\omega_k))$ which, in turn, is $(2, 1)$-approximated by $CMLF(\omega_k)$.

is optimal on the reliability index. Using chunks of exponentially increasing size, $\left\lceil \log_{1+\frac{\epsilon}{2}} \left(\frac{C_{\max}^{\max}}{C_{\max}^{\min}} \right) \right\rceil$ steps have to be performed.

Remark that the cardinality of the generated set is polynomial: the algorithm generates less than $\left\lceil \log_{1+\frac{\epsilon}{2}} \frac{C_{\max}^{\max}}{C_{\max}^{\min}} \right\rceil \leqslant \left\lceil \log_{1+\frac{\epsilon}{2}} m \right\rceil$ solutions which is polynomial in $\frac{1}{\epsilon}$ and in the size of the instance.

We now show that the set of solutions constructed by Algorithm 9.1 approximates the Pareto set within $(2 + \epsilon, 1)$.

THEOREM 9.6
Algorithm 9.1 is a $(2 + \epsilon, 1)$ approximation algorithm of the Pareto set of the EFFICIENTRELIABLE *problem.*

PROOF Let π^* be a Pareto-optimal schedule. Then, there exists $k \in \mathbb{N}$ such that $(1 + \frac{\epsilon}{2})^{k-1} C_{\max}^{\min} \leqslant C_{\max}(\pi^*) \leqslant (1 + \frac{\epsilon}{2})^k C_{\max}^{\min}$. We show that π_k is an $(2 + \epsilon, 1)$-approximation of π^*. This is illustrated by Figure 9.4.

- Reliability. $rel(\pi_k) \leqslant rel^*((1 + \frac{\epsilon}{2})^k C_{\max}^{\min})$ (by Theorem 9.5). π^* is Pareto optimal, hence $rel(\pi^*) = rel^*(C_{\max}(\pi^*))$. But, $C_{\max}(\pi^*) \leqslant (1 + \frac{\epsilon}{2})^k C_{\max}^{\min}$. Since rel^* is a decreasing function, we have: $rel(\pi_k) \leqslant rel(\pi^*)$.

- Makespan. $C_{\max}(\pi_k) \leqslant 2(1 + \frac{\epsilon}{2})^k C_{\max}^{\min} = (2 + \epsilon)(1 + \frac{\epsilon}{2})^{k-1} C_{\max}^{\min}$ (by Theorem 9.5) and $C_{\max}(\pi^*) \geqslant (1 + \frac{\epsilon}{2})^{k-1} C_{\max}^{\min}$. Thus, $C_{\max}(\pi_k) \leqslant (2 + \epsilon) C_{\max}(\pi^*)$.

□

9.6 Fairness as Multi-Objective Optimization

In this section, we investigate the problem of scheduling jobs for several agents. The problem is not only to schedule jobs efficiently but also to allocate resources in a fair way.

The TWOAGENTMINSUM problem is used to illustrate all concepts. Let us denote by n_A (resp. n_B) the number of jobs of agent A (resp. agent B). p_i^A denotes the processing time of J_i^A, the i-th job of agent A. Recall that in this problem, the machine is a single processor.

9.6.1 The Meaning of Fairness

A Pareto-optimal solution can be perceived as unfair for some agents. Consider the following instance of TWOAGENTMINSUM: each agent has three jobs with sizes $p_1^A = p_1^B = 1, p_2^A = p_2^B = 3, p_3^A = p_3^B = 5$ (later in this section we will assume that the jobs are numbered according to increasing execution times). The solution $(J_1^A, J_2^A, J_3^A, J_1^B, J_2^B, J_3^B)$ that schedules first the jobs of A, achieves $(\Sigma C_i^A, \Sigma C_i^B) = (14, 41)$. Despite this solution is Pareto-optimal (any other schedule would increase ΣC_i^A), B may "feel" treated unfairly. Consequently, the notion of fairness should put further restrictions on the set of Pareto-optimal solutions.

There is no widely accepted agreement about the meaning of *fairness* in multi-objective scheduling. The Webster dictionary defines the term *fair* as "free from self-interest, prejudice, or favoritism". Consequently, a solution $(14, 41)$ should be considered as being fair, as long as the scheduler does not take into account the ownership of jobs while producing a schedule. However, if, for instance, A produces definitely more jobs than B, and the scheduler treats all the jobs equally, A's jobs will be favored. Moreover, really selfish agents would produce many copies of their jobs, which not only disfavors other agents, but also worsens the performance of the whole system.

Therefore, in order to be fair, the scheduler should take into account the gains of the agents from the system. *Minimax* (Rawlsian social welfare function) [31] is one of the most popular approaches to ensure fairness. In *minimax*, the scheduler minimizes the maximum of the objectives. Assuming that each job has a different owner, *minimax* corresponds to the classic makespan minimization. *Leximax* [13] is a refinement of *minimax*, if two solutions have the same values of the maximum (worst-off) objectives, the scheduler compares them according to values of the second-worst objectives (and, if they are also equal, to the third-worst, and so on). As a consequence, a non-*leximax*-dominated solution is Pareto-optimal. Another related concept is the min-max fairness [8]. However, there are problems in which *leximax*-optimal solutions are not min-max fair [29]. In the remainder of this chapter, we will use *leximax*, since it is more popular in the scheduling community.

Note that *leximax* fairness does not favor *a priori* one user. Each job is scheduled considering all the jobs and changing the instance by reordering the users keeps a similar *leximax* optimal solution. Thus, an important property of a fair scheduler is to guarantee the symmetry of the users.

In the problems addressed in this section, all agents share the same objective function. The values of agents' objective functions can be directly compared in order to measure the fairness of the resulting distribution. This property is usually called *distributive fairness* [23]. It is discussed in Section 9.6.4.

9.6.2 Axiomatic Theory of Fairness

In this section, we consider a multi-objective approach to fairness that returns several possible solutions. Generally, this approach imposes further restrictions on Pareto-optimal solutions. The concept of optimality of a solution is modified to also contain fairness. This enables the decision-maker to explicitly choose among these solutions, in contrast with choosing between values of parameters of the optimization algorithm that will produce "the" fair solution. Consequently, the balance between fairness and efficiency is made explicitly by the decision-maker.

Distributive fairness can be precisely described by axioms, which define a relation of *equitable dominance* [23]. Similarly to the relation of Pareto-dominance, equitable dominance characterizes a set of solutions to a multi-objective optimization problem.

DEFINITION 9.6 *The* relation of equitable dominance *is the asymmetric, irreflexive and transitive relation that verifies the three following properties:*

symmetry *it ignores the ordering of the objective values; all permutations of the objective functions are equitably-equivalent.*

monotony *a solution which is Pareto-dominated is also equitably-dominated.*

principle of transfers *A transfer of any small amount from one objective to any other relatively better-off objective results is more equitable, i.e.,* $f_i > f_j \Rightarrow (f_1, \ldots, f_i, \ldots, f_j, \ldots, f_k)$ *is equitably-dominated by* $(f_1, \ldots, f_i - \epsilon, \ldots, f_j + \epsilon, \ldots, f_k)$, *for a maximization problem.*

Equitable-efficiency is defined similarly as Pareto-optimality: solution σ^* is equitably-efficient or equitably-optimal if it is not equitably-dominated by any other solution.

The axioms of symmetry, monotony, and principle of transfers have straightforward interpretation (see Figure 9.5 for graphical reference).

Symmetry states that no objective is preferred *a priori*, *e.g.*, the fictive solution $B = (73, 76)$ is equitably-equivalent to $A = (76, 73)$. Monotony

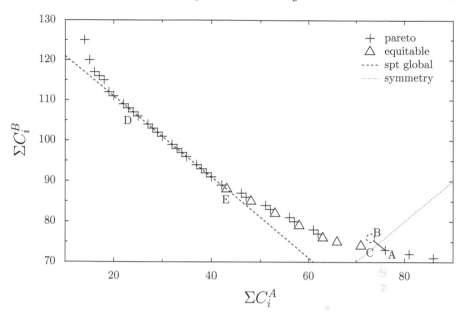

FIGURE 9.5: Relation of equitable dominance. (Equitably-optimal solutions are denoted by triangles.)

states that equitably-optimal solutions must be Pareto-optimal, *e.g.*, $C = (71, 74)$ equitably-dominates $B = (73, 76)$. Therefore, C equitably-dominates A. Finally, the principle of transfers states that a solution improving a worse-off objective, at the expense of deteriorating a better-off objective is preferred, *e.g.*, $E = (43, 88)$ is preferred to any solution of D, for instance, $(33, 98)$.

Principle of transfers is analogous to optimization of the Gini index [32], *i.e.*, the minimization of the surface between the line of perfect equality and the Lorenz curve. Note also that the relation of Definition 9.6 is analogous to the Generalized Lorentz Dominance, used in the theory of majorization [26].

Remind that all equitably efficient solutions are also Pareto-optimal solutions. Thus, the enumeration of the equitably-efficient solutions sounds easier and should help the decision maker.

9.6.3 Application to TWOAGENTMINSUM

In the following, we study three different instances of TWOAGENTMINSUM on which the axiomatic theory of fairness is proved to be useful. To get a better understanding of the instance structure, we give for each of them a graphical representation of the solution space in $(\Sigma C_i^A, \Sigma C_i^B)$ coordinates (Figures 9.6, 9.7, and 9.8). Equitably-optimal solutions are denoted by "$\triangle$". Solutions that are Pareto-optimal, but not equitably-optimal, are denoted by "$+$".

We also show the line of the perfect equity $\Sigma C_i^A = \Sigma C_i^B$ (dotted line), and the line of system efficiency $\min(\Sigma C_i^A + \Sigma C_i^B)$ (dashed line). All the solutions on the dashed line of system efficiency result in optimal *global* sum of completion times $\Sigma C_i = \Sigma C_i^A + \Sigma C_i^B$. A solution that is closest to the perfect equity line is the *leximax* solution. The solutions above the perfect equity line favor agent A, as the resulting $\Sigma C_i^A < \Sigma C_i^B$. Similarly, solutions below the perfect equity line favor agent B.

Axiomatic Fairness as *Leximax*

The easiest situation occurs when jobs produced by agents are similar (Figure 9.6). Here, both agents produced jobs of sizes $1, 3$, and 5. The resulting Pareto set crosses the equity line close to the efficiency line. Consequently, there are only two equitably-efficient schedules, $(J_1^B, J_1^A, J_2^B, J_2^A, J_3^A, J_3^B)$ with $(\Sigma C_i^A, \Sigma C_i^B) = (23, 24)$ and $(J_1^A, J_1^B, J_2^A, J_2^B, J_3^B, J_3^A)$ with $(\Sigma C_i^A, \Sigma C_i^B) = (24, 23)$.

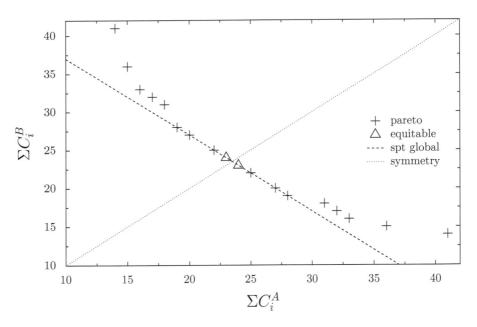

FIGURE 9.6: When agents produce similar loads (here, both agents produced three jobs of sizes $\{1, 3, 5\}$), the axioms of fairness chose the min-max solutions.

Axiomatic Fairness Balancing Efficiency and Equity

When one of the agents produces longer jobs, it is disfavored by the globally-optimal **SPT** solution (Figure 9.7). Here, agent A (resp. B) produced three jobs of size 1 (resp. 2). Globally-optimal SPT schedule $(J_1^A, J_2^A, J_3^A, J_1^B,$

J_2^B, J_3^B) results in $(\Sigma C_i^A, \Sigma C_i^B) = (6, 21)$. Yet, the min max solution is $(\Sigma C_i^A, \Sigma C_i^B) = (16, 16)$ and it can be interpreted as inefficient since it leads to a significant increase of the global sum of completion times ($\Sigma C_i = 32$, comparing to $\Sigma C_i = 27$ in the previous case). According to axiomatic theory of fairness, all the solutions between SPT and min max are equitable. Thus, the decision maker can balance the equity of the system and its efficiency. Note that the solutions below the equity line are not equitable, as they are Pareto-dominated by solutions symmetric to the solutions above the equity line.

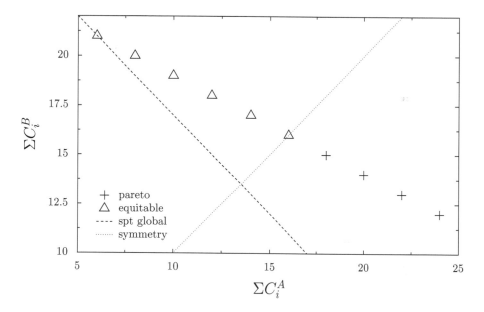

FIGURE 9.7: When agent loads differ (here, agent A produced three jobs of length 1, and agent B three jobs of length 2), axiomatic fairness balances the equity and efficiency.

In a more complex example (Figure 9.8) the notion of fairness substantially reduces the number of equitable solutions. Here, agent A produced jobs of sizes $1, 3$, and 5, and agent B jobs of sizes $1, 3, 5, 5, 5$, and 5. There are 7 equitably-optimal schedules, compared to 44 Pareto-optimal ones.

9.6.4 Problems with Different Objective Functions

It is difficult to extend the previous analysis for problems with different objective functions (non distributive fairness).

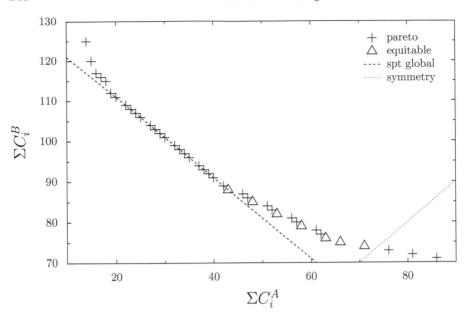

FIGURE 9.8: A more complex example, in which only some Pareto-optimal solutions between SPT and min max are equitably-optimal

First, the objectives are not always expressed in the same unity and thus no arithmetic operation on them is meaningful. Second, even if the objectives are the same, they can have different scales. If one of the agents uses the makespan, while the other one uses the sum of completion times, *leximax* approach will almost always favor the sum of completion times user, even in a multi-processor setting.

9.6.5 Aggregative Fairness

A commonly used mean to solve a multi-objective optimization problem is to aggregate objectives of individual agents $f_1, \ldots, f_k$ by some functions, e.g., $\sum_i \lambda_i f_i$, and then to solve the resulting mono-objective optimization problem with ordinary means. A similar approach is used in aggregative fairness. The goal is to chose a function that provides fair aggregation of the individual objectives.

An infinite number of functions can be used as such fair aggregations [27]. Here, we provide a brief survey of the most used functions in the literature. Note that it is possible to choose an aggregation, whose optimal value is equitable in the sense of axioms from Section 9.6.2 [23]. Without loss of generality, we assume that all the objective functions are minimized.

$\sum_i f_i$ optimizes the sum of objectives. The main drawback is that an improvement of an objective f_i that is already better-off than some other objective f_l ($f_i < f_l$) is indistinguishable from the improvement of f_l of the same quantity.

$\sum_i f_i^2$ –the L2 norm– avoids the aforementioned problem. An ϵ decrease in f_l decreases the value of the aggregation more than the same decrease in f_i.

$\min \max_i f_i$ optimizes the worst-off objective, at the cost of increasing values of other objectives uncontrollably. $\min \max_i f_i$ is generalized by *leximax* operator.

Aggregative fairness suffers from many drawbacks. Firstly, similarly to aggregative multi-objective optimization, not all Pareto-optimal solutions can be produced, even when weights are added.

Secondly, those approaches return only one fair solution. Thus, the decision-maker cannot measure the drop in the system's efficiency, resulting from the proposed solution. For instance, in TWOAGENTMINSUM, let us assume that A produced three jobs of size 1, and B three jobs of size 2. The min max solution, $(16, 16)$, results in system-wide $\Sigma C_i = 32$. The system-wide optimal solution, resulting from ordering jobs in Shortest Processing Time (SPT) order, is $(6, 21)$, leading to $\Sigma C_i = 28$. When given only one of these solutions, the decision-maker does not have enough information to balance the fairness (*i.e.*, the difference between ΣC_i^A and ΣC_i^B) with the efficiency of the system (ΣC_i).

9.7 Conclusion

In this chapter, we presented a survey of recent techniques used in multi-objective optimization. This subject is more and more popular and motivates today a lot of studies. Many interesting results have been obtained for solving specific multi-objective problems with *ad hoc* methods (mostly bi-objective ones).

We can mainly distinguish between three kinds of techniques. The first one is to find a reasonable way to aggregate the various objectives into a single one and solve the corresponding problem with a standard method. We did not really focus on this way in this chapter, because the problem is more in the model than in the design of resolution methods. The second one is to determine a global zenith approximation of the Pareto curve. It is not always adequate to reduce the problem in a single point, and moreover, as we discussed in this chapter, determining the zenith point is not always tractable. However, in many cases parametrized variants of the original problem are possible; the zenith solution of each variant is then a feasible solution of the original problem. The third one is what we consider as the most promising way. It consists in approximating the Pareto curve. Efficient techniques can be found for providing a polynomial number of good trade-offs in order to help the decision-maker. There are many promising research problems in this area: In particular, we would like to deal with problems that have many objectives. It sounds possible to handle such problems using for instance similar techniques as those presented in this chapter as fairness. Otherwise, the scheduling problem can be considered as a game where the users are the players. Game theory would be a nice tool to investigate this problem when the scheduler becomes distributed.

References

[1] A. Agnetis, P. Mirchandani, D. Pacciarelli, and A. Pacifici. Scheduling problems with two competing agents. *Operations Research*, 52(2):229–242, 2004.

[2] S. Albers and H. Fujiwara. Energy-efficient algorithms for flow time minimization. In *23rd International Symposium on Theoretical Aspects of Computer Science (STACS)*, volume 3884 of *LNCS*, pages 621–633. Springer, 2006.

[3] E. Angel, E. Bampis, and A. Kononov. A FPTAS for approximating the Pareto curve of the unrelated parallel machines scheduling problem with costs. In *Algorithms - ESA 2001*, volume 2161 of *LNCS*, pages 194–205. Springer, 2001.

[4] J. Aslam, A. Rasala, C. Stein, and N. Young. Improved bicriteria existence theorems for scheduling. In *SODA '99: Proceedings of the Tenth Annual ACM-SIAM Symposium on Discrete Algorithms*, pages 846–847, 1999.

[5] I. Assayad, A. Girault, and H. Kalla. A bi-criteria scheduling heuristic for distributed embedded systems under reliability and real-time constraints. In *2004 International Conference on Dependable Systems and Networks (DSN'04)*, pages 347–356, June 2004.

[6] K. Baker and J. Smith. A multiple-criterion model for machine scheduling. *Journal of Scheduling*, 6:7–16, 2003.

[7] M. Bender, S. Muthukrishnan, and R. Rajaraman. Improved algorithms

for stretch scheduling. In *Proceedings of the 13th Annual ACM-SIAM Symposium on Discrete Algorithms (SODA)*, pages 762–771, 2002.

[8] D. Bertsekas and R. Gallager. *Data Networks*. Prentice-Hall, 1987.

[9] P. Brucker. *Scheduling Algorithms*. Springer-Verlag, Berlin, 5th edition, 2007.

[10] D. Bunde. Power-aware scheduling for makespan and flow. In *18th Annual ACM Symposium on Parallel Algorithms and Architectures*, pages 190–196, 2006.

[11] Y. Caniou and E. Jeannot. Experimental study of multi-criteria scheduling heuristics for GridRPC systems. In *Euro-Par 2004 Parallel Processing*, volume 3149 of *LNCS*, pages 1048–1055. Springer, 2004.

[12] C. Chekuri, R. Motwani, B. Natarajan, and C. Stein. Approximation techniques for average completion time scheduling. In *SODA '97: Proceedings of the Eighth Annual ACM-SIAM Symposium on Discrete Algorithms*, pages 609–618, 1997.

[13] M. Chen. Individual monotonicity and the leximin solution. *Economic Theory*, 15(2):353–365, 2000.

[14] M. Detyniecki. *Mathematical aggregation operators and their application to video querying*. PhD thesis, Université Paris VI, 2000. `http://www.lip6.fr/reports/lip6.2001.002.html`.

[15] J. Dongarra, E. Jeannot, E. Saule, and Z. Shi. Bi-objective scheduling algorithms for optimizing makespan and reliability on heterogeneous systems. In *SPAA '07: Proceedings of the Nineteenth Annual ACM Symposium on Parallelism in Algorithms and Architectures*, pages 280–288. ACM Press, June 2007.

[16] P.-F. Dutot, L. Eyraud, G. Mounié, and D. Trystram. Bi-criteria algorithm for scheduling jobs on cluster platforms. In *Symposium on Parallel Algorithms and Architectures*, pages 125–132. ACM Press, 2004.

[17] D. Feitelson. Workload modeling for computer systems performance evaluation. `http://www.cs.huji.ac.il/~feit/wlmod/`.

[18] M. R. Garey and D. S. Johnson. *Computers and Intractability. A Guide to the Theory of NP-Completeness*. W. H. Freeman & Co, San Francisco, 1979.

[19] R. Graybill and R. Melhem, editors. *Power Aware Computing*. Series in Computer Science. Kluwer Academic/Plenum Publishers, New York, May 2002.

[20] K. Ho. Dual criteria optimization problems for imprecise computation tasks. In Leung [25], chapter 35.

[21] H. Hoogeveen. Multicriteria scheduling. *European Journal of Operational Research*, 167(3):592–623, Dec. 2004.

[22] E. Jeannot, E. Saule, and D. Trystram. Bi-objective approximation scheme for makespan and reliability optimization on uniform parallel machines. In *Euro-Par 2008*, volume 5168 of *LNCS*. Springer, Aug. 2008.

[23] M. Kostreva, W. Ogryczak, and A. Wierzbicki. Equitable aggregations and multiple criteria analysis. *European Journal of Operational Research*, 158:362–377, 2004.

[24] C. Laforest. A tricriteria approximation algorithm for Steiner tree in graphs. Technical Report 69-2001, LaMI, 2001.

[25] J. Y.-T. Leung, editor. *Handbook of Scheduling: Algorithms, Models, and Performance Analysis*. Chapman & Hall/CRC, 2004.

[26] A. Marshall and I. Olkin. *Inequalities: Theory of Majorization and Its Applications*. Academic Press, 1979.

[27] J. Mo and J. Warland. Fair end-to-end window-based congestion control. In *Proc. of SPIE '98: International Symposium on Voice, Video and Data Communications*, 1998.

[28] C. Papadimitriou and M. Yannakakis. On the approximability of trade-offs and optimal access of web sources. In *41st Annual Symposium on Foundations of Computer Science*, pages 86–92, 2000.

[29] B. Radunovic and J. Le Boudec. A Unified Framework for Max-Min and Min-Max Fairness with Applications. *IEEE/ACM Transactions on Networking*, 15(5):1073–1083, Oct. 2007.

[30] A. Rasala, C. Stein, E. Torng, and P. Uthaisombut. Existence theorems, lower bounds and algorithms for scheduling to meet two objectives. In *SODA '02: Proceedings of the Thirteenth Annual ACM-SIAM Symposium on Discrete Algorithms*, pages 723–731, 2002.

[31] J. Rawls. *The Theory of Justice*. Harvard Univ. Press, 1971.

[32] P. Samuelson and W. Nordhaus. *Economics*. McGraw-Hill, 16th edition, 1998.

[33] E. Saule, P.-F. Dutot, and G. Mounié. Scheduling with storage constraints. In *IEEE International Symposium on Parallel and Distributed Processing*, Apr. 2008.

[34] S. Shatz and J. Wang. Task allocation for maximizing reliability of distribued computer systems. *IEEE Tansactions on Computer*, 41(9):1156–1169, Sept. 1992.

[35] D. Shmoys and E. Tardos. Scheduling unrelated machines with costs. In *Proceedings of the Fourth Annual ACM/SIGACT-SIAM Symposium on Discrete Algorithms*, pages 448–454, 1993.

[36] D. B. Shmoys, J. Wein, and D. P. Williamson. Scheduling parallel machines on-line. *SIAM Journal on Computing*, 24(6):1313–1331, Dec. 1995.

[37] M. Skutella. Approximation algorithms for the discrete time-cost trade-off problem. *Mathematics of Operations Research*, 23(4):909–929, 1998.

[38] S. Srinivasan and N. Jha. Safety and reliability driven task allocation in distributed systems. *IEEE Transactions on Parallel and Distributed Systems*, 10(3):238–251, Mar. 1999.

[39] C. Stein and J. Wein. On the existence of schedules that are near-optimal for both makespan and total weighted completion time. *Operational Research Letters*, 21(3):115–122, Oct. 1997.

[40] R. Steueur. *Multiple Criteria optimization: theory, computation and application*. John Wiley, 1986.

[41] W. Stromquist. How to cut a cake fairly. *The American Mathematical Monthly*, 87(8):640–644, 1980.

[42] V. T'kindt and J. Billaut. *Multicriteria Scheduling*. Springer, 2007.

[43] V. T'kindt, K. Bouibede-Hocine, and C. Esswein. Counting and enumeration complexity with application to multicriteria scheduling. *Annals of Operations Research*, Apr. 2007.

[44] M. Voorneveld. Characterization of Pareto dominance. *Operations Research Letters*, 31:7–11, 2003.

Chapter 10

Comparisons of Stochastic Task-Resource Systems

Bruno Gaujal

INRIA and Université de Grenoble

Jean-Marc Vincent

INRIA and Université de Grenoble

Abstract   In this chapter, we show how to compare the performance of stochastic task resource systems. Here, task resource systems are modeled by dynamic systems whose inputs are the task and resource data, given under the form of random processes, and whose outputs are the quantities of interest (response times, makespan). The approach presented here is based on stochastic orders. Several typical examples (some simple, other rather sophisticated) are used to show how orders can help the system designer to compare several options but also to compute bounds on the performance of a given system.

10.1   Motivation

The main object of this chapter is to present several methods that can be used to study stochastic task-resource models through comparisons.

Using random variables in task-resource systems can be useful to model uncertainties: one uses statistical data on tasks and resources to infer a distribution of task sizes, or arrival times, for example. The first drawback of this approach is the fact that inferring a distribution from a finite sample can be very difficult and hazardous. Actually, in most cases, one chooses a distribution beforehand and only infers parameters. This may not be very satisfactory from the system designer point of view. Another difficulty with the stochastic approach is the fact that when the distributions of the input processes of task-resource system are complex, it is often very difficult, or even impossible (see Section 10.4.2, for example) to compute the distribution of its output (or even to compute its expectation).

However, these difficulties are often counterbalanced by the great analytical power of probability theory. Indeed, a random process can also be used as a *simplification* of a complex deterministic process. For example the process made of the superposition of the release dates of many independent tasks can often be simplified as a single Poisson process, being the limit of the superposition of a large number of independent arbitrary processes [17]. In fact limit theorems (see Section 10.5.3) are of paramount importance in probability theory. The law of large numbers or the central limit theorem may constitute the most well-known tools used to easily analyze very complex dynamic systems.

Another useful property which helps to overcome the technical difficulties of distribution inference is the insensibility properties. In some cases (see Section 10.5.1), the task-resource system average behavior only depends on the expectation (or the first two moments) of the input data and not on the whole distribution. This makes distribution inference less critical.

Finally, stochastic structural properties can also be used to derive qualitative properties based on the structure of the system (see Section 10.5) valid for all distributions. Task-resource system analysis is very sensitive to the initial assumptions. An easy polynomial problem may become NP-hard in the strong sense when a small assumption is changed. The complexity of classical questions such as computing the makespan of the system jumps from one class of complexity to the next, sometimes in a counter-intuitive way (see Chapter 2 for an illustration of this). However, when one focuses on bounds and qualitative properties (such as monotonicity with respect to the input parameters), then the classification of task-resource systems becomes much more stable. Furthermore, there exist several efficient techniques that can be used to tackle them. Actually, this is the object of this chapter.

We will present three methods through several examples that may help system designers assess qualitative properties (and quantitative bounds) of task-resource systems. These methods are all based on a model of task-resource systems as input/output operators using initial exogenous random input processes and transforming them into random processes, resulting from the action of resources over tasks.

The first technique presented here is the use of appropriate *orders* between random variables and processes. The second one is called the *coupling* technique and allows one to compare two systems over a single trajectory. The third one is called *association* and will help us deal with dependencies.

10.2 Task-Resource Models

Tasks (or jobs) are characterized by their arrival times (or release dates) that form a point process $0 \leqslant r_1 \leqslant r_2 \leqslant r_3 \leqslant \cdots$ over the positive real line, and by the sizes of tasks that form a sequence of real numbers $\sigma_1, \sigma_2, \ldots$. We also use the inter-arrival times, $\tau_1 = r_1, \tau_i = r_i - r_{i-1}, \quad i \geqslant 2$.

In the following we will consider both finite and infinite cases (with finite or infinite number of tasks) for which arrival times as well as the sizes are real random processes.

Resources are characterized by their number $m \in \mathbb{N} \cup \{+\infty\}$ and their respective speeds $v_1, \ldots, v_m$. Therefore, the size of the tasks is given is seconds (the time for a resource of speed 1 to treat a task).

Once the tasks and the resources are specified, the system is still not completely defined. The system designer may also want to take synchronizations between resources and communications between tasks into account. Additionally, tasks and resources may be constrained by dependencies, availability conditions and matchings between tasks and resources.

Once the system is specified, our goal will be to analyse the best allocation policy (which task is executed on which machine) and the best scheduling policy (what is the sequence of execution of tasks on each resource) with respect to a given criterion (typically completion times or response times).

10.2.1 Static Systems

A static task-ressource system is made of a finite number of tasks (N) and ressources (m). Hence, most systems studied in Chapters 2, 3, 4, and 9 belong to the class of static task-ressource systems. The main object of this section is to abstract a task-resource system into a functional operator transforming inputs into outputs.

We can view a static model as an operator, ϕ, whose input is the exogenous data (arrival times of the tasks, sizes, processor speeds) $X = \phi(Z_1, \cdots, Z_N)$. Its output X corresponds to the state of the system and the quantity of interest to the system designer—makespan, departure times of all tasks, response times, stretch, etc.—is a function h of X. It is often the case that one aggregates the performance h into the dynamics ϕ by computing directly $h \circ \phi(Z_1, \cdots, Z_N)$.

Here are examples of static systems viewed as input/output operators:

- $1||C_{max}$ schedule problem: $X = C_{max} = \phi(\sigma_1 \ldots \sigma_N) \stackrel{\text{def}}{=} \sum_i \sigma_i$. This case is straightforward and no scheduling actually takes place here.

- $1||\sum C_i$ schedule problem: $X = \phi(\sigma_1 \ldots \sigma_N) \stackrel{\text{def}}{=} \min_{\alpha \in perm\{1...N\}} \sum_{i=1}^{N}(N - i+1)\sigma_{\alpha(i)}$. This is a well-known problem which can also be solved under appropriate stochastic assumptions on task sizes (see Section 10.4.1).

- $P_\infty|prec|C_{max}$ schedule problem: this system is also called a PERT graph. Here, $C_{max} = \phi(\sigma_1 \ldots \sigma_N) \stackrel{\text{def}}{=} \max_{c \in \mathcal{P}(G)} \sum_{i \in c} \sigma_i$, where G is the acyclic precedence graph of the tasks and $\mathcal{P}(G)$ is the set of paths in G. Here again, no scheduling takes place, the goal being to compute the makespan. This case is studied in Section 10.4.2.

10.2.2   Dynamic Systems

The other main class is when the system has an infinite horizon (infinite number of tasks). This is the case for the systems studied in Chapters 5 and 7.

In that case we consider task-resource systems with an infinite number of tasks arriving in the system, according to the sequence of arrival times $r_1 \leqslant \cdots \leqslant r_n \leqslant \cdots$ that go to infinity with n (the sequence does not have an accumulation point).

Dynamic models can be modeled by an operator, indexed by n. $X_n = \phi_n(X_{n-1}, Z_n)$, $\forall n \geqslant 0$. Z_n corresponds to a point of the exogenous process (arrival times, sizes, resource availability), X_n is the state at system at step n, and the quantity of interest is a sequence of functions of $X_1 \ldots X_n$.

Here are some examples of an operator definition for infinite task graph systems.

We use the same notation as for classical scheduling problems. However one must keep in mind that the tasks are infinitely many here.

- Consider a system with one resource and an infinite number of tasks, processed in a FCFS manner, and one wants to compute the response time of the tasks. This problem can be written as: $1|r_j|W_j = C_j - r_j$. Here W_j corresponds to the response time of task j: $W_j = C_j - r_j$, often denoted T_j in the scheduling literature. This quantity satisfies $W_j = \phi(W_{j-1}, \tau_j, \sigma_j) = \max(W_{j-1} + \sigma_j - \tau_j, \sigma_j)$. This recurrence formula is similar to Lindley's formula [3] and is illustrated by Figure 10.1. It can be solved (by induction) as $W_j = \max_{k=1\ldots j} \sum_{i=k}^{j} \sigma_i - (r_i - r_{i-1})$ or $W_j = \sigma_j$, if the previous quantity is smaller than σ_j. This problem is addressed in Section 10.5.1.

- More complicated systems fit the operator model. For example, computing the best average response time over unit tasks ($\sigma_i = 1$) over m processors with speeds $v_1, \cdots, v_n$ summing to 1. This is a scheduling problem that can be denoted as $Q_m|r_j|\sum_i W_i$ and can be modeled by a Markov Decision Process. The best average response time (denoted J) is the solution of the well-known Bellman equation. More precisely, the triple $(J^*, V(.), \pi^*(.))$ is the unique finite solution of the following equation where $\tau = \mathbb{E}\tau_j$ and e_k is a vector whose coordinates are all null except coordinate k, equal to one. ($V(.)$ is an unknown function

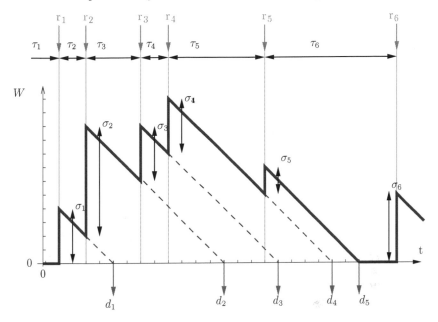

FIGURE 10.1: The evolution of the response times and Lindley's formula.

of $x \in \mathbb{N}^k$, $\pi^*(.)$ is the best allocation of the next task under state x, corresponding to the arg-min in the equation).

$$V(x) = J + \min_{\pi} \left(\tau V(x + e_{\pi(x)}) + (1 - \tau) \sum_{i=1}^{K} v_i V(x - e_i) \right).$$

These kinds of scheduling problems where the function ϕ is implicit are usually more difficult to solve and are out of the scope of this chapter. However, several structural results in the same flavor as Theorems 10.3 and 10.5 can be found in [13]. Several computational issues in this context are addressed in [5].

10.3   Stochastic Orders

This section deals with the comparison of stochastic variables and processes. The goal is to introduce the most common concepts that are used in the literature while keeping in mind the task-resource model. Therefore, the presentation may not always be as general as possible and has a task-resource bias. Indeed, many other stochastic orders have been defined and used in many different applications (in the book *Comparison methods for stochastic models and risks* [14], Muller and Stoyan define 49 different orders). The presentation used here is more or less based on this book.

10.3.1 Orders for Real Random Variables

This section is devoted to the presentation of the most classical orders for real random variables. In the following, we consider that all random variables are defined on a common probability space $(\Omega, \mathcal{A}, P)$ with values in $\mathbb{R}$, unless an explicit construction is done. The distribution function of X is defined as $F_X(x) = P(X \leqslant x)$ for all $x \in \mathbb{R}$. Its density (if it exists) is $f_X(x) = dF(x)/dx$. The expectation of X is $\mathbb{E}X \overset{\text{def}}{=} \int_{x \in \mathbb{R}} x dF_X(x) = \int_{x \in \mathbb{R}} x f_X(x) dx$, if X admits a density.

There are two obvious ways to compare two random variables.

One may say that X is smaller than Y if $P(X \leqslant Y) = 1$ (denoted $X \leqslant_{as} Y$). This way to compare random variables may be too strong since it involves the distributions of X and Y but also their joint distribution. In particular this means that two independent random variables can never be compared this way, no matter their distribution on a common domain. This impairs the usefulness of this order.

On the other extreme, one may say that X is smaller than Y if $\mathbb{E}X \leqslant \mathbb{E}Y$ (denoted $X \leqslant_\mu Y$). This crude comparison works well in a linear context but as soon as non-linear dynamics are used, the comparisons based on expectations may collapse.

Example 10.1
Here is a small example showing that comparing the expectations of the inputs does not help to compare the expected outputs. Consider a system made of periodic tasks and one resource. In a first case, one has arrivals every 4 seconds of tasks of size 4. Using our notations, this means that $r_i = 4i + X$ and $\sigma_i = 4$, for all $i \in \mathbb{N}$, where X is a real random variable with a finite expectation corresponding to the instant when the system starts. The load per second up to time t is $L(t) = \sum_{i=0}^{\lfloor (t-X)/4 \rfloor} \sigma_i / t$. By using Wald's Theorem [17], the expected load per time unit verifies

$$4/t((t - \mathbb{E}X)/4 - 1) \leqslant \mathbb{E}L(t) \leqslant (t - \mathbb{E}X)/t.$$

Since X has a finite expectation the expected load per second goes to one when t goes to infinity: $\mathbb{E}L(t) \to 1 = \mathbb{E}L$.

Now, consider a second system with arrivals of tasks of size 4 at times $10n + Y$ and $10n + 1 + Y$, for all n (again Y is a random initial time with a finite expectation). Using the current notations, $\sigma_i' = 4$ for all i and $r_{2i}' = 10i + Y, r_{2i+1}' = 10i + 1 + Y$. Now, the expected load per second up to time t is $L'(t) = \sum_{i=0}^{N(t)} \sigma_i'/t$ with $N(t) = \lfloor (t-Y)/10 \rfloor + \lfloor (t-Y-1)/10 \rfloor$. By using

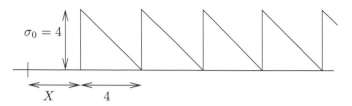

FIGURE 10.2: Case 1: The tasks never wait for the resource.

Wald's Theorem,

$$4/t((t - \mathbb{E}Y)/10 - 1 + (t - \mathbb{E}Y - 1)/10 - 1)$$
$$\leqslant \quad \mathbb{E}L'(t) \quad \leqslant$$
$$4(t - \mathbb{E}Y)/10 + 4(t - \mathbb{E}Y - 1)/10.$$

Since Y has a finite expectation the expected load per second goes to $8/10$ when t goes to infinity: $\mathbb{E}L'(t) \to 8/10 = \mathbb{E}L'$. Therefore, $\mathbb{E}L \geqslant \mathbb{E}L'$.

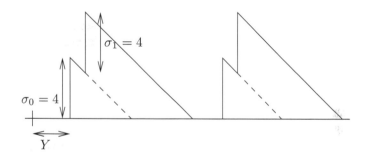

FIGURE 10.3: Case 2: One task out of two waits for the resource.

Now, let us consider the expected response times W of the tasks, i.e., the time elapsed between the arrival of the task and the end of its execution by the resource. In the first case, no task ever waits for the resource to be available, therefore, $W_i = 4$ for all i, so that $\mathbb{E}W = 4$ (see Figure 10.2).

In the second case, even tasks never wait for the resource $W_{2i} = 4$ while odd tasks wait for one time unit for the previous task to finish its execution: $W_{2i+1} = 5$ (see Figure 10.3). Therefore the expected response time in the second system is $\mathbb{E}W' = 9/2$. So that $\mathbb{E}W' \geqslant \mathbb{E}W$. □

This kind of behavior will be studied in full details in section 10.5.1.1, where we will show that response times are increasing functions of loads (as intuition tells), when the right order is used.

10.3.1.1   The Strong Order

The most classical stochastic order is called the stochastic order (or strong order). It was introduced by Karlin in the 1960s and many others [12] and has many equivalent definitions.

DEFINITION 10.1 (strong order)   *The real random variable X is smaller that the real random variable Y for the strong order (denoted $X \leqslant_{st} Y$) if for all $t \in \mathbb{R}$, $F_X(t) \geqslant F_Y(t)$ (or $P(X \leqslant t) \geqslant P(Y \leqslant t)$, or $P(X > t) \leqslant P(Y > t)$, for all $t \in \mathbb{R}$).*

The strong order has several characterizations, which will be useful in the following.

THEOREM 10.1
The following propositions are equivalent.
(i) $X \leqslant_{st} Y$
(ii) Coupling: There exist two variables X' and Y' in (Ω, A, P) with the same distribution as X and Y respectively such that $X'(\omega) \leqslant Y'(\omega)$ for all $\omega \in \Omega$.
(iii) Integral definition: For all increasing function f, $\mathbb{E}(f(X)) \leqslant \mathbb{E}(f(Y))$, whenever the expectations exist.

PROOF   Here are the main ideas of this proof. Let us show that $(i) \Rightarrow (ii)$. Denote by $F_X^{-1}(u) \stackrel{\text{def}}{=} \inf\{x : F_X(x) \geqslant u\}$ for $0 < u < 1$, the inverse distribution function of X. Let U, defined on the common probability space $(\Omega, \mathcal{A}, P)$, be uniformly distributed on $(0,1)$. Let $X' = F_X^{-1}(U)$ and $Y' = F_Y^{-1}(U)$. By definition of F_X^{-1} and F_Y^{-1}, X' and Y' have the same distributions as X and Y, respectively. Since according to (i), $F_X \geqslant F_Y$ point-wise, then $F_X^{-1} \leqslant F_Y^{-1}$ point-wise. This implies $X'(\omega) \leqslant Y'(\omega)$ for all $\omega \in \Omega$. Let us show that $(ii) \Rightarrow (iii)$. Since $X'(\omega) \leqslant Y'(\omega)$ for all $\omega \in \Omega$, then the comparison also holds for all increasing function f: $f(X'(\omega)) \leqslant f(Y'(\omega))$ for all $\omega \in \Omega$. Taking expectations on both sides, $\mathbb{E}f(X) = \mathbb{E}f(X') \leqslant \mathbb{E}f(Y') = \mathbb{E}f(Y)$. Finally, $(iii) \Rightarrow (i)$ because the indicator function $f_t(x) = \mathbb{1}_{(t,\infty)}(x)$ is increasing and $P(X > t) = \mathbb{E}f_t(X) \leqslant \mathbb{E}f_t(Y) = P(Y > t)$.   ☐

The strong order has many interesting properties (see [14] for a more exhaustive list).

- The strong order compares all moments for positive variables: using the integral definition with $f(x) = x^n$, $X \leqslant_{st} Y \Rightarrow \mathbb{E}X^n \leqslant \mathbb{E}Y^n$ for all n.

- Moreover, $X \leqslant_{st} Y$ and $\mathbb{E}X = \mathbb{E}Y$ imply $F_X = F_Y$.

- If Z is independent of X and Y and $X \leqslant_{st} Y$ then $f(X, Z) \leqslant_{st} f(Y, Z)$ for any function f non-decreasing in the first argument. In particular,

$X + Z \leqslant_{st} Y + Z$ and $\max(X, Z) \leqslant_{st} \max(Y, Z)$.

- If $(X_n)_{n\in\mathbb{N}}$ and $(Y_n)_{n\in\mathbb{N}}$ are sequences of random variables such that $X_n \leqslant_{st} Y_n$ for all n, then $\lim_n X_n \leqslant_{st} \lim_n Y_n$.

10.3.1.2 Stronger Orders

Some orders are stronger than the strong order and may be useful in task-resource problems as seen in Section 10.4.

Example 10.2

Consider the following problem with one task of unknown size σ and two resources that may break down. To execute the task, one can choose between the two resources with respective random lifetimes X and Y. If $X \leqslant_{st} Y$ then one should choose Y, which has a longer lifetime according to the stochastic order. Doing so, one maximizes chances that the task of size σ is fully executed before breakdown: $P(X > \sigma) \leqslant P(Y > \sigma)$. Now what happens if the task has a release time r and both ressources are still alive at time r. Is Y still a better choice? In other words, is it true that $P(X > \sigma + r | X > r) \leqslant P(Y > \sigma + r | Y > r)$? Well, not necessarily. Assume for example that Y is uniform over $[0, 3]$ (with distribution F) and X has a distribution G with densities $2/3, 0, 1/3$ on $[0, 1], (1, 2], (2, 3]$, respectively. Then, $X \leqslant_{st} Y$ ($G \geqslant F$). However $X_1 = (X | X > 1)$ and $Y_1 = (Y | Y > 1)$ are such that Y_1 is uniform over $[0, 2]$ (with density $1/2$) and X_1 has a distribution with densities 0 and 1, on $[0, 1], (1, 2]$, which means that $Y_1 \leqslant_{st} X_1$. ⬛

So a natural question is: what order is preserved under conditioning?

DEFINITION 10.2   *The hazard rate (or failure rate) of X is defined by:*

$$r_X(t) \stackrel{\text{def}}{=} \lim_{\varepsilon \to 0} \frac{P(X < t + \varepsilon | X > t)}{\varepsilon} = \frac{f_X(t)}{1 - F_X(t)} = -\frac{d}{dt} \ln(1 - F_X(t)).$$

DEFINITION 10.3 (hazard rate order) $X \leqslant_{hr} Y$ *if* $r_X(t) \geqslant r_Y(t)$, *for all* $t \in$.

The main properties [14] of the hazard rate order are:

- The *hr* order is preserved under aging: $X \leqslant_{hr} Y$ implies $(X | X > t) \leqslant_{hr} (Y | Y > t)$, $\forall t \in \mathbb{R}$ (aging property).

- The *hr* order is stronger than the *st* order.

- If $X \leqslant_{hr} Y$ then there exists two random variables X^* and Y^* with the same distribution as X and Y, respectively, such that $\mathbb{E}g(X^*, Y^*) \leqslant$

$\mathbb{E}g(Y^*, X^*)$ for all g s.t. $g(x, y) - g(y, x)$ is increasing in $x, \forall x \geqslant y$ (coupling property).

Another order which is even stronger than hr is the likelihood ratio which preserves st under any conditioning.

DEFINITION 10.4 (likelihood ratio order)   *Let $U = [a, b], V = [c, d]$ with $a < c$ and $b < d$. Then $X \leqslant_{lr} Y$ if $P(X \in V)P(Y \in U) \leqslant P(X \in U)P(Y \in V)$ or equivalently $(X|X \in U) \leqslant_{st} (Y|Y \in U)$.*

Example 10.3
Here is a case where the likelihood ratio is useful. Consider a task resource problem with one task of unknown size σ, but known release date r and deadline d. Two processors can execute the task. They may both breakdown and have a respective life span of X and Y (which are random variables). Which processor should be used to maximize chances to get the task properly executed, provided both processors are still alive at time r? It turns out that one should always choose Y over X if and only if $X \leqslant_{lr} Y$.   □

Here are the main properties of the likelihood ratio ordering.

- The lr order is stronger than the hr order.

- If $X \leqslant_{lr} Y$ then there exist two random variables X^* and Y^* with the same distribution as X and Y, respectively, such that $\mathbb{E}g(X^*, Y^*) \leqslant \mathbb{E}g(Y^*, X^*)$ for all functions g verifying $g(x, y) - g(y, x) \geqslant 0, \quad \forall x \geqslant y$ (coupling property).

10.3.1.3   Convex Orders

The convex orders are used to compare the variability of stochastic variables.

DEFINITION 10.5 (convex order)   *$X \leqslant_{cx} Y$ if $\mathbb{E}f(X) \leqslant \mathbb{E}f(Y)$ for all convex functions f, whenever the expectations exist.*

Note that $X \leqslant_{cx} Y$ implies that $\mathbb{E}X = \mathbb{E}Y$ by applying the definition with the convex functions $f(x) = x$ and $f(x) = -x$. However, $varX \leqslant varY$, which was not true for the previous orders.

To compare variables with different expected values, one can combine the strong order with the convex order.

DEFINITION 10.6 (increasing convex order)   *$X \leqslant_{icx} Y$ if $\mathbb{E}f(X) \leqslant \mathbb{E}f(Y)$ for all increasing convex functions f, whenever the expectations exist.*

The most important property of the convex order is the Strassen Representation Theorem.

THEOREM 10.2 (Strassen Representation Theorem [19])
$X \leqslant_{cx} Y$ *if and only if there exist two variables X' and Y' with the same distribution as X and Y, respectively, such that $X' = \mathbb{E}(Y'|X')$.*
$X \leqslant_{icx} Y$ *if and only if there exists Z_1 (resp. Z_2) such that $X \leqslant_{st} Z_1 \leqslant_{cx} Y$ (resp. $X \leqslant_{cx} Z_2 \leqslant_{st} Y$).*

Note that this theorem is a coupling result: one can find versions of X and Y that are coupled so that they can be compared for almost all ω, in the projection over all events measurable by X.

Here is an example that illustrates the fact that the convex order is related with the variability of stochastic variables. This example also shows the usefulness of the Strassen representation.

Example 10.4
Consider a task-resource system with a single task and a single processor. In one version of the system, the task has size σ, which is a random variable and a resource with speed 1. In the second version, the task is the same as previously but the resource has a random speed v independent of σ and $\mathbb{E}(1/v) = 1$ then, the completion times are, respectively, σ and σ/v, with the same expectation: $\mathbb{E}\sigma = \mathbb{E}\sigma \mathbb{E}(1/v) = \mathbb{E}(\sigma/v)$.

However, $\sigma \leqslant_{cx} \sigma/v$. This is a direct application of the Strassen theorem: consider $\mathbb{E}(\sigma/v|\sigma) = \mathbb{E}(\sigma|\sigma)\mathbb{E}(1/v|\sigma) = \sigma\mathbb{E}1/v = \sigma$. □

The following graphs give the relations between stochastic orders ($A \to B$ means that A implies B).

$$\leqslant_{as} \to \leqslant_{lr} \to \leqslant_{hr} \to \leqslant_{st} \to \leqslant_{icx} \to \leqslant_{\mu}$$

$$\leqslant_{as} \to \leqslant_{cx} \to \leqslant_{icx} \to \leqslant_{\mu} .$$

10.3.2 Orders for Multidimensional Random Variables

When one wants to compare finite or infinite sequences of real random variables, things become more complicated because the joint distribution of the variables will play a role in the comparisons. Recall that in the univariate case, the integral definition coincides with the distribution condition: $\mathbb{E}f(X) \leqslant \mathbb{E}f(Y)$ for all increasing f is equivalent to $P(X \leqslant x) \geqslant P(Y \leqslant x)$, and equivalent to $P(X > x) \leqslant P(Y > x)$. This is not the case anymore in the multidimensional case because upper and lower orthants are not the only sets that can be compared.

Therefore, we introduce three orders:

DEFINITION 10.7   *Let X and Y be random vectors of size n.*

- *The usual stochastic (strong) order is $X \leqslant_{st} Y$ if $\mathbb{E}f(X) \leqslant \mathbb{E}f(Y)$ for all increasing f from $\mathbb{R}^n$ to $\mathbb{R}$, as long as expectations exists.*

- *The upper orthant order is: $X \leqslant_{uo} Y$ if $P(X > x) \leqslant P(Y > x)$, for all $x \in \mathbb{R}^n$.*

- *The lower orthant order is: $X \leqslant_{lo} Y$ if $P(X \leqslant x) \geqslant P(Y \leqslant x)$, for all $x \in \mathbb{R}^n$.*

It should be obvious by using multidimensional indicator functions that $X \leqslant_{st} Y$ implies $X \leqslant_{uo} Y$ as well as $X \leqslant_{lo} Y$.

In the following, we will also be mainly concerned with the following integral orders.

DEFINITION 10.8 $(Z_1, \ldots, Z_n) \leqslant_{cx} (Z_1', \ldots, Z_n')$ *if for all convex function f from $\mathbb{R}^n$ to $\mathbb{R}$, $\mathbb{E}f(Z_1, \ldots, Z_n) \leqslant \mathbb{E}f(Z_1', \ldots, Z_n')$, whenever the expectations exist.*

$(Z_1, \ldots, Z_n) \leqslant_{icx} (Z_1', \ldots, Z_n')$ *if for all increasing convex function f from $\mathbb{R}^n$ to $\mathbb{R}$, $\mathbb{E}f(Z_1, \ldots, Z_n) \leqslant \mathbb{E}f(Z_1', \ldots, Z_n')$, whenever the expectations exist.*

10.3.3   Association

Association is used to strengthen the concept of positive correlation. Two variables X and Y are positively correlated if $cov(X, Y) \stackrel{\text{def}}{=} \mathbb{E}(X - \mathbb{E}X)(Y - \mathbb{E}Y) \geqslant 0$. In words, whenever X is big, Y is big as well. Association says the same for arbitrary functions of X and Y. In particular, independent variables are associated.

DEFINITION 10.9   *A random vector (resp. a sequence) X is associated if $cov(g(X), f(X)) \geqslant 0$ for all increasing f and g from $\mathbb{R}^n$ (resp. $\mathbb{R}^\mathbb{N}$) to $\mathbb{R}$.*

In dynamic systems, association is more useful than the simple positive correlation because it propagates through monotonic operators.

LEMMA 10.1
Consider a dynamic system $X_{n+1} = \phi(X_n, Z_n)$. If ϕ is non-decreasing in both variables and if $(Z_n)_{n \in \mathbb{N}}$ are associated then $(X_n)_{n \in \mathbb{N}}$ are associated.

Finally, association implies a comparison with independent versions of the vectors.

LEMMA 10.2

If the vector $(X_1, \ldots, X_n)$ is associated then $(X_1, \ldots, X_n) \leqslant_{lo} (X_1^*, \ldots, X_n^*)$ and $(X_1, \ldots, X_n) \leqslant_{uo} (X_1^*, \ldots, X_n^*)$, where $(X_1^*, \ldots, X_n^*)$ is an independent version of $(X_1, \ldots, X_n)$.

PROOF (Sketch) First, note that $cov(g(X), f(X)) \geqslant 0$ is equivalent to $\mathbb{E}(f(X)g(X)) \geqslant \mathbb{E}f(X)\mathbb{E}g(X)$ which implies by induction that $\mathbb{E}(\prod_i f_i(X_i)) \geqslant \prod_i \mathbb{E}f_i(X_i) = \mathbb{E}(\prod_i f_i(X_i^*))$, by using independence of $(X_1^*, \ldots, X_n^*)$. Finally, using increasing indicator functions for the f_i implies upper and lower orthant comparisons. □

This ends the presentation of stochastic orders, that will be put into practice in the following sections.

10.4 Applications to Static Problems

In this section we will focus on static task-resource problems. We recall that we consider a static model as an operator, ϕ, whose input is the exogenous data (arrival times of the tasks, sizes, processor speeds) $X = \phi(Z_1, \cdots, Z_N)$. Its output X corresponds to the state of the system or to the quantity of interest to the system designer: makespan, departure times of all tasks, response times, stretch.

For a static system, considering the task-resource system as a rather abstract operator is very useful to design a method to assess comparisons, as shown by the following theorem.

THEOREM 10.3

If, for $\omega = st$, cx, or icx, $(Z_1, \ldots, Z_n) \leqslant_\omega (Z_1', \ldots, Z_n')$ and if ϕ is non-decreasing, then $\phi(Z_1, \ldots, Z_n) \leqslant_\omega \phi(Z_1', \ldots, Z_n')$.

PROOF The proof of this theorem is direct by using a coupling argument and the integral definition of the orders. Let $(Z_1, \ldots, Z_n) \leqslant_w (Z_1', \ldots, Z_n')$. Then for any increasing function f, one has $\mathbb{E}f \circ \phi(Z_1, \ldots, Z_n) \leqslant \mathbb{E}f \circ \phi(Z_1', \ldots, Z_n')$. This means that $X_n = \phi(Z_1, \ldots, Z_n) \leqslant_{st} \phi(Z_1', \ldots, Z_n') = X_n'$. The proof for cx and icx is similar since the composition of convex (resp. increasing convex) functions is convex (resp. increasing convex). □

10.4.1   The $1||\sum C_i$ Problem, Revisited

This is one of the most classical problem from scheduling theory in the task resource framework. One resource with no scheduling restriction executes N tasks, all arriving at time 0. The objective is to minimize the sum of the completion times (or the average completion time).

We consider all tasks to be of independent sizes $\sigma_1, \ldots, \sigma_N$.

For a given schedule (or permutation) α, the objective function is $X_\alpha = \sum_{i=1}^{N} C_i = \sum_{i=1}^{N}(N-i+1)\sigma_{\alpha(i)}$ (this is an instance of the operator ϕ defined above).

We consider two particular schedules: SEPT (shortest expected processing time) and LEPT (largest expected processing time).

It should be clear that for any permutation α, $\mathbb{E}X_{SEPT} \leqslant \mathbb{E}X_\alpha \leqslant \mathbb{E}X_{LEPT}$. Indeed, $\mathbb{E}X_\alpha = \mathbb{E}\sum_{i=1}^{N} C_i = \sum_{i=1}^{N}(N-i+1)\mathbb{E}\sigma_{\alpha(i)}$. This is a well-known result that SEPT is the best policy on average and that LEPT is the worse. But can we say more?

Yes, indeed! Using hr and lr orders on the variables σ_i, one can compare the distributions of X, and not only their expectations.

THEOREM 10.4 Shanthikumar and Yao [18]

If $\sigma_i \leqslant_{lr} \sigma_{i+1}$ for all i, then $X_{SEPT} \leqslant_{st} X_\alpha \leqslant_{st} X_{LEPT}$.
If $\sigma_i \leqslant_{hr} \sigma_{i+1}$ for all i, then $X_{SEPT} \leqslant_{icx} X_\alpha \leqslant_{icx} X_{LEPT}$.

PROOF   First, the proof uses a coupling technique and then an interchange argument (the proof is only done for hr, the other case being very similar).

For any permutation $\alpha \neq SEPT$ there exists k such that $\alpha(k) = j, \alpha(k+1) = i$ with $\sigma_i < \sigma_j$. Let $\gamma = \alpha$ except $\gamma(k) = i, \gamma(k+1) = j$ (γ is closer to $SEPT$ than α).

Now, consider that the sizes of the tasks are the same under both execution sequences (this is a coupling between both systems). Under this coupling, $X_\alpha = \sigma_j + (N-k)(\sigma_i+\sigma_j)+Y$ and $X_\gamma = \sigma_i + (N-k)(\sigma_i+\sigma_j)+Y$ where Y is the contribution of the other jobs, independent of σ_i and σ_j. The next point is to consider the function $g(x,y) \stackrel{\text{def}}{=} f(x+(N-k)(x+y))$. It should be clear that $g(x,y) - g(y,x)$ is increasing as long as f is convex and increasing. Therefore, $\sigma_i \leqslant_{hr} \sigma_j$ implies $\mathbb{E}f(\sigma_j + (N-k)(\sigma_i + \sigma_j)) \geqslant \mathbb{E}f(\sigma_i + (N-k)(\sigma_i + \sigma_j))$ for all increasing convex f, by using the coupling bivariate characterization of $\leqslant_{hr}$.

Finally, $\sigma_i + (N-k)(\sigma_i+\sigma_j) \leqslant_{icx} \sigma_j + (N-k)(\sigma_i+\sigma_j)$ implies $X_\gamma \leqslant_{icx} X_\alpha$.

⬜

10.4.2   PERT Graphs

A PERT graph is a more general static model: N tasks of sizes $\sigma_1, \ldots, \sigma_N$ are to be executed over a large number of resources ($m > N$) and are constrained by an acyclic communication graph G. The set of paths in G is denoted by $\mathcal{P}(G)$.

PERT graphs are impossible to solve (compute the makespan) analytically in general [11]. However, one can use comparisons to prove several results. Here are the ingredients that can be used to compare (and compute bounds for) PERT graphs.

The first step is to construct the operator ϕ that computes the makespan of the system from the input processes. Since the number of resources is unlimited, the makespan is the length of the critical path in $\mathcal{P}(G)$, that is the path over which the sum of sizes of tasks is the largest. $C_{max} = \phi(\sigma_1, \ldots, \sigma_N) \overset{\text{def}}{=} \max_{c \in \mathcal{P}(G)} \sum_{i \in c} \sigma_i$. Computing the distribution of C_{max} when the distributions of $\sigma_1, \ldots, \sigma_N$ are general is a very difficult problem [11]. This is due to the fact that the paths in G share nodes so that their execution times are not independent. Hence, the distribution of the maximum of such variables is very tricky to obtain.

On the other hand, it should be clear at this point that ϕ is convex and increasing. Therefore a direct application of Theorem 10.3 yields the following result:

$$\sigma_i \leqslant_\omega \sigma_i' \text{ implies } C_{max} \leqslant_\omega C_{max}' \text{ with } \omega = st \text{ or } \omega = icx.$$

Next, one can use Lemma 10.1 to assess that if task sizes are independent (or associated), then the paths are all associated and are therefore bounded by independent versions.

LEMMA 10.3
For any path $c \in \mathcal{P}(G)$, let $\sigma_c = \sum_{i \in c} \sigma_i$. Then $\sigma_c \leqslant_{st} \sigma_c^* \overset{\text{def}}{=} \sum_{i \in c} \sigma_i^*$, where $(\sigma_1^*, \ldots, \sigma_n^*)$ are independent versions of $(\sigma_1, \ldots, \sigma_n)$. Furthermore, $C_{max} \leqslant_{st} C_{max}^* \overset{\text{def}}{=} \max_{c \in \mathcal{P}(G)} \sigma_c^*$.

Finally, one may introduce additional properties on the size distributions to get further results.

DEFINITION 10.10   New Better than Used in Expectation (NBUE)   *The variable X is NBUE if for all $t \in \mathbb{R}$, $\mathbb{E}(X - t | X > t) \leqslant \mathbb{E}X$.*

Assuming that the tasks sizes are all NBUE is not a strong assumption since, in most cases, a partial execution of a task should not increase the expected remaining work.

An immediate corollary is that, if σ is NBUE, then $\sigma \leqslant_{cx} X^e(\mathbb{E}\sigma)$, where

$X^e(\mathbb{E}\sigma)$ is a random variable exponentially distributed with expectation $\mathbb{E}\sigma$: $P(X^e(\sigma) > t) = \exp(-t/\mathbb{E}\sigma)$.

Combining this with the previous results implies that

$$\max_{c \in \mathcal{P}(G)} \sum_{i \in c} \mathbb{E}\sigma_i \leqslant_{icx} \mathbb{E}C_{max} \leqslant_{icx} \max_{c \in \mathcal{P}(G)} \sum_{i \in c} X^e(\mathbb{E}\sigma_i) \leqslant_{icx} \max_{c \in \mathcal{P}(G)} X^e(\sum_{i \in c} \mathbb{E}\sigma_i).$$

Then, this last expression has a closed form distribution:

$$P\left(\max_{c \in \mathcal{P}(G)} X^e(\sum_{i \in c} \mathbb{E}\sigma_i) \leqslant t \right) = \prod_{c \in \mathcal{P}(G)} \left(1 - exp(-t/\sum_{i \in c} \mathbb{E}\sigma_i) \right),$$

as soon as σ_i are all NBUE and associated (or independent).

The lower bound $\max_{c \in \mathcal{P}(G)} \sum_{i \in c} \mathbb{E}\sigma_i$ is deterministic and easy to compute (it is obtained by replacing the sizes of all tasks by their expected value). The second bound has a distribution which can be sampled efficiently by inverting each member of the product. This means that these bounds can be used to get numerical upper and lower estimates of the completion time of arbitrary PERT graphs.

In particular, when the task graphs have regular structures (tree, diamond graphs, level graphs,...), the set of paths $\mathcal{P}(G)$ has a particular structure, for example, the number of paths with fixed length is known. When task execution times are NBUE, and independent and identically-distributed (iid), the previous expression has a simpler analytic form [22]:

$$P\left(\max_{c \in \mathcal{P}(G)} X^e(\sum_{i \in c} \mathbb{E}\sigma_i) \leqslant t \right) = \prod_{k}(1 - exp(-t/k\mathbb{E}\sigma))^{n_k},$$

where n_k is the number of paths with length k.

If the task graph has a serie-parallel recursive structure, the upper bound is tightened by recursively applying the technique in the parallel part by bounding the maximum of exponential independent random variables and in the series part by bounding sums of exponential which are NBUE [7].

10.5 Applications to Dynamic Systems

In this section, we consider systems with an infinite number of tasks arriving in the system, according to an arrival sequence $r_1 \leqslant \cdots \leqslant r_n \leqslant \cdots$ that goes to infinity with n.

Dynamic models can be modeled by an operator, indexed by n. $X_n = \phi_n(X_{n-1}, Z_n)$, $\forall n \geqslant 0$, where Z_n corresponds to the input exogenous process (arrival times, sizes, resource availability, speeds), X_n is the state of the system at step n, and the quantity of interest is a function of $X_1 \ldots X_n \ldots$

DEFINITION 10.11 monotonic systems   *A dynamic system is* time-monotonic *for order ω if $X_n \leqslant_\omega X_{n-1}$.*
A dynamic system is ω-isotonic if $\forall k, Z_k \leqslant_\omega Z'_k \Rightarrow \forall n, X_n \leqslant_\omega X'_n$.

Whenever a dynamic system has monotonicity or isotonicity properties, then one may establish comparisons by comparing the inputs of functionals.

THEOREM 10.5
if $(Z_1, \ldots, Z_n) \leqslant_\omega (Z'_1, \ldots, Z'_n)$ and all ϕ_n are increasing, (resp. increasing and convex) then $X_n \leqslant_\omega X'_n$, with $\omega = st$ (resp. $\omega = icx$).

PROOF   The proof is similar to the proof of Theorem 10.3, using an induction on n. ⬚

10.5.1   Single Queues

Queues may be the simplest task-resource systems falling into the category of dynamic systems. They are made of an infinite number of tasks and two resources: one is a server executing the tasks and one is a buffer (with a finite or infinite capacity) storing the arriving tasks before their execution by the server. In spite of their apparent simplicity, queues are still the source of many open problems.

Let us recall that W_n denotes the response time of the n-th task. We recall that FIFO queues are dynamic systems of the form $W_n = \phi(W_{n-1}, Z_n)$ with $Z_n \stackrel{\text{def}}{=} \sigma_n - \tau_n$ and ϕ defined by Lindley's equation:

$$W_n = \max(W_{n-1} + Z_n, \sigma_n) .$$

THEOREM 10.6
For all n, $W_n \leqslant_{st} W_{n+1}$ in a queue, initially empty.

PROOF   The proof is done by a backward coupling argument known as Loynes' scheme [3]. First, let us construct (on the common probability space) two trajectories by going backward in time: $\sigma^1_{i-n} = \sigma^2_{i-n-1}$ with the same distribution as σ_i and $r^1_{i-n} = r^2_{i-n-1}$, with the same distribution as $r_i - r_{n+1}$ for all $0 \leqslant i \leqslant n+1$ and $\sigma^1_{-n-1} = 0$. By construction, $W^1_0 = W_n$ and $W^2_0 = W_{n+1}$. Also, it should be clear that $0 = W^1_{-n+1} \leqslant W^2_{-n+1}$ almost surely.

This implies $W^1_{-i} \leqslant W^2_{-i}$ so that $W_n \leqslant_{st} W_{n+1}$. This proof is illustrated in Figure 10.4. ⬚

This has many consequences in terms of existence and uniqueness of a

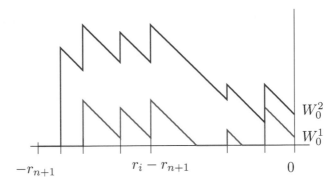

FIGURE 10.4: Backward coupling of release times and sizes.

stationary (or limit) regime for general queues, because increasing sequences must converge. For more on this, see for example [3].

10.5.1.1   Comparisons Based on the Input Process

In queues, the input load plays a major role in the performance measures. When $\mathbb{E}\tau \leqslant \mathbb{E}\sigma$, then the queue is unstable and all performance measures go to infinity. However, even when $\mathbb{E}\tau$ and $\mathbb{E}\sigma$ are fixed, the performance measure may still vary because of the variability of both processes. There exists a folk theorem (Ross conjecture [16]) that says: *things work better when the input traffic has less variability.*

Here is a precise version of Ross conjecture that derives directly from Lindley's equation.

THEOREM 10.7
If $(W_0, Z_1, \ldots, Z_n) \leqslant_\omega (W'_0, Z'_1, \ldots, Z'_n)$ then $W_n \leqslant_\omega W'_n$ (with $\omega = st$ or icx).

PROOF $\phi(z, w) = max(w + z, 0)$ is convex and increasing for both variables. The proof follows by a direct application of Theorem 10.5.   ⬜

Note that the proof is very short and elegant once the framework has been built.

In this context, it is rather straightforward to derive other interesting properties such as the following. Once the traffic intensity is fixed in a single queue, then the average waiting time is smallest when the arrivals are periodic.

Another consequence is that if the arrival process is Poisson and the task sizes are NBUE, then the average waiting time can be bounded by the waiting time in a well known queue (M/M/1), with Poisson arrivals and exponential task sizes: $\mathbb{E}W \leqslant \frac{1}{\mathbb{E}(1/\sigma_1) - \mathbb{E}(1/r_1)}$. The proof uses the same ingredients as those used to derive bounds for PERT graphs (Section 10.4.2).

Several extensions of this theorem are possible.

THEOREM 10.8 Altman, Gaujal, and Hordijk [2]

If the release sequence $(r_n)_{n \in \mathbb{N}}$ is fixed in a FIFO queue then $\sigma_1, \ldots, \sigma_n \leqslant_{cx}$ $\sigma'_1, \ldots, \sigma'_n$ implies $(W_1, \ldots, W_n) \leqslant_{icx} (W_1, \ldots, W_n)$.

The proof of this theorem (not reported here) is again based on a convexity of the waiting time with respect to the input sequence, called multimodularity [2].

10.5.1.2 Service Discipline

In a queue with one or more servers, tasks may be served according to disciplines (the most common ones are FIFO, LIFO, and PS (processor sharing)).

The policy PS has insensibility, reversibility, and product form properties [17]. Namely, when tasks are processed using processor sharing then expected response times only depend on the expectation of the sizes and not on the whole distribution (when the inter-release dates are exponentially distributed).

As for the FIFO policy (denoted (F) in the following), it has optimality properties, in terms of response times:

THEOREM 10.9

In a single queue, consider two scheduling policies, FIFO (F) and any arbitrary service discipline π, not based on job sizes. Then $f(W_1^F, \cdots, W_n^F) \leqslant_{st}$ $f(W_1^\pi, \cdots, W_n^\pi)$ for all n and all convex increasing and symmetric function f.

PROOF   The proof is done using a coupling technique and majorization.

Since task sizes and release dates are independent, we can index the sizes in the order of service (and not in the order of arrivals) under policy π as well as under policy F. Let D_i be the ith departure time. If the task sizes are indexed according to their departure dates, then $D_i^\pi = D_i^F$ for all task i.

Then $W_i^F = D_i - r_i$ and assume that π interchanges the departure of j and $j+1$: $W_j^\pi = D_{j+1} - r_j$ and $W_{j+1}^\pi = D_j - r_{j+1}$

Now, it should be obvious that $W_j^\pi + W_{j+1}^\pi = W_j^F + W_{j+1}^F$, $W_j^\pi > W_j^F$ and $W_{j+1}^\pi < W_{j+1}^F$. Therefore, if f is increasing convex and symmetric (or more generally if f is Schur convex) $f(W_j^\pi, W_{j+1}^\pi) \geqslant f(W_j^F, W_{j+1}^F)$.

In general, consider all tasks (n) within a busy period of the system, then, $W_1^\pi + \cdots + W_{j+1}^\pi = W_1^F + \cdots + W_n^F$ and interchanging a pair of customers out of order under π, reduces the value of $f(W_1^\pi, \ldots, W_n^\pi)$ down to the value of $f(W_1^F, \ldots, W_n^F)$ for any Schur convex function f.

If the first n tasks do not form a busy period, then $W_1^\pi + \cdots + W_{j+1}^\pi \geqslant W_1^F + \cdots + W_n^F$ and again $f(W_1^\pi, \ldots, W_n^\pi) \geqslant f(W_1^F, \ldots, W_n^F)$ for any Schur convex function f. ▯

10.5.2 Networks of Queues

To model complex networks or parallel computers, one needs to take into account the fact that resources are interconnected and that tasks may need treatment by several resources. This kind of system is often modeled by networks of queues, in which the tasks leaving one server may be queued into the buffer of another server for further treatment.

An open network Q consists of m queues $Q_1, \ldots, Q_m$ interconnected by a communication network. Each queue Q_i has a buffer, whose capacity is denoted by $c_i \in \mathbb{N} \cup \{\infty\}$, $i = 1, \ldots, m$. The state space (number of tasks) in the queue Q_i is $\mathcal{X}_i = \{0, \ldots, c_i\}$. Hence, the state space $\mathcal{X}$ of the network is $\mathcal{X} = \mathcal{X}_1 \times \cdots \times \mathcal{X}_m$. The state of the system is described by a vector $x = (x_1, \ldots, x_m)$ with x_i the number of tasks in queue Q_i.

An event $e \in \mathcal{E}$ in this network corresponds to a state change. The events are of three possible types: arrival of a new task, completion of a task, or transfer of one task (after completion) from one queue to another.

The whole dynamic is an operator over the state space of the system

$$\phi : \mathcal{X} \times \mathcal{E} \to \mathcal{X}.$$

Now, the evolution of the network is described by the recursive sequence

$$X_{n+1} = \phi(X_n, e_{n+1}), \tag{10.1}$$

where X_n is the state of the network at time n and $\{e_n\}_{n \in \mathbb{N}}$ is a sequence of random variables corresponding to events.

10.5.2.1 Polling from Several Queues: Balanced Sequences

This is a queuing system with a single processing resource and m queues buffering tasks, coming from m exogenous processes (see Figure 10.5).

Here is the scheduling problem to be solved: one must choose the sequence $a = (a_1, \cdots, a_i, \cdots)$ of queues that the server must visit in order to minimize the average expected response time of all tasks. a_i indicates which queue (say j) is visited by the server at step i. For mathematical purposes, a_i is a vector of size m whose components are all null except the j-th one, equal to 1: $a_i(j) = 1$.

Here R^j is the infinite sequence of release dates into queue j, $R^j = (r_i^j)_{i \in \mathbb{N}}$ and the task size sequence entering queue j is $\sigma^j = (\sigma_i^j)_{i \in \mathbb{N}}$. The frequencies of visits to all queues, $\alpha = (\alpha_1, \ldots, \alpha_m)$, is:

$$\alpha = \limsup_{N \to \infty} \frac{1}{N} \sum_{i=1}^{N} a_i.$$

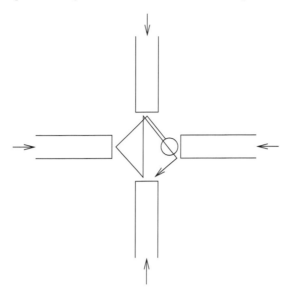

FIGURE 10.5: A polling system.

The average response time of all tasks can be written:

$$\phi_a(R^1, \cdots, R^m, \sigma^1, \cdots, \sigma^m) = \limsup_{n \to \infty} \frac{1}{n} \sum_{i=1}^{n} \alpha_j \sum_{j=1}^{m} (\mathbb{E}W_1^j + \cdots + \mathbb{E}W_n^j),$$

where W_i^j is the response time of task i in queue j.

The problem is difficult to solve in general. However, one may again use structural properties of the function ϕ_a to get comparisons. The function ϕ_a is a convex increasing function of $(a_1 \cdots, a_i, \cdots)$ for fixed R (see [1]). This has a direct consequence using Theorem 10.5.

LEMMA 10.4 Altman, Gaujal, and Hordijk [2]
If $(\sigma^1, \cdots, \sigma^m) \leqslant_{icx} (\beta^1, \cdots, \beta^m)$ then

$$\phi_a(R^1, \cdots, R^m, \sigma^1, \cdots, \sigma^m) \leqslant \phi_a(R^1, \cdots, R^m, \beta^1, \cdots, \beta^m).$$

A more precise result can be obtained. Consider a periodic polling sequence with period T, and frequency $\alpha_1, \cdots, \alpha_m$. Let us introduce the unbalance in queue j, $U_j(a)$ as follows:

$$U_j(a) \stackrel{\text{def}}{=} \max_{k} \frac{1}{T} \sum_{i=k}^{T+k} (a_i(j) - i\alpha_j).$$

The total unbalance is $U(a) = \sum_{j=1}^{m} U_j(a)$.

For a given queue j, there exists a most balanced polling sequence b_j: $U_j(b_j)$ is minimal over all periodic polling sequence with period T, and frequency α_j. See Figure 10.6 for an illustration of this.

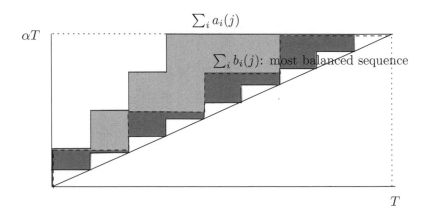

FIGURE 10.6: Here is an example for sequences with period 9 and $\alpha_j = 4/9$. The solid broken line corresponds to $a(j) = (1,1,1,1,0,0,0,0,0)$. The dashed line is the most balanced sequence with frequency 4/9: $b(j) = 1,0,1,0,1,0,1,0,0$. The unbalance of $b(j)$ is the dark area, the unbalance of $a(j)$ is the light area plus the dark one.

THEOREM 10.10 Altman, Gaujal, and Hordijk [2]
If the release sequences and the sizes of all tasks are stationary independent sequences, then $U(a) \leqslant U(a')$ implies $\phi_a \leqslant \phi_{a'}$. Furthermore, if b is a virtual polling sequence composed by the most balanced sequences for each queue (b may not be a real polling sequence since two most regular sequences may be in conflict), there exists a constant C such that

$$\phi_b \leqslant \phi_a \leqslant \phi_b + CU(a).$$

PROOF   The proof is again based on convexity results, and is not reported here. See [2] for a detailed proof.   ▯

Basically, this theorem says that once the frequencies are given, choosing the best polling sequence for the server corresponds to choosing the most regular visit pattern in each queue. This has the same flavor as Ross' conjecture, and has relations with the results presented in Chapter 5 taken in a stochastic context.

The main remaining difficulty is to compute the optimal frequencies of the visits to each queue. This can be done for deterministic polling systems with two queues [10]. In that case, let $\lambda_1 = 1/\tau_1$ be the input intensity in the first queue and λ_2 be the input intensity in the second queue. The optimal frequency of visits to the first queue is a function of λ_1 and λ_2 represented in Figure 10.7. Each zone corresponds to a constant rational value for the frequency. Frontiers correspond to jumps from one value of the frequency to the next. Note that the structure of this graph is highly complex with a fractal behavior.

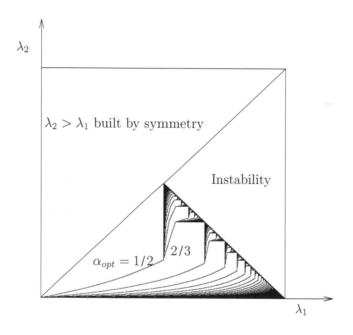

FIGURE 10.7: The best frequency of the server allocations w.r.t. input intensities.

10.5.3   Stochastic Comparisons and Simulation Issues

Simulation approaches are alternative methods to estimate the behavior of stochastic queuing systems. Simulations also benefit from the use of orders and monotonicity properties.

Propp and Wilson used a backward coupling [15] to derive a simulation algorithm providing perfect sampling (i.e., whose distribution is exactly stationary) of the state of *discrete time finite Markov chains*. Here, we adapt their algorithm and show that backward coupling can be used to simulate queuing networks very efficiently.

Here, we will assume that capacities are finite, the release times of all task processes have a Poisson distribution. Their sizes are independent and exponentially distributed. The stochastic process X_n becomes a finite Markov chain.

Let $\phi^n : \mathcal{X} \times \mathcal{E} \to \mathcal{X}$ denote the function whose output is the state of the network after n events: $\phi^n (s, e_{1 \to n}) = \phi (\dots \phi (\phi (s, e_1), e_2), \dots, e_n)$. This notation can be extended to set of states. So, for a set of states $A \subset \mathcal{X}$, we denote $\phi^n (A, e_{1 \to n}) = \{\phi^n (s, e_{1 \to n}), s \in A\}$.

THEOREM 10.11
For all sequences of independently generated events, there exists $\ell \in \mathbb{N}$ such that $\lim_{n \to \infty} |\phi^n (\mathcal{X}, e_{-n+1 \to 0})| = \ell$ almost surely. The system couples if $\ell = 1$. In that case, the value of $X_0 = \phi^n(\mathcal{X}, e_{-n+1 \to 0})$ is steady-state distributed (i.e., X_0 and $X_1 = \phi(X_0, e_1)$ have the same distribution, and so forth).

This existence theorem has an algorithmic counterpart, called the Perfect Simulation Algorithm (PSA).

Algorithm 10.1: Perfect Simulation Algorithm (PSA) of networks

---

Data: A dynamic representation of the system $X_{n+1} = \phi(X_n, e_{n+1})$, and an infinite sequence $e_0, e_{-1}, e_{-2}, \dots$ of independent identical events.

Result: A state $x^* \in \mathcal{X}$ generated according to the limit distribution of the queuing network.

begin
    $m := 1$;
    foreach *state* $x \in \mathcal{X}$ **do**
        $\lfloor\ y(x) := x$;
    repeat
        $y(\cdot) := y(\phi(\cdot, e_{-m}))$;
        $m := m + 1$;
    until $\{y(x), x \in \mathcal{X}\}$ *are all equal (x^* denotes this common value)*
    return x^*
end

The main drawback of Algorithm 10.1 is the fact that one needs to simulate the system starting with all states in $\mathcal{X}$, which could be too large for Algorithm 10.1 to be used in practice.

Several approaches have been used to overcome this problem. The main one is already present in [15]. When the state space $\mathcal{X}$ is ordered and when the function $\phi(\cdot, e)$ is non-decreasing for all e, then it is possible to generate a steady state by starting Algorithm 1 with maximal and minimal states only. This technique has been successfully used in [20] to construct PSA for network of queues. When $\phi(\cdot, e)$ is not monotonic, one can still use monotonic bounds, as in [9]. In [6], it is shown that extremal states can also be found for perfect simulations of a special class of closed non-monotonic Petri nets.

10.5.3.1 Envelopes

Given an order $\preceq$, for which $\mathcal{X}$ is a lattice, then one can consider a new transition function,

$$\Gamma \stackrel{\text{def}}{=} (\overline{\phi}, \underline{\phi}) : \mathcal{X} \times \mathcal{X} \times \mathcal{E} \rightarrow \mathcal{X} \times \mathcal{X}$$

with

$$\overline{\phi}(M, m, e) \stackrel{\text{def}}{=} \sup_{m \preceq x \preceq M} \phi(x, e),$$

$$\underline{\phi}(M, m, e) \stackrel{\text{def}}{=} \inf_{m \preceq x \preceq M} \phi(x, e),$$

whenever $m \preceq M$.

Let us call $Top \stackrel{\text{def}}{=} \sup \mathcal{X}$ (resp. $Bot \stackrel{\text{def}}{=} \inf \mathcal{X}$) the top (resp. bottom) element of $\mathcal{X}$ with respect to order $\preceq$.

From this, one can construct a sequence of ordered pairs of states $S(n) = (Y(n), Z(n))$, defined by iterating Γ starting with (Top, Bot) and using a sequence of independent random events e_{-n}:

$$(Y(-n+1), Z(-n+1)) = \Gamma(Y(-n), Z(-n), e_{-n}),$$

up to time 0.

THEOREM 10.12

Assume that the random sequence $S(n) = (Y(n), Z(n))$ hits the diagonal (i.e., states of the form (x, x)) when starting at some finite time in the past:

$$K \stackrel{\text{def}}{=} \min \{n : Y(0) = Z(0)\}.$$

Then K is a backward coupling time of X. In other words, $\Gamma(Top, Bot, e_{-K}, \dots, e_0)$ has the steady-state distribution of X.

PROOF   The function Γ has been constructed such that once the event sequence is chosen, $Z(-n) \preceq X(-n) \preceq Y(-n)$ for all initial conditions for the real system $X(-K)$. Consider a stationary initial condition $X(-K)$. Then, $X(0) = \phi(X(-K), e_{-K+1}, \dots, e_{-1})$ is also steady-state distributed by stationarity. It just happens that $Y(0) = X(0) = Z(0)$ by definition of K so that $Z(0) = Y(0)$ is also steady-state distributed.   ☐

Now, Algorithm 10.1 can be adapted to the envelope simulation: start the simulation with only two states: *Top* and *Bot* and iterate using the bidimensional function Γ.

In general, this approach may not gain over the general coupling techniques because of three problems:

(P_1) The assumption that $S(n)$ hits the diagonal may not be verified.

(P_2) Even if Theorem 10.12 applies, the coupling time can become prohibitively large.

(P_3) The time needed to compute $\overline{\phi}(M, m, u)$ and $\underline{\phi}(M, m, u)$ might depend on the number of states between m and M which could affect the efficiency of this approach.

Algorithm 10.2: Backward simulation of a queuing system using envelopes.

Data: A function Γ and an infinite sequence $e_0, e_{-1}, e_{-2}, \dots$ of independently generated events.

Result: A state $X \in \mathcal{X}$ generated according to the limit distribution of the network

begin
    $n := 1; T := Top; B := Bot$
    repeat
        for $i = n$ *downto* 1 **do**
            $T := \overline{\phi}(T, B, e_i)$ $B := \underline{\phi}(T, B, e_i)$
        $n := 2n$
    until $T = B$
    return T
end

Some Applications.

When the function $\phi(., e)$ is monotone then the functions $\overline{\phi}(M, m, e)$ and $\underline{\phi}(M, m, e)$ degenerate into $\phi(M, e)$ and $\phi(m, e)$, respectively. In this case, problems P_1 and P_3 always have a positive answer [15]. As for problem P_2, the coupling time is shown to have an exponentially decaying distribution and an expectation which is polynomial is the parameters of the system [8].

When the queuing network is not monotone, the envelopes can still be used to simulate efficiently rather general classes of queueing networks. For example networks of m finite queues (of capacity c) with general index routing (as in [21]) and batch arrivals (which break the monotonicity property). In that case, envelopes always couple with probability one (Problem (P_1)). Problem (P_2) is solved by using a partial split of the trajectories when the states reached by the lower and upper envelopes get close in a queue. Problem (P_3) is solved by constructing an algorithm computing ϕ with complexity $O(m \log(c))$ [4].

Other examples where problems $(P_1), (P_2), (P_3)$ can be solved in a satisfying manner are networks of N finite queues with negative customers and/or with fork and join nodes, which are not monotonic.

This technique has been used to assess the performance of several grid scheduling policies for monotonic as well as non-monotonic grid systems (see for example [5]).

References

[1] E. Altman, B. Gaujal, and A. Hordijk. Optimal open-loop control of vacations, polling and service assignment. *Queueing Systems*, 36:303–325, 2000.

[2] E. Altman, B. Gaujal, and A. Hordijk. *Discrete-Event Control of Stochastic Networks: Multimodularity and Regularity*, volume 1829 of *LNM*. Springer-Verlag, 2003.

[3] F. Baccelli and P. Brémaud. *Elements of Queueing Theory*. Springer-Verlag, 1994.

[4] V. Berten, A. Busic, B. Gaujal, and J.-M. Vincent. Can we use perfect simulation for non-monotonic markovian systems? In *Roadef*, pages 166–167, Clermont-Ferrand, 2008.

[5] V. Berten and B. Gaujal. Brokering strategies in computational grids using stochastic prediction models. *Parallel Computing*, 33(4-5):238–249, 2007. Special Issue on Large Scale Grids.

[6] A. Bouillard and B. Gaujal. Backward coupling in bounded free-choice nets under markovian and non-markovian assumptions. *Journal of Discrete Event Dynamics Systems: Theory and Applications*, 18(4):473–498, 2008. Special issue of selected papers from the Valuetools conference.

[7] B. Dodin. Bounding the project completion time distributions networks. *Operations Research*, 33(4):862–881, 1985.

[8] J. Dopper, B. Gaujal, and J.-M. Vincent. Bounds for the coupling time in queueing networks perfect simulation. In *Numerical Solutions for*

Markov Chain (NSMC06), pages 117–136, Charleston, June 2006. Celebration of the 100th anniversary of Markov.

[9] J.-M. Fourneau, I. Y. Kadi, N. Pekergin, J. Vienne, and J.-M. Vincent. Perfect Simulation and Monotone Stochastic Bounds. In *2nd International Conference Valuetools'07*, Nantes, France, Oct. 2007.

[10] B. Gaujal, A. Hordijk, and D. V. der Laan. On the optimal open-loop control policy for deterministic and exponential polling systems. *Probability in Engineering and Informational Sciences*, 21:157–187, 2007.

[11] J. Kamburowski. Bounding the distribution of project duration in PERT networks. *Operation Research Letters*, 12:17–22, 1992.

[12] S. Karlin. Dynamic inventory policy with varying stochastic demands. *Management Sciences*, 6:231–258, 1960.

[13] G. Koole. Structural results for the control of queueing systems using event-based dynamic programming. Technical Report WS-461, Vrije Universiteit Amsterdam, 1996.

[14] A. Müller and D. Stoyan. Comparison methods for stochastic models and risks. *Wiley Series in Probability and Statistics*. Wiley, 2002.

[15] J. G. Propp and D. B. Wilson. Exact sampling with coupled Markov chains and applications to statistical mechanics. *Random Structures and Algorithms*, 9(1-2):223–252, 1996.

[16] S. M. Ross. Average delay in queues with non-stationary Poisson arrivals. *Journal of Applied Probability*, 15:602–609, 1978.

[17] S. M. Ross. *Introduction to Probability Models*. Academic Press, 2003.

[18] J. Shanthikumar and D. D. Yao. Bivariate characterization of some stochastic order relations. *Advances in Applied Probability*, 23(3):642–659, 1991.

[19] V. Strassen. The existence of probability measures with given marginals. *Annals of Mathematical Statistics*, 36:423–439, 1965.

[20] J.-M. Vincent. Perfect Simulation of Monotone Systems for Rare Event Probability Estimation. In *Winter Simulation Conference*, Orlando, Dec. 2005.

[21] J.-M. Vincent and J. Vienne. Perfect simulation of index based routing queueing networks. *Performance Evaluation Review*, 34(2):24–25, 2006.

[22] N. Yazici-Pekergin and J.-M. Vincent. Stochastic bounds on execution times of parallel programs. *IEEE Trans. Software Eng.*, 17(10):1005–1012, 1991.

Chapter 11

The Influence of Platform Models on Scheduling Techniques

Lionel Eyraud-Dubois

INRIA and Université de Bordeaux

Arnaud Legrand

CNRS and Université de Grenoble

Abstract In this chapter, we first review a wide representative panel of platform models and topology used in the literature. We explain why such models have been used and why they may be irrelevant to current applications and current platforms. Then we present through a few simple examples the influence of these models on the complexity of the corresponding problems.

11.1   Introduction

The object of this chapter is to discuss different models used in the field of scientific computing. The large need of computational resource motivates the use of parallel systems, in which several processors communicate to perform a given task. The optimization of these communications is an important part of the development of parallel applications, and it requires an appropriate model of the platform. The choice of a model is the first step for solving a practical problem, and it influences greatly the obtained results. However, the best model is not necessarily the most accurate one. For example, the well-known PRAM models are very simple (synchronous, unit-time, and contention-free communications), but they have led to very interesting results about parallel complexity and algorithm analysis.

We review in detail in Section 11.2 the most important characteristics of common models used in the scheduling literature, and the situations in which they should be considered. In the next sections, we illustrate this characterization with three different problems (namely divisible load scheduling in Section 11.3, placement for iterative algorithms in Section 11.4, and data redistribution in Section 11.5). For each of them, we study several modeling options as well as the influence of the choice of a model in terms of complexity and performance of the solutions.

11.2 Platform Modeling

In this section, we describe several characteristics of the main communication models used in the literature. It is possible to adjust these characteristics almost independently, allowing to model a wide variety of situations. Indeed, the appropriate model for a study is always a compromise between realism and tractability (and if it is meant for practical use, the instantiation of its parameters from the real world is also an important issue). It is thus important to carefully select which aspects of the system needs to be modeled precisely, and which aspects can have a more approximate model.

In the following, we will first present how to model the *topology* of the platform, and its implications on the problems of routing. Then we will see how to model *point-to-point* communications, and say a word about the *heterogeneity* of the platform. The next important point is the modeling of *contentions*, for which there are several interesting and complementary approaches. Lastly, we will study the *interactions* between computations and communications.

Then, through various examples, we study how these models impact the complexity of corresponding scheduling problems.

11.2.1 Modeling the Topology

Link-level topology – Graphs are a common and quite natural way of modeling a communication network, in which vertices represent computing nodes, or the internal routers and switches, and edges represent cables or communication links. This yields a very precise model, which is actually used by network simulators to evaluate the performance of communication protocols. It is also useful for dedicated parallel platforms, since they often have a well-known, simple and regular interconnection scheme that makes it possible to consider such a precise model. Topologies which have been quite successful include grids (two- or three-dimensional), torus, and hypercubes.

More off-the-shelf platforms operating on more standard networks (which may be local or wide-area networks) can however not really be modeled in such a way. Acquiring and managing all the required information would actually be too costly, and it is unclear whether the precision gained by such a model would be worth the cost. It is thus necessary to consider coarser models, adapted to the application and situation, while keeping in mind the real underlying topology.

Organization of the application – Most applications have a structured communication scheme, and it is sometimes possible to use this structure to simplify the model of the topology. Master/Slave applications are a good example of this situation. In that case, one master node holds all the data of the problem, and needs to distribute them to a collection of slave nodes, which will perform the computations. Therefore, the network links between two slaves do not need to be taken into account, since all communications take place between the master and one of the slaves. Of course, the precise model to use may depend on the platform: it can be a star if the contention is assumed to happen only at the master node, or a tree (with the master as a root) to model the contention more precisely.

A variety of applicative topology are commonly used; they include *chains* or *rings*, where each node only communicates with two neighbors, *grids* or *torus*, which are basically multi-dimensional versions of chains or rings, and *hypercubes*, which are becoming widespread in peer-to-peer applications.

Application-level topology – It is also possible to consider a simplified view of the link-level topology, that would only include the *visible* aspects for the application. Routing protocols on most platforms allow two nodes to seamlessly communicate, even if the path connecting them crosses several routers. It is then natural to exclude these routers from the network model, and propose a model for the whole network connection between these two hosts (the actual models for this situation are discussed in Section 11.2.2).

On platforms with more than two hosts, directly extending this view leads to the *clique* model, in which communications between two different pairs of nodes do not interfere. This model can be adapted for some (for example, latency-bound) applications, or in situations where the nodes' bandwidths are small compared to the capacity of the backbone links connecting them (as would be the case in a peer-to-peer setting). In many cases, however, this hypothesis is not applicable. There are two main approaches to give more precise models. The first approach is to consider that the platform model is a general graph, so as to be sure to capture all possible situations. However, this often leads to untractable problems, and is also quite difficult to instantiate (i.e., find the input graph and values that represent a given practical situation that we want to model). The second approach is to restrict the class of graphs considered; popular choices are trees or hierarchical structures like clusters of clusters. Since they are not as general, the corresponding problems are easier to solve, and the models can be reasonably instantiated.

Modeling the routing – More precise models often mean that there are several possible routes between two hosts. There are two ways for the choice of the route: either the communication algorithm has to choose (and optimize) which route to use, or it is specified in the instance, together with the network model. The more classical solution used in academic works is the first one, but it yields very difficult problems on general graphs, and it is most often unrealistic, since applications cannot select which route the messages will use.

11.2.2   Modeling Point-to-Point Communication Time

A simple affine model – Let us assume that a node, P_i, wishes to send a message M of size m to another node, P_j. The time needed to transfer a message along a network link is roughly linear in the message length. Therefore, the time needed to transfer a message along a given route is also linear in the message length. This is why the time $C_{i,j}(m)$ required to transfer a message of size m from P_i to P_j is often modeled as

$$C_{i,j}(m) = L_{i,j} + m/bw_{i,j} = L_{i,j} + m.c_{i,j},$$

where $L_{i,j}$ is the start-up time (also called latency) expressed in seconds and $bw_{i,j}$ is the bandwidth expressed in bytes per second. For convenience, we also define $c_{i,j}$ as the inverse of $bw_{i,j}$. $L_{i,j}$ and $bw_{i,j}$ depend on many factors: the length of the route, the communication protocol, the characteristics of each network link, the software overhead, the number of links that can be used in parallel, whether links are half-duplex or full-duplex, etc. This model was proposed by Hockney [13] to model the performance of the Intel Paragon and is probably the most commonly used.

The LogP family – Many aspects are not taken into account in the Hockney model. One of them is that at least three components are involved in a communication: the sender, the network and the receiver. There is *a priori* no reason for all these components to be fully synchronized and busy at the same time. It may thus be important to evaluate precisely when each one of these components is active and whether it can perform some other task when not actively participating in the communication.

The LogP [10] model, and other models based on it, have been proposed as more precise alternatives to the Hockney model. More specifically, these models account for the fact that the node may partially overlap some of the communications. Messages, say of size m, are split in small packets whose size is bounded by the Maximum Transmission Unit (MTU), w. In the LogP model, L is an upper bound on the latency. o is the *overhead*, defined as the length of time for which a node is engaged in the transmission or reception of a packet. The gap g is defined as the minimum time interval between consecutive packet transmissions or consecutive packet receptions at a node. During this time, the node cannot use the communication coprocessor, i.e., the network card. The reciprocal of g thus corresponds to the available per-node communication bandwidth. Lastly, P is the number of nodes in the platform. Figure 11.1 depicts the communication between nodes each equipped with a

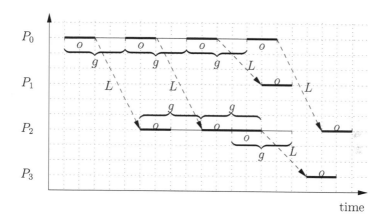

FIGURE 11.1: Communication under the LogP model.

network card under the LogP model. In this model, sending m bytes with packets of size w takes time

$$\mathcal{C}(m) = 2o + L + \left\lfloor \frac{m-w}{w} \right\rfloor \cdot g \, .$$

The processor occupation time on the sender and on the receiver is equal to

$$o + \left(\left\lfloor \frac{m}{w} \right\rfloor - 1 \right) \cdot g \, .$$

One goal of this model was to summarize in a few numbers the characteristics of parallel platforms. It would thus have enabled the easy evaluation of complex algorithms based only on the previous simple parameters. However, parallel platforms are very complex and their architectures are often hand-tuned with many optimizations. For instance, often, different protocols are used for short and long messages. LogGP [1] is an extension of LogP where G captures the bandwidth for long messages. One may however wonder whether such a model would still hold for average-size messages. pLogP [16] is thus an extension of LogP when L, o and g depends on the message size m. This model also introduces a distinction between the sender and receiver overhead (o_s and o_r).

A complex affine model –   One drawback of the LogP models is the use of floor functions to account for explicit MTU. The use of these non-linear functions causes many difficulties for using the models in analytical/theoretical studies. As a result, many fully linear models have also been proposed. We summarize them through the general scenario depicted in Figure 11.2.

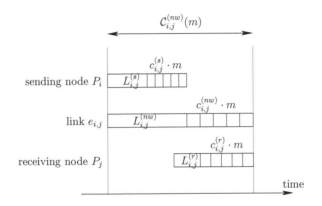

FIGURE 11.2: Sending a message of size m from P_i to P_j.

We define a few notations to model the communication from P_i to P_j:

- The time during which P_i is busy sending the message is expressed as an affine function of the message size: $\mathcal{C}_{i,j}^{(s)}(m) = L_{i,j}^{(s)} + m \cdot c_{i,j}^{(s)}$. The start-up time $L_{i,j}^{(s)}$ corresponds to the software and hardware overhead paid to initiate the communication. The quantity $c_{i,j}^{(s)}$ corresponds to the inverse of the transfer rate that can be achieved by the processor when sending data to the network (say, the data transfer rate from the main memory of P_i to a network card able to buffer the message).

- Similarly, the time during which P_j is busy receiving the message is expressed as an affine function of the message length m, namely $\mathcal{C}_{i,j}^{(r)}(m) = L_{i,j}^{(r)} + m \cdot c_{i,j}^{(r)}$.

- The total time for the communication, which corresponds to the total occupation time of the link $e_{i,j} : P_i \rightarrow P_j$, is also expressed as an affine function $\mathcal{C}_{i,j}^{(nw)}(m) = L_{i,j}^{(nw)} + m \cdot c_{i,j}^{(nw)}$. The parameters $L_{i,j}^{(nw)}$ and $c_{i,j}^{(nw)}$ correspond, respectively, to the start-up cost and to the inverse of the link bandwidth.

Simpler models (e.g., Hockney) do not make the distinction between the three quantities $\mathcal{C}_{i,j}^{(s)}(m)$ (emission by P_i), $\mathcal{C}_{i,j}^{(nw)}(m)$ (occupation of $e_{i,j} : P_i \rightarrow P_j$) and $\mathcal{C}_{i,j}^{(r)}(m)$ (reception by P_j). Such models use $L_{i,j}^{(s)} = L_{i,j}^{(r)} = L_{i,j}^{(nw)}$ and $c_{i,j}^{(s)} = c_{i,j}^{(r)} = c_{i,j}^{(nw)}$. This amounts to assuming that the sender P_i and the receiver P_j are blocked throughout the communication. In particular, P_i cannot send any message to another neighbor P_k during $\mathcal{C}_{i,j}^{(nw)}(m)$ time units. However, some system/platform combinations may allow P_i to proceed to another send operation before the entire message has been received by P_j. To account for this situation, more complex models would use different functions for $\mathcal{C}_{i,j}^{(s)}(m)$ and $\mathcal{C}_{i,j}^{(nw)}(m)$, with the obvious condition that $\mathcal{C}_{i,j}^{(s)}(m) \leqslant \mathcal{C}_{i,j}^{(nw)}(m)$ for all message sizes m (which implies $L_{i,j}^{(s)} \leqslant L_{i,j}^{(nw)}$ and $c_{i,j}^{(s)} \leqslant c_{i,j}^{(nw)}$). Similarly, P_j may be involved only at the end of the communication, during a time period $\mathcal{C}_{i,j}^{(r)}(m)$ smaller than $\mathcal{C}_{i,j}^{(nw)}(m)$.

Here is a summary of the general framework. Assume that P_i initiates a communication of size m to P_j at time $t = 0$:

- Link $e_{i,j} : P_i \rightarrow P_j$ is busy from $t = 0$ to $t = \mathcal{C}_{i,j}^{(nw)}(m) = L_{i,j}^{(nw)} + L \cdot c_{i,j}^{(nw)}$

- Processor P_i is busy from $t = 0$ to $t = \mathcal{C}_{i,j}^{(s)}(m) = L_{i,j}^{(s)} + m \cdot c_{i,j}^{(s)}$, where $L_{i,j}^{(s)} \leqslant L_{i,j}^{(nw)}$ and $c_{i,j}^{(s)} \leqslant c_{i,j}^{(nw)}$

- Processor P_j is busy from $t = \mathcal{C}_{i,j}^{(nw)}(m) - \mathcal{C}_{i,j}^{(r)}(m)$ to $t = \mathcal{C}_{i,j}^{(nw)}(m)$, where $\mathcal{C}_{i,j}^{(r)}(m) = L_{i,j}^{(r)} + m \cdot c_{i,j}^{(r)}$, $L_{i,j}^{(r)} \leqslant L_{i,j}^{(nw)}$ and $c_{i,j}^{(r)} \leqslant c_{i,j}^{(nw)}$

Banikazemi et al. [3] propose a model that is very close to the general model presented above. They use affine functions to model the occupation time of the processors and of the communication link. The only minor difference is that they assume that the time intervals during which P_i is busy sending (of duration $\mathcal{C}_{i,j}^{(s)}(m)$) and during which P_j is busy receiving (of duration $\mathcal{C}_{i,j}^{(r)}(m)$) do not overlap. Consequently, they write

$$\mathcal{C}_{i,j}(m) = \mathcal{C}_{i,j}^{(s)}(m) + \mathcal{C}_{i,j}^{(nw)}(m) + \mathcal{C}_{i,j}^{(r)}(m) .$$

In [3] a methodology is proposed to instantiate the six parameters of the affine functions $\mathcal{C}_{i,j}^{(s)}(m)$, $\mathcal{C}_{i,j}^{(nw)}(m)$ and $\mathcal{C}_{i,j}^{(s)}(m)$ on a heterogeneous platform. The authors show that these parameters actually differ for each processor pair and depend upon the CPU speeds.

A simplified version of the general model was proposed by Bar-Noy et al. [4]. In this variant, the time during which an emitting processor P_i is blocked does not depend upon the receiver P_j (and similarly the blocking time in reception does not depend upon the sender). In addition, only fixed-size messages are considered in [4], so that one obtains

$$\mathcal{C}_{i,j} = \mathcal{C}_i^{(s)} + \mathcal{C}_{i,j}^{(nw)} + \mathcal{C}_j^{(r)} . \tag{11.1}$$

Back on the relevance of the simple affine model − In parallel machines, point-to-point communications are often implemented with more or less simple protocols (store-and-forward, cut-through). Depending on the size of the message a protocol may be preferred to another but in most cases, the resulting communication time is still an affine function of the message size. On the dedicated networks of such parallel platforms, with good buffer management, there is almost no message loss and thus no need for flow-control mechanisms. By contrast, in large networks (e.g., a subset of the Internet) there is a strong need for congestion avoidance mechanisms. Protocols like TCP split messages just like wormhole protocols but do not send all packets at once. Instead, they wait until they have received some acknowledgments from the destination node. To this end, the sender uses a sliding window of pending packet transmissions whose size changes over time depending on network congestion. The size of this window is typically bounded by some operating system parameter W_{max} and the achievable bandwidth is thus bounded by $W_{max}/($Round Trip Time$)$. Such flow-control mechanisms involve a behavior called *slow start*: it takes time before the window gets full and the maximum bandwidth is thus not reached instantly. One may thus wonder whether the Hockney model is still valid in such networks. It turns out that the model remains valid in most situations but $L_{i,j}$ and $B_{i,j}$ cannot be determined via simple measurements.

11.2.3   Heterogeneity

In the most general models, each of the values associated to a node (computing power, communication overhead, etc.) or to a link (bandwidth, latency, etc.) are different, in order to model the *heterogeneity* of the platform. This is relevant when modeling large networks, in which these capacities can vary greatly from one component to the other. However, this variety often leads to intractable problems.

In contrast, in the *homogeneous* models, these values are assumed to be equal. This assumption is valid for many parallel systems (like clusters for example, which are made of strictly identical machines), and problems of this sort can often be solved polynomially.

It is also possible to use intermediate approaches, depending on the situation. Models with homogeneous communications, but heterogeneous computational speeds, for example, can be appropriate for *networks of workstations*, made of many standard workstations connected by a local network. The *cluster of clusters* architecture is another example, in which several homogeneous clusters are connected together to form a large-scale platform. All the machines inside a cluster are identical, but may be different from one cluster to the other. Furthermore, any kind of network model may be used for the interconnection of the clusters.

A classical approach to handle heterogeneity is to bound the ratio between the speed of the fastest and the slowest resource, and to express the quality of the solution, or the running time of the algorithm, as a function of this bound. This is quite useful when the heterogeneity is too large to be neglected, but remains small enough for the results to be interesting.

11.2.4 Modeling Concurrent Communications

As we have seen in Section 11.2.2 there is a wide variety of interconnection networks. There is an even wider variety of network technology and protocols. All these factors greatly impact the performance of point-to-point communications. They impact even more the performance of concurrent point-to-point communications. The LogP model could probably be used to evaluate the performance of concurrent communications but it has clearly not been designed to this end. In this section, we present a few models that account for the interference between concurrent communications.

Multi-port − Sometimes a good option is simply to ignore contention. This is what the multi-port model does. All communications are considered to be independent and do not interfere with each others. In other words, a given processor can communicate with as many other processors as needed without any degradation in performance. This model is often associated to a clique interconnection network to obtain the macro-dataflow model. This model is widely used in scheduling theory since its simplicity makes it possible to prove that problems are hard and to know whether there is some hope of proving interesting results.

The major flaw of the macro-dataflow model is that communication resources are not limited. The number of messages that can simultaneously circulate between processors is not bounded, hence an unlimited number of communications can simultaneously occur on a given link. In other words, the communication network is assumed to be contention-free. This assumption is of course not realistic as soon as the number of processors exceeds a few units. Some may argue that the continuous increase in network speed favors the relevance of this model. However, with the growing popularity of multi-core multi-processor machines, contention effects can clearly no longer be neglected. Note that the previously discussed models by Banikazemi et al. [3] and by Bar-Noy et al. [4] are called *multi-port* because they allow a sending processor to initiate another communication while a previous one is still taking place. However, both models impose an overhead to pay before engaging in another operation. As a results these models do not allow for fully simultaneous communications.

Bounded multi-port − Assuming an application that uses threads and possibly a node that uses multi-core technology, the network link is shared by incoming and outgoing communications. Therefore, the sum of the bandwidths allotted by the operating system to all communications cannot exceed the bandwidth of the network card. The bounded multi-port model proposed by Hong and Prasanna [14] is an extension of the multi-port model with a bound on the sum of the bandwidths of concurrent communications at each node. An unbounded number of communications can thus take place simultaneously, provided that they share the total available bandwidth.

Note that in this model, there is no degradation of the aggregate throughput. Such a behavior is typical for protocols with efficient congestion control mechanisms (e.g., TCP). Note, however, that this model does not express how the bandwidth is shared among the concurrent communications. It is generally assumed in this model that the application is allowed to define the bandwidth allotted to each communication. In other words, the bandwidth sharing is performed by the application and not by the operating system, which is not the case in many real-world scenarios.

1-port (unidirectional or half-duplex) – To avoid unrealistically optimist results obtained with the multi-port model, a radical option is simply to forbid concurrent communications at each node. In the 1-port model, a processor can either send data or receive data, but not simultaneously. This model is thus very pessimistic as real-world platform can achieve some concurrency of computation. On the other hand, it is straightforward to design algorithms that follow this model and thus to determine their performance *a priori*.

1-port (bidirectional or full-duplex) – Nowadays most network cards are full-duplex, which means that emissions and receptions are independent. It is thus natural to consider a model in which a processor can be engaged in a single emission *and* in a single reception at the same time.

The bidirectional 1-port model is used by Bhat et al. [6, 7] for fixed-size messages. They advocate its use because "current hardware and software do not easily enable multiple messages to be transmitted simultaneously." Even if non-blocking multi-threaded communication libraries allow for initiating multiple send and receive operations, they claim that all these operations "are eventually serialized by the single hardware port to the network." Experimental evidence of this fact has recently been reported by Saif and Parashar [23], who report that asynchronous sends become serialized as soon as message sizes exceed a few tens of kilobytes. Their results hold for two popular implementations of the MPI message-passing standard, MPICH on Linux clusters and IBM MPI on the SP2.

The one-port model fully accounts for the heterogeneity of the platform, as each link has a different bandwidth. It generalizes a simpler model studied by Banikazemi et al. [2], Liu [19], and Khuller and Kim [15]. In this simpler model, the communication time $\mathcal{C}_{i,j}(m)$ only depends on the sender, not on the receiver: in other words, the communication speed from a processor to all its neighbors is the same.

k-ports –   A node may have $k > 1$ network cards and a possible extension of the 1-port model is thus to consider that a node cannot be involved in more than an emission and a reception on each network card.

Bandwidth sharing –   The main drawback of all these models is that they only account for contention on nodes. However, other parts of the network can be limiting performance, especially when multiple communications occur concurrently. It may thus be interesting to write constraints similar to the bounded multi-port model on each network link.

Consider a network with a set $\mathcal{L}$ of network links and let bw_l be the bandwidth of resource $l \in \mathcal{L}$. Let a route r be a non-empty subset of $\mathcal{L}$ and let us consider $\mathcal{R}$ the set of active routes. Set $A_{l,r} = 1$ if $l \in r$ and $A_{l,r} = 0$ otherwise. Lastly let us denote by ρ_r the amount of bandwidth allotted to connection r. Then, we have

$$\forall l \in \mathcal{L} : \sum_{r \in \mathcal{R}} A_{l,r}\rho_r \leqslant bw_l \,. \tag{11.2}$$

We also have

$$\forall r \in \mathcal{R} : \rho_r \geqslant 0 \,. \tag{11.3}$$

The network protocol will eventually reach an equilibrium that results in an allocation ρ such that $A\rho \leqslant bw$ and $\rho \geqslant 0$. Most protocols actually optimize some metric under these constraints. Recent works [22, 20] have successfully characterized which equilibrium was reached for many protocol. For example ATM networks recursively maximize a weighted minimum of the ρ_r whereas some versions of TCP optimize a weighted sum of the $\log(\rho_r)$ or even more complex functions.

Such models are very interesting for performance evaluation purposes but as we will see later, even when assuming that bandwidth sharing is performed by the application and not by the operating system, they often prove too complicated for algorithm design purposes.

11.2.5   Interaction between Communication and Computation

Computation times are also an important part of most of the models. Sometimes (for example when optimizing collective communications, like a broadcast), computations can be ignored to place all the focus on communications. But most of the time, communications are required to perform computations on distant nodes, and it is then necessary to model the interactions between these two activities.

In sequence –   In the classical model, communications and computations are performed *in sequence*: it is necessary to wait for the end of a communication before starting an associated computation, and vice-versa. This make perfect sense, since it is not possible to start working on data that has not arrived yet. Several variations can be proposed in this setting, to give more precisions on these interactions.

The first precision concerns the *overlapping* of communications and computations, and is quite connected to the previous discussion about contention. Most of the models (like the macro-dataflow model for example) assume that it is possible for a node to communicate *and* compute at the same time (since the communication coprocessor does all the communication-related work); in other models, the communication takes up all of the computing power. In practice, communications use the memory bus and hence interfere with computations. Such interferences have for example been observed and measured in [17]. Taking such interferences into account may become more and more important with multi-core architectures but these interferences are not well understood yet. The choice of one or the other model may depend on the actual programming or hardware environment, since both behaviors can be found in real-world applications

Another variation of this aspect concerns the computation times. In the macro-dataflow model, they are fixed by the input instance, and the only choice of the scheduling algorithm is to select the starting time and the computing resource that will process each task. In contrast, the *divisible* model allows to divide the computations in chunks of arbitrary size, and the computation time of a chunk depends on its size. The most general model for this dependency is an affine function, with a latency and a processing speed. However, the latency is often assumed to be zero, since this greatly simplifies the problem.

As flows – Some applications, like video encoding for example, process data as a *flow*. In that case, it can make sense to relax the previous assumption, and to allow the computation to start before the communication is over. Of course, this is strongly related to the divisible case, which when considered with a *steady-state* point of view, reduces to this kind of model. This relaxation often leads to much simpler problems, which can be solved by maximum flow techniques.

11.3   Scheduling Divisible Load

In the problem of divisible load scheduling, we consider a large amount of computations that need to be performed by a certain number of "slave" nodes. The "master" node originally holds all the data, and the goal is to organize the communications to distribute this data among the participating slaves. The model is called "divisible" because the computations can be divided in chunks

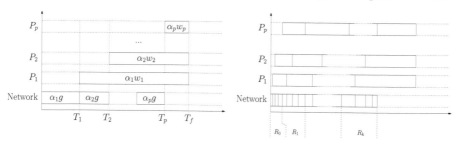

FIGURE 11.3: Single round versus multi-round for one-port divisible load scheduling.

of arbitrary sizes. It is thus suited for large "bag-of-tasks" applications, in which the granularity is low enough to be neglected. Readers interested in the details of the results presented here are invited to read Chapter 8, in which the divisible load model is presented in details.

11.3.1 Single Round

In the single round version of this problem, the master only communicates once with each slave. Consequently, each slave first receives all the data alloted to it before performing the computation. The difference between single and multi-round is depicted on Figure 11.3.

Notations – We consider p slave processors P_1, ..., P_p. Processor P_i executes a unit of work in time w_i and will be alloted a share α_i of the total workload W (we assume $\alpha_i \geqslant 0$ and $\sum_{j=1}^{p} \alpha_j = W$). Therefore, the time needed by processor P_i to process its share is $\alpha_i.w_i$. The time to communicate a unit of data to processor P_i is denoted by c_i, so the communication of its share of data will take time $\alpha_i.c_i$. The problem is to find a good allocation of the total workload (the values of the α_i), and when appropriate a good ordering of the communications toward the slaves, so as to minimize the total completion time. The dual version of this problem is to maximize the workload W that can be processed within a given time bound T, which we consider in the rest of this section to simplify the notations and sketches of the proofs as both problems have the same complexity [24]. Note that we assume that the master node does not perform any computation. Last, in the whole section, we will not consider return messages: a processor completes as soon as its computation is finished.

In the following, we consider different contention models, with increasing difficulty.

Unbounded multi-port – In the unbounded multi-port model, contention is ignored: the master can communicate in parallel with all the slaves. All communications thus start at time 0, and the completion time of processor i is simply $\alpha_i.(c_i + w_i)$. It is then very simple to ensure that all slaves finish at the same time T by setting $\alpha_i = \frac{T}{(c_i+w_i)}$. With the additional constraint that all α_i sum up to 1, we can find the optimal allocation with

$$\alpha_i = \frac{\frac{1}{c_i+w_i}}{\sum_j \frac{1}{c_j+w_j}}$$

With this simple model, it is also possible to take latencies into account. Let us assume that the completion time of a participating processor i is now $L_i + \alpha_i.(c_i + w_i)$. To find an optimal solution, we can simply sort the processors by increasing L_i, and "fill" the W units of load. More precisely, if $L_1 \leqslant L_2 \leqslant \cdots \leqslant L_p$, the procedure is to assign load to P_1 until its completion time reaches L_2, then assign load to both P_1 and P_2 (the respective shares are computed as previously) until the completion time reaches L_3, and so on until the deadline T is reached.

1-port model – In the 1-port model, the master can only communicate with *one* slave at a time. It is thus necessary to specify the ordering of the slaves in addition to the workload distribution. If we renumber the processors by their order of communication, then slave P_i starts its communication at time $\sum_{j<i} \alpha_j.c_j$, and its completion time is $\sum_{j\leqslant i} \alpha_j.c_j + \alpha_i.w_i$. It is still true that all processors participate in an optimal solution, since the outgoing port of the master node always has some idle time at the end of any solution. It is thus always possible to improve a partial solution by scheduling the communication of an additional processor at the end. Furthermore, all processors finish their computation at the same time. The best way to see this is to write the corresponding linear program (for a fixed ordering):

$$\text{MAXIMIZE} \quad W = \sum_i \alpha_i \quad \text{SATISFYING THE (IN)EQUATIONS}$$

$$\begin{cases} \forall i, & \alpha_i \geqslant 0 \\ \forall i, & \sum_{j\leqslant i} \alpha_j c_j + \alpha_i w_i \leqslant T \end{cases}$$

Notice now that an optimal solution is necessarily on the border of the corresponding polytope, and that the first type of constraints can not be equalities, because all processors participate. Counting the remaining constraints and variables implies that all constraints of the second type have to be equalities, and thus all processors finish at the same time T.

The problem is then completely solved by an exchange argument, which proves that it is always best to schedule a processor with lower c_i first. The optimal solution thus orders processors by increasing c_i, and allocates the workload so that all complete at time T. In this case it is still possible to write closed-form formulas for the α_i.

Considering latencies makes the problem more complicated. It is still polynomial if latencies are homogeneous ($L_i = L$ for all i), but it is no longer the case that all processors participate. However, the problem becomes NP-hard with heterogeneous latencies, even when all $c_i = 0$ [24].

Bounded multi-port – In the bounded multi-port model, the master is once again allowed to communicate with several slaves at a time, but with a maximum bandwidth bw_0. Thus, a complete specification of the communication shedule requires to specify the instantaneous bandwidth $bw_i(t)$ associated with processor i at time t, with the constraint that their sum does not exceed bw_0 at any time, and of course that each $bw_i(t)$ does not exceed the maximum input bandwidth[1] bw_i of processor i. Fortunately, it can be proved [5] that the functions $bw_i(t)$ can be chosen piecewise constant with no more than p different values.

Given a bandwidth allocation $(bw_i(t))_{i \leq p}$, we can define the time t_i at which the communication towards processor i is finished with $t_i = \max\{t | bw_i(t) > 0\}$. We also define the share α_i of work it has received with $\alpha_i = \int_0^{t_i} bw_i(t)dt$. The completion time of processor i is then $t_i + \alpha_i.w_i$. The problem can then be formulated by:

$$\text{MAXIMIZE} \quad W = \sum_i \alpha_i \quad \text{SATISFYING THE (IN)EQUATIONS}$$

$$\begin{cases} \forall i, t & bw_i(t) \leq bw_i \\ \forall t & \sum_i bw_i(t) \leq bw_0 \\ \forall i & \int_0^{t_i} bw_i(t) = \alpha_i \\ \forall i & t_i + \alpha_i.w_i \leq T \end{cases}$$

Surprisingly, unlike all other divisible problems without latencies, this problem turns out to be NP-hard [5]. It is indeed possible to prove a reduction from 2-PARTITION, by building an instance where all $bw_i.w_i = 1$. If a partition of the bw_i in two sets of sum bw_0 exists, the optimal solution is to allocate their maximum bandwidth to all processors of the first set (their t_i is then equal to $\frac{T}{2}$), then similarly with the second set (with a t_i equal to $\frac{3T}{4}$). The total work performed is then $\frac{3}{4}T.bw_0$, and any solution that does not perfectly partition the processors performs less work.

Introducing this more realistic model makes the problem more difficult, but it is still worthy to be considered. Indeed, these solutions allow to use more efficiently the output link of the master node, which is the only critical resource. Furthermore, even if finding an optimal solution is difficult, it is fairly easy to give a "good" solution. With two processors for example, the *worst* greedy solution performs at least $\frac{8}{9}$ as much work as the optimal solution.

---

[1] For this model, it is more convenient to introduce the bandwidth bw_i instead of the time c_i to communicate one unit of data used in the previous models. Of course, both are related by $bw_i = \frac{1}{c_i}$.

11.3.2 Multi-Round

In single-round schedules, as soon as some contention is introduced, the processors that receive their data last remain idle for all the beginning of the schedule. It seems thus natural to perform several rounds in sequence, so that all processors can start to compute as soon as possible, and this is why the multi-round model was introduced. However, without latencies the problem is not correctly formulated, because there is no finite optimal solution: as proven in Theorem 8.3, a linear communication model allows to improve any solution by adding a sufficiently small round at the beginning of the schedule. Consequently, multi-round models always include latencies, since this physical parameter makes infinite solutions unusable in practice. We first consider the more classical one-port version of the problem, and then explore a bounded multi-port approach.

Problem statement –   A full specification of the allocation requires to provide an *activation sequence*: σ_k is the processor used in the kth communication from the master, and $\alpha^{(k)}$ is the workload sent to processor σ_k in activation k. The starting time of the kth communication is $\sum_{l<k} L_{\sigma_l} + \alpha^{(l)}.c_{\sigma_l}$, and after the end of this communication, the corresponding processor needs to compute a workload of $\sum_{l \geq k, \sigma_l = \sigma_k} \alpha^{(l)}$ for all the subsequent activations. The feasibility constraint for this activation k is thus:

$$\sum_{l \leq k} \left(L_{\sigma_l} + \alpha^{(l)}.c_{\sigma_l} \right) + \sum_{l \geq k, \sigma_l = \sigma_k} \alpha^{(l)}.w_{\sigma_k} \leq T$$

As seen previously, the one-round problem with one-port communication and latencies is NP-complete. This problem is a generalization, and as such is NP-hard. It is actually unclear if it is in $\mathcal{NP}$, because there is no polynomial bound on the number of activations, and so encoding a solution to check for feasibility might not be possible in polynomial time.

With a fixed activation sequence (*i.e.*, with given σ_k), the previous constraint allows to write a linear program that gives the values of the $\alpha^{(k)}$ and thus solves the problem. If only the number N_{act} of activations is fixed, it is possible to introduce (0,1)-variables $\chi_i^{(j)}$ with the convention that $\chi_i^{(j)} = 1$ if and only if $\sigma_j = i$ and write a mixed integer linear program to solve the problem, with branch-and-bound methods [24]. However, this takes exponential time and becomes untractable when N_{act} grows.

The size of the solution is also an issue for practical considerations, and this motivated the search for regular schedules that can be encoded in a compact way.

Uniform Multi-Round –   The Uniform Multi-Round heuristic [25] imposes that in a given round, all participating processors have the same computation time. For round j, this means that all $\alpha_i^{(j)}.w_i$ are equal to the round size S_j (here a round is composed of n activations, where n is the number of participating processors. $\alpha_i^{(j)}$ is thus the share of work received by slave i in round j). To ensure that there are no idle times in the communications, the total communication time of round $j + 1$ has to be equal to the size of round j. This yields a geometrical increase in the round sizes. The selection of the participating processors is performed heuristically, by sorting them by decreasing bandwidth and ensuring that the increase of the round sizes is large enough.

This algorithm has no worst-case performance guarantee, but exhibits very good performance in practice.

Periodic schedule – Searching for periodic schedules is another natural way for encoding schedules in a compact way. The approach to build such schedules is to write constraints on one period for communication and computation capacities, without specifying an ordering of the operations. This gives a *steady-state* formulation of the problem, which can be solved optimally since all variables are rational. The solution of this formulation specifies how much work each processor receives during a period, and it can be turned into a valid schedule by ensuring that the data necessary for the computations of period k are sent in period $k - 1$.

In consequence, the returned solution has an optimal throughput, except for a relatively small number of initialization and finalization periods. It is possible to choose the period length so that this algorithm is asymptotically optimal, *i.e.*, its worst-case performance ratio tends to 1 when the optimal makespan grows to infinity. These solution are thus very good in practice for application with large datasets. Furthermore, the algorithm has a very low time complexity, and can easily be extended to more elaborate platform graphs.

Bounded Multi-Port – Let us now consider briefly the multi-port model. In a multi-round multi-port setting, it is natural to imagine that slaves can start computing as soon as they have received their first part of data. This means that we are in a *flow* setting, as described in Section 11.2.5. This gives a totally relaxed problem, whose formulation is exactly the same as the previous steady-state approach. It is thus fairly easy to solve, and does actually not need rounds: once the allocation is determined, all participating processors follow the allocation until all work has been performed.

This shows how an underlying application and software that allows such a fine control of communications can greatly simplify the problem and allow for much more performance than one-port implementations (the proof of theorem 8.5 shows that this performance gap is at most two on master-slave platforms).

11.4   Iterative Algorithms on a Virtual Ring

In this section, we study the mapping of iterative algorithms onto hetero-geneous clusters. Such algorithms typically operate on a large collection of application data, which will be partitioned over the processors. At each it-eration, some independent calculations will be carried out in parallel, and then some communications will take place. This scheme is very general, and encompasses a broad spectrum of scientific computations, from mesh based solvers (e.g., elliptic PDE solvers) to signal processing (e.g., recursive convo-lution), and image processing algorithms (e.g., mask-based algorithms such as thinning). The application data is partitioned over the processors, which are arranged along a virtual ring. At each iteration, independent calculations are carried out in parallel, and some communications take place between con-secutive processors in the ring. The question is to determine how to slice the application data into chunks, and to assign these chunks to the processors, so that the total execution time is minimized. Note that there is no reason *a priori* to restrict to a uni-dimensional partitioning of the data, and to map it onto a uni-dimensional ring of processors: more general data partitionings, such as two-dimensional, recursive, or even arbitrary slicings into rectangles, could be considered. But uni-dimensional partitionings are very natural for most applications, and, as will be shown in this section, the problem to find the optimal one under a realistic communication model is already very diffi-cult.

The target architecture is a fully heterogeneous cluster, composed of diffe-rent-speed processors that communicate through links of different bandwidths. On the architecture side, the problem is twofold: (i) select the processors that will participate in the solution and decide for their ordering, that will represent the arrangement into a ring; (ii) assign communication routes from each participating processor to its successor in the ring. One major difficulty of this ring embedding process is that some of the communication routes will (most probably) have to share some physical communication links: indeed, the communication networks of heterogeneous clusters typically are sparse, i.e., far from being fully connected. If two or more routes share the same physical link, we have to decide which fraction of the link bandwidth is to be assigned to each route.

There are three main sources of difficulties in this problem:

- the heterogeneity of the processors (computational power, memory, etc.);

- the heterogeneity of the communications links;

- the irregularity of the interconnection network.

In Section 11.4.1, we model this problem and state our objective function and in Section 11.4.2, 11.4.3, and 11.4.4 we study this load-balancing problem under various assumptions to illustrate the impact of these hypothesis on the hardness of the problem. The original formulation of this problem and a detailed presentation of these results can be found in [18].

11.4.1   Problem Statement

Our target applications generally comprise a set of data (typically, a matrix) and at each step, each processor performs a computation on its chunk of data and exchanges the border of its chunk of data with its neighbor processors. To determine the execution time of such an application, we need to determine 1) which processors should be used, 2) the amount of data to give them, and 3) how to cut the set of data. As we have previously explained, it is natural to restrict to a unidimensional slicing of a large matrix. Therefore, the borders and neighbors are easily defined, a constant volume of data D_c is exchanged between neighbors at each step, and processors are virtually organized into a ring. Let $\operatorname{succ}(i)$ and $\operatorname{prec}(i)$ denote the successor and the predecessor of P_i in the virtual ring. Last, we denote by D_w the overall amount of work to process at each step.

Let us denote by P_1, ..., P_p our set of processors. Processor P_i executes a unit of work in a time w_i and will be alloted a share α_i of the matrix (we assume $\alpha_i \geqslant 0$ and $\sum_{j=1}^{p} \alpha_j = D_w$). Therefore, the time needed by processor P_i to process its share is $\alpha_i.w_i$. Let us denote by $c_{i,j}$ the cost of a unit of communication from P_i to P_j. At each step, the cost of sending from P_i to its successor is thus $D_c.c_{i,\operatorname{succ}(i)}$. Last, we assume that all processors operate under the 1-port full-duplex model. Our problem can then be stated as follows:

DEFINITION 11.1 Ring Load-Balancing   *Select q processors among p, order them as a ring and distribute the data among them so as to minimize*

$$\max_{1 \leqslant i \leqslant p} \mathbb{I}\{i\}[\alpha_i.w_i + D_c.(c_{i,\operatorname{prec}(i)} + c_{i,\operatorname{succ}(i)})],$$

where $\mathbb{I}\{i\}[x] = x$ if P_i participates in the computation, and 0 otherwise.

11.4.2   Complete Homogeneous Platform

In this section, we restrict ourselves to the situation where the platform graph is a complete graph (i.e., there is a communication link between any two processors) and all links have the same capacity (*i.e.*, $\forall i, j \ c_{i,j} = c$). These hypotheses have many consequences. First, either the most powerful processor performs all the work, or all the processors participate. Then, if all processors participate, all end their share of work simultaneously. These properties can easily be shown by exchange arguments. Therefore, there is a

τ such that $\alpha_i.w_i = \tau$ for all i and thus $D_w = \sum_i \alpha_i = \sum_i \frac{\tau}{w_i}$. The time of the optimal solution is thus

$$T_{\text{step}} = \min\left\{ D_w.w_{\min}, D_w.\frac{1}{\sum_i \frac{1}{w_i}} + 2.D_c.c \right\}$$

The heterogeneity of the processors is thus not a real issue in this problem and, as expected, most of the difficulty comes from the network modeling and communication optimization.

11.4.3 Complete Heterogeneous Platform

In this section, we still restrict ourselves to the situation where the platform graph is a complete graph (i.e., there is a communication link between any two processors) but the capacity of network links may differ from one link to another. Again, using exchange arguments, it is easy to prove that all participating processors end their share of work simultaneously. However, it may be the case that not all processors participate to the computation. Therefore, we have $T_{\text{step}} = \alpha_i.w_i + D_c.(c_{i,\text{succ}(i)} + c_{i,\text{prec}(i)})$ for any participating processor P_i.

Let us assume that we have determined the set of participating processors (or that all processors participate to the computation). Then, from $\sum \alpha_i = D_w$, we get

$$\sum_{i=1}^{p} \frac{T_{\text{step}} - D_c.(c_{i,\text{succ}(i)} + c_{i,\text{prec}(i)})}{w_i} = D_w,$$

and thus

$$\frac{T_{\text{step}}}{D_w.w_{\text{cumul}}} = 1 + \frac{D_c}{D_w} \sum_{i=1}^{p} \frac{c_{i,\text{succ}(i)} + c_{i,\text{prec}(i)}}{w_i}, \quad \text{where } w_{\text{cumul}} = \frac{1}{\sum_i \frac{1}{w_i}}.$$

Therefore, T_{step} is minimal when $\sum_{i=1}^{p} \frac{c_{i,\text{succ}(i)} + c_{i,\text{prec}(i)}}{w_i}$ is minimal and our problem is actually to find an Hamiltonian cycle of minimal weight in a graph where the edge from P_i to P_j has a weight of $d_{i,j} = \frac{c_{i,j}}{w_i} + \frac{c_{j,i}}{w_j}$. It is thus possible to prove that the ring load-balancing problem on a heterogeneous complete graph is NP-complete even when the processor selection has already been done.

Many classical techniques can however be used to solve this problem. For example, one can write a mixed integer linear program for this problem when restricting to a ring of size q:

MINIMIZE T SATISFYING THE (IN)EQUATIONS

$$
\begin{cases}
(1)\ x_{i,j} \in \{0,1\} & 1 \leqslant i,j \leqslant p \\
(2)\ \sum_{i=1}^{p} x_{i,j} \leqslant 1 & 1 \leqslant j \leqslant p \\
(3)\ \sum_{i=1}^{p} \sum_{j=1}^{p} x_{i,j} = q & \\
(4)\ \sum_{i=1}^{p} x_{i,j} = \sum_{i=1}^{p} x_{j,i} & 1 \leqslant j \leqslant p \\[4pt]
(5)\ y_i \in \{0,1\} & 1 \leqslant i \leqslant p \\
(6)\ \sum_{i=1}^{p} y_i = 1 & \\
(7)\ u_i \text{ integer}, u_i \geqslant 0 & 1 \leqslant i \leqslant p \\
(8)\ -p.y_i - p.y_j + u_i - u_j + q.x_{i,j} \leqslant q-1 & 1 \leqslant i,j \leqslant p, i \neq j \\
(9)\ \alpha_i \geqslant 0 & 1 \leqslant i \leqslant p \\
(10)\ \sum_{i=1}^{p} \alpha_i = D_w & \\
(11)\ \alpha_i \leqslant D_w \sum_{j=1}^{p} x_{i,j} & 1 \leqslant i \leqslant p \\
(12)\ \alpha_i.w_i + D_c \sum_{j=1}^{p}(x_{i,j}c_{i,j} + x_{j,i}c_{j,i}) \leqslant T & 1 \leqslant i \leqslant p
\end{cases}
$$

Constraints (1) to (8) are a classical ILP formulation for selecting a ring of size q in a graph inspired from the traveling salesman problem formulation. $x_{i,j}$ is a binary variable indicating whether processor i immediately follows processor j on the ring and the non-zero y_i represents the "origin" of the q-ring. u_i represents the position on the tour at which P_i is visited and the constraint (8) on u_i ensures that the tour does not split into sub-tours.

No relaxation in rationals seems possible here but heuristics inspired from those used to solve the traveling salesman problem (e.g., Lin-Kernighan) can be adapted. It is also possible to build the ring using a greedy algorithm: initially we take the best pair of processors, and for a given ring we try to insert any unused processor in between any pair of neighbor processors in the ring and select the best solution.

11.4.4   Arbitrary Heterogeneous Platform

In this section, we do not impose any restriction on the platform graph. The communication networks of heterogeneous clusters being typically sparse (i.e., far from being fully connected), it is very likely that some communication routes will have to share some physical communication links. It may thus be important to take the link sharing into account. We will first show that when the processor selection and ordering has been done, even the load-balancing (i.e., computing the α_i) becomes a complex problem.

Let us assume that our platform graph comprises a set of communications links: $e_1, ..., e_n$. We denote by bw_{e_m} the bandwidth of link e_m and we denote by $\mathcal{S}_i$ the path from P_i to $P_{\mathrm{succ}(i)}$ in the network. $\mathcal{S}_i$ uses a fraction $s_{i,m}$ of the bandwidth bw_{e_m} of link e_m. P_i needs a time $D_c/\min_{e_m \in \mathcal{S}_i} s_{i,m}$ to send to its successor a message of size D_c. Note that unlike in the previous sections,

$c_{i,\mathrm{succ}(i)}$ is thus no more a given constant: we need to enforce $c_{i,\mathrm{succ}(i)}.s_{i,m} \geqslant 1$ for all $e_m \in \mathcal{S}_i$.

Symmetrically, there is a path $\mathcal{S}_i$ from P_i to $P_{\mathrm{prec}(i)}$ in the network, which uses a fraction $p_{i,m}$ of the bandwidth b_{e_m} of link e_m. The $s_{i,m}$ and $p_{i,m}$ have to respect the following constraints: $\sum_{1 \leqslant i \leqslant p} s_{i,m} + p_{i,m} \leqslant bw_{e_m}$

Therefore, the problem sums up to a quadratic system (with terms like $c_{i,\mathrm{succ}(i)}.s_{i,m}$ in the constraints) if the processors are selected, ordered into a ring and the communication paths between the processors are known.

The greedy heuristic of the previous section can easily be adapted but is much more expensive to compute than earlier. Note also that extending heuristics like those used to solve the traveling salesman problem is much harder in this context than it was for complete interconnexion graphs.

Unsurprisingly, using more realistic models makes the corresponding problem more and more intractable. However, one may wonder whether a good solution (derived from an optimal algorithm or from an excellent heuristic) on a poor model should be preferred to a poor solution (derived from a simple heuristic) on a good model. There is no general answer to this question but as shown in [18], the latter seems to be preferable to the former in the context of iterative algorithms on a virtual ring: ignoring contention leads to important "unexpected" performance degradation and the corresponding rings have thus really poor performance (see [18] for more details).

11.5   Data Redistribution

Our third problem arises naturally when coupling code on a grid computing platform. In such a context, data often have to be redistributed from one cluster to another. Launching all communications simultaneously and hoping that the low-layer network flow control mechanisms will produce a good solution is generally not a good idea. Consider for example the platform depicted on Figure 11.4(a) and the redistribution on Figure 11.4(b). We assume that at a given time-step, any given network resource is equally shared between competing flows. The so-called "Brute-force" approach [8] starts all communications at the same time and thus completes the redistribution at time 2.5 (see Figure 11.5(a)). It should be noted that between time 0 and 1.5, the backbone link is the bottleneck for flow C whereas between time 1.5 and 2.5 it is not a bottleneck anymore. As C is the last finishing flow, if C had been alloted more bandwidth between time 0 and 1.5, it would have finished earlier. That is why the schedule depicted in Figure 11.5(b) completes the redistribution at time 2.

The Brute-force redistribution has thus poor performance when compared to redistribution where communications are scheduled. The problem of mini-

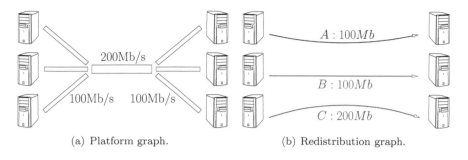

(a) Platform graph.   (b) Redistribution graph.

FIGURE 11.4: Platform graph and redistribution graph

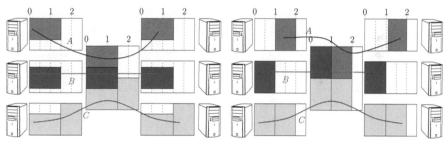

(a) Brute-force schedule: completion at time   (b) Clever schedule: completion at time 2.
2.5.

FIGURE 11.5: Brute-force and optimal schedule.

mizing the redistribution time can be modeled in various ways and the complexity of the corresponding problem changes drastically depending on the hypothesis and constraints one imposes. The goal of this section is to illustrate how these various hypothesis and constraints may affect the modeling and the resolution of the corresponding problem.

Let V_1 denote the set of nodes of the first cluster and V_2 denote the set of nodes of the second cluster. Our platform graph (see Figure 11.6) is defined by bw_1, bw_2 and bw_b, which denote respectively the bandwidth of the network cards of the first cluster, the bandwidth of the network cards of the second cluster, and the bandwidth of the backbone. A redistribution is defined by a set $E \subseteq V_1 \times V_2$ and an application $S : E \mapsto \mathbb{Q}$ indicating the amount of data that should be transferred from a node of V_1 to a node of V_2.

DEFINITION 11.2 Redistribution Problem   *Given a platform graph* $G_P = (V_1, V_2, bw_1, bw_2, bw_b)$ *and a set of redistributions* $G_A = (E, S)$, *minimize the redistribution time.*

The previous general problem is not fully specified and many modeling

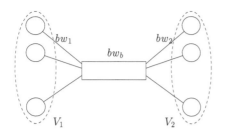 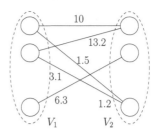

FIGURE 11.6: Platform graph.   FIGURE 11.7: Bipartite graph.

options can be considered. In Section 11.5.1, we formally model and state our problem as a matching decomposition problem. We state some complexity results and emphasize the practical and theoretical advantages and drawbacks of such a model. In Section 11.5.2, we use a different approach using elastic flows and also list the pros and cons.

11.5.1 The Matching Approach

We can first note that any flow from V_1 to V_2 is limited either by the network of a cluster or by the backbone link. Therefore the time needed to complete a single partial redistribution $e = (v_1, v_2)$ is defined by

$$\tau(v_1, v_2) = \frac{S(v_1, v_2)}{\min(bw_1, bw_2, bw_b)}$$

$G = (V_1 \cup V_2, E, \tau)$ defines a weighted bipartite graph where the topology and the characteristics of the platform graph does not appear anymore (see Figure 11.7). Note that other partial redistributions may occur in parallel if they do not exceed the capacity of the bottleneck link and use distinct nodes. At most $k = \left\lfloor \frac{bw_b}{\min(bw_1, bw_2, bw_b)} \right\rfloor$ such redistributions can occur in parallel therefore, a matching of size smaller than k is thus a valid set of redistributions. Therefore the redistribution problem amounts to the following matching problem:

DEFINITION 11.3 K-Preemptive Bipartite Scheduling (no latency)   *Given a bipartite graph $G = (V_1, V_2, \tau)$ and a bound k, decompose G into a set of weighted matchings $(\tau_1, E_1), \ldots, (\tau_h, E_h)$ whose sizes are smaller than k and such that:*

$$\forall (v_1, v_2) \in E : \tau(v_1, v_2) \leqslant \sum_{l \text{ s.t } (v_1, v_2) \in E_l} \tau_l$$

and $\sum_{l=1}^{h} \tau_l$ is minimum.

This KPBS problem without latency amounts to a classical fractional matching problem (even with the constraint on the size of the matchings) and can be solved for example using linear programming, the ellipsoid method or flow techniques. In the original problem defined in [8], there is however an additional latency term β in the objective function accounting for the reconfiguration time. Instead of minimizing $\sum_{l=1}^{h} \tau_l$, the general KPBS problem seeks to minimize $h.\beta + \sum_{l=1}^{h} \tau_l$.

The general KPBS problem is strongly NP-hard [8]. There are actually two main difficulties in this problem. First, there is a trade-off to find between the number of steps and the latency. Second, we restrict our search to bounded-size matchings. This constraint on the size of the matchings cannot easily be removed (it was introduced when reducing the redistribution problem to KPBS) but it is not what makes the problem hard. The PBS problem [8], which is the exact same problem where the bound on the size of matchings is removed, is also strongly NP-hard [11, 12] and cannot be approximated with a ratio smaller than $\frac{7}{6}$ unless $\mathcal{P} = \mathcal{NP}$ [9]. Actually, only the β term is an issue.

One can thus see that the choice of modeling the latency is crucial in this context! In the original problem formulation by Wagner *et al.* [8], the aim was to design an MPI-based library with such optimizations. There was a barrier between each set of redistributions and it was important to minimize the number of such barriers. In this context, the bound on the size of the matchings was also important to avoid overloading the network and not degrade too much other user's experiences of the network. Such a bound may however be too strict and allowing slightly larger matchings can lead to better results in practice.

Actually, by reducing the redistribution problem to the KPBS problem (and thus losing some information about the platform topology), we restrict ourselves to particular schedules where some resource are somehow wasted:

- Assume for example that $bw_1 \ll bw_2 \leqslant bw_b$. Then there could be multiple communications toward a given v_2 without experiencing a degradation of the communication time (it would require to remove the 1-port constraint that underlies many MPI implementations though). It is easy to introduce the maximum number of incoming communications at node v_2 as $\lfloor bw_2/bw_1 \rfloor$. Then, the problem reduces to finding *factors* in the bipartite graph G with constraints on the size of the factors. As previously, the problem becomes NP-hard as soon as β is introduced.

  Note that there are also problems for which the latency is not an issue but where the hardness really comes from the bound on the number of simultaneous connections [21].

- We reduce to factor or matching problems by ensuring that all our communications do not interfere with each other. Doing so, some bandwidth may be wasted. For example, if $bw_1 = bw_2 = 100$ and $bw_b = 199$, we will have $k = 1$ and the full capacity of the backbone will never be used. If we assume that the communication speed of each flow can be controlled (e.g., using kernel modules implementing QoS flow control mechanisms) then the available capacity could be used more efficiently. We detail the problem under such assumptions in the next section.

11.5.2 The Elastic Flows Approach

For a given redistribution schedule σ, let us denote by T_σ the time needed to complete all flows. For any link l of type $i(l)$, with $i(l) \in \{1, 2, b\}$, $T_\sigma.bw_{i(l)}$ is an upper bound on the amount of data transmitted through link l. Thus, if we denote by $\mathcal{L}$ the set of links, we have:

$$\forall l \in \mathcal{L}: \sum_{(v_1,v_2)|l\in v_1 \to v_2} S(v_1, v_2) \leqslant T_\sigma bw_{i(l)}.$$

Therefore, if we consider a schedule σ' such that each flow (v_1, v_2) is alloted a bandwidth $S(v_1, v_2)/T_\sigma$, we get a feasible schedule for the "Bandwidth sharing" model of Section 11.2.4 with the same completion time as σ. Therefore, the optimal T is simply:

$$\max_{l \in \mathcal{L}} \frac{\sum_{(v_1,v_2)|l\in v_1 \to v_2} S(v_1, v_2)}{bw_{i(l)}}.$$

The problem is thus much simpler under such assumptions than it was in the previous model. Note that introducing latencies is not a problem when we consider such a schedule as all flows start and finish at the same time. It can be seen as a *regulated* "brute force" approach. Hence, such a schedule is generally not usable in practice. Indeed, there is generally a bound per machine on the number of concurrent open connections. Moreover, having a large number of concurrent flows is very intrusive to other users that may want to use the backbone link. That is why it may be useful to reorganize this schedule so as to minimize the number of concurrent flows. This minimization can be seen as a packing problem and is thus NP-hard. However, there are simple and efficient approximations and one can thus get very good solutions with this approach. Therefore, by using a *borderline* model (that does not really make sense anymore when too many communications occur in parallel), we can derive a *borderline* solution that can be re-optimized afterward into a realistic and efficient solution.

11.6   Conclusion

Through these various examples, we have seen that there are always several ways of modeling the same situation, and that the choice *does* matter. Modeling is an art, and deciding what is important and what can be neglected requires a good knowledge of the application and of the underlying platform.

The focus of this chapter has been on the influence of the model on the complexity class of the problem; however the fact that a problem is NP-hard does not mean that the model should not be considered. It is often better to provide an approximate solution of a good model than an optimal solution of an imperfect model, which does not take into account an important characteristic of the system and thus may prove less efficient in practice.

References

[1] A. Alexandrov, M. Ionescu, K. Schauser, and C. Scheiman. LogGP: Incorporating long messages into the LogP model for parallel computation. *Journal of Parallel and Distributed Computing*, 44(1):71–79, 1997.

[2] M. Banikazemi, V. Moorthy, and D. K. Panda. Efficient collective communication on heterogeneous networks of workstations. In *Proceedings of the 27th International Conference on Parallel Processing (ICPP'98)*. IEEE Computer Society Press, 1998.

[3] M. Banikazemi, J. Sampathkumar, S. Prabhu, D. Panda, and P. Sadayappan. Communication modeling of heterogeneous networks of workstations for performance characterization of collective operations. In *HCW'99, the 8th Heterogeneous Computing Workshop*, pages 125–133. IEEE Computer Society Press, 1999.

[4] A. Bar-Noy, S. Guha, J. S. Naor, and B. Schieber. Message multicasting in heterogeneous networks. *SIAM Journal on Computing*, 30(2):347–358, 2000.

[5] O. Beaumont, L. Eyraud-Dubois, and N. Bonichon. Scheduling divisible workload on heterogeneous platforms under bounded multi-port model. In *Heterogeneous Computing Workshop HCW'2008*. Proceedings of IPDPS 2008, IEEE Computer Society Press, 2008.

[6] P. Bhat, C. Raghavendra, and V. Prasanna. Efficient collective communication in distributed heterogeneous systems. In *ICDCS'99 19th*

International Conference on Distributed Computing Systems, pages 15–24. IEEE Computer Society Press, 1999.

[7] P. Bhat, C. Raghavendra, and V. Prasanna. Efficient collective communication in distributed heterogeneous systems. *Journal of Parallel and Distributed Computing*, 63:251–263, 2003.

[8] J. Cohen, E. Jeannot, N. Padoy, and F. Wagner. Message Scheduling for Parallel Data Redistribution between Clusters. *IEEE Transactions on Parallel and Distributed Systems*, 17:1163–1175, 2006.

[9] P. Crescenzi, D. Xiaotie, and C. Papadimitriou. On approximating a scheduling problem. *Combinatorial Optimization*, 5:287–297, 2001.

[10] D. Culler, R. Karp, D. Patterson, A. Sahay, E. Santos, K. Schauser, R. Subramonian, and T. von Eicken. LogP: a practical model of parallel computation. *Communication of the ACM*, 39(11):78–85, 1996.

[11] S. Even, A. Itai, and A. Shamir. On the complexity of timetable and multicommodity flow problem. *SIAM Journal on Computing*, 5:691–703, 1976.

[12] I. Gopal and C. Wong. Minimizing the number of switching in an SS/TDMA system. *IEEE Transactions on Communications*, 33(6):497–501, 1985.

[13] R. W. Hockney. The communication challenge for MPP: Intel Paragon and Meiko CS-2. *Parallel Computing*, 20:389–398, 1994.

[14] B. Hong and V. Prasanna. Distributed adaptive task allocation in heterogeneous computing environments to maximize throughput. In *International Parallel and Distributed Processing Symposium IPDPS'2004*. IEEE Computer Society Press, 2004.

[15] S. Khuller and Y. A. Kim. On broadcasting in heterogenous networks. In *Proceedings of the Fifteenth Annual ACM-SIAM Symposium on Discrete Algorithms*, pages 1011–1020. Society for Industrial and Applied Mathematics, 2004.

[16] T. Kielmann, H. E. Bal, and K. Verstoep. Fast measurement of LogP parameters for message passing platforms. In *Parallel and Distributed Processing: 15 IPDPS 2000 Workshops*, volume 1800 of *LNCS*. Springer, 2000.

[17] B. Kreaseck, L. Carter, H. Casanova, and J. Ferrante. On the interference of communication on computation in java. In *IPDPS*, 2004.

[18] A. Legrand, H. Renard, Y. Robert, and F. Vivien. Mapping and load-balancing iterative computations on heterogeneous clusters with shared links. *IEEE Transactions on Parallel Distributed Systems*, 15(6):546–558, 2004.

[19] P. Liu. Broadcast scheduling optimization for heterogeneous cluster systems. *Journal of Algorithms*, 42(1):135–152, 2002.

[20] S. H. Low. A duality model of TCP and queue management algorithms. *IEEE/ACM Transactions on Networking*, 2003.

[21] L. Marchal, Y. Yang, H. Casanova, and Y. Robert. Steady-state scheduling of multiple divisible load applications on wide-area distributed computing platforms. *International Journal of High Performance Computing Applications*, 20(3):365–381, 2006.

[22] L. Massoulié and J. Roberts. Bandwidth sharing: Objectives and algorithms. In *INFOCOM (3)*, pages 1395–1403, 1999.

[23] T. Saif and M. Parashar. Understanding the behavior and performance of non-blocking communications in MPI. In *Proceedings of Euro-Par 2004: Parallel Processing*, LNCS 3149, pages 173–182. Springer, 2004.

[24] Y. Yang, H. Casanova, M. Drozdowski, M. Lawenda, and A. Legrand. On the complexity of multi-round divisible load scheduling. Research Report 6096, INRIA, Jan. 2007.

[25] Y. Yang, K. van der Raadt, and H. Casanova. Multiround algorithms for scheduling divisible loads. *IEEE Transactions on Parallel and Distributed Systems*, 16(11):1092–1102, 2005.

Index

Allocation, 164
Application model
   Bag of tasks, 162
   Independent tasks, 188
   PERT graph, 267
   Series of broadcasts, 163
   Series of multicast, 163
   Task graph, 172, 182, 267
Approximation algorithm, 23, 24, 34,
   41, 93, 233, 234, 237, 238
Approximation scheme
   FPTAS (Fully polynomial-time
    approximation scheme), 28,
    82
   PTAS (Polynomial-time approx-
    imation scheme), 24, 82
Average cycle time, 105, 109

Bellman-Ford algorithm, 112

Certificate, 13
Classification scheme
   Job characteristics, 10
   Optimality creiterion, 10
   Processor environment, 10
Communication delays, 32
   Large communication delays, 44
Communication model
   Affine cost, 201, 209–214, 284–
    288
   Linear cost, 189–209, 214
   Multi-port, 289
    Bounded, 162, 167, 178, 295
    Unbounded, 293
   One-port, 294
    Bidirectional, 162, 167, 178,
     179, 290

    Unidirectional, 162, 166, 178,
     188, 290
Competitive ratio, 51, 53, 56–59, 71
Completion time, 81, 91
   Sum of completion times, 219–
    224, 231, 233, 234, 244–246,
    266
   Total weighted completion time,
    81
Complexity classification, 18
Counting problems, 232
Critical circuit, 111
Cyclic schedule, 103, 129, 169

Deadline, 4, 73
   Agreeable deadline, 72
Decomposed Software Pipelining, 121
Dependence, *see* Precedence con-
    straints
Difference constraint system, 112
Divisible load scheduling, 187–216,
   292–297
   Multi-round strategies, 204–215,
    296
   Return messages, 214–215
   Single-round strategies, 191–
    204, 214–215, 293
Duplication, 42
Dynamic programming, 190
Dynamic schedule, 124
Dynamic speed scaling, 67
Dynamic system, 256, 268

Earliest schedule, 116
Edge-coloring problem, 26
Ellipsoid method, 175
Embedded system, 129
Energy consumption, 62, 68, 71

311